003770

SOUTH HILL
EDUCATION CENTRE
6010 Fraser Street
Vancouver, B.C. V5W 2Z7
Ph: 324-2414 Fax: 324-9476

EARTH DYNAMICS
Studies in Physical Geography

Ron Chasmer

Toronto
Oxford University Press Canada

Oxford University Press Canada, 70 Wynford Drive,
Don Mills, Ontario, Canada M3C 1J9

Oxford   New York
Athens   Auckland   Bangkok   Bombay
Calcutta   Cape Town   Dar es Salaam   Delhi
Florence   Hong Kong   Istanbul   Karachi
Kuala Lumpur   Madras   Madrid   Melbourne
Mexico City   Nairobi   Paris   Singapore
Taipei   Tokyo   Toronto

and associated companies in
Berlin   Ibadan

OXFORD is a trademark of Oxford University Press.

Copyright © Oxford University Press Canada 1995. All rights reserved. No part of this book may be reproduced in any form without the written permission of the publisher. Photocopying or reproducing mechanically in any other way parts of this book without the written permission of the publisher is an infringement of the copyright law.

Every effort has been made to trace the original source of material and photographs contained in this book. Where the attempt has been unsuccessful the publisher would be pleased to hear from copyright holders to rectify any omissions.

This book is printed on permanent (acid-free) paper. ∞

Canadian Cataloguing in Publication Data

Chasmer, Ron
   Earth dynamics: studies in physical geography

Includes index.
ISBN 0-19-540984-1

1. Physical geography. 2. Earth sciences.
I. Title.

GB55.C53 1995   910'.02   C94-931883-3

Editor: Loralee Case
Editorial assistant: Micaela Gates
Design and illustrations: VISUT*ronX*
Cover design: Heather Delfino
Photo research: Natalie Pavlenko-Lomago
Printed in Canada by The Bryant Press

5 99

ACKNOWLEDGEMENTS

The publisher and author wish to thank the following people for reviewing the manuscript of this text.

Ron Bonham
Education Consultant
Cookstown, Ontario

John Chalk
Templeton Secondary School
Vancouver, British Columbia

Don Farquharson
Port Perry High School
Port Perry, Ontario

Ken O'Connor
Geography Consultant
Scarborough Board of Education
Scarborough, Ontario

James Taylor
Cornwallis District High School
Canning, Nova Scotia

Dedicated to Laura and Ryan

Cover photographs: Wave background:
   Peter Cade/Tony Stone Images

   Volcano:
   K.Krafft/Explorer/Publiphoto

Maurice and Katia Krafft (shown on the cover) bravely confronted nature's deadly fury during their field studies of erupting volcanoes.
While photographing Japan's Unzen volcano shortly after its eruption in June 1991, the lives of the renowned French volcanologists were extinguished by a massive rush of searing ash and deadly gases.

CONTENTS

Key
📖 Traditional Approach
✱ Thematic (Disasters) Approach
⚙ Integrative Studies Approach
🍁 Field Studies Approach

Introduction 1

**PART 1 Physical Geography:
An Overview** 3

📖 CHAPTER 1: *The Physical Environment:
An Overview* 4
    Introduction..................................4
    Elements of Physical Geography 5

📖 CHAPTER 2: *Maps: Important Tools
for the Geographer* 6
    Introduction..................................6
    Scale 7
    Direction......................................7
    Grids..8
    Legends..9
    Projections 9
🍁 Field Study...................................14

**PART 2 Energy Systems: What Makes
the World Go Round?** 15

CHAPTER 3: *Energy from the Sun* 16
    Introduction................................16
    The Sun as an Energy Source 17
    Solar Energy and Natural Processes...17
    Balancing Energy Flows 18
    Other Forms of Solar Energy 18
    Non-solar Energy Sources..............19

CHAPTER 4: *Energy Flows
in the Geosphere* 20
    Introduction................................20
    Energy in the Lithosphere 21
    Energy in the Atmosphere.............21
    Energy in the Hydrosphere 21
    Energy in the Biosphere 21

📖 CHAPTER 5: *Energy Systems in the
Lithosphere*....................................22
    Introduction................................22
    Nuclear Energy 23
    Potential Energy..........................24
    Kinetic Energy 24
    Chemical Energy..........................25

📖 CHAPTER 6: *Energy Systems in the
Atmosphere* 26
    Introduction................................26
    Radiant Energy 27
    Earth-Sun Relationships................27
    Thermal Energy 30
    Kinetic Energy 34

📖 CHAPTER 7: *Energy Systems in
the Hydrosphere* 40
    Introduction................................40
    A Storehouse of Solar Energy.........41
    Kinetic Energy 42
    Thermal Energy 46
    Potential Energy..........................54

📖 CHAPTER 8: *Energy Systems in
the Biosphere* 56
    Introduction................................56
    Kinetic Energy 56
    Chemical Energy..........................58
    Potential Energy..........................63

📖 CHAPTER 9: *Landscape Systems* 65
    Introduction................................65
    An Alpine Landscape System 65
🍁 Field Study...................................68

**PART 3 Studying the Atmosphere:
Weather and Climate** 69

📖 CHAPTER 10: *The Atmosphere* 70
    Introduction................................70
    Properties of the Atmosphere..........71

⚙ CHAPTER 11: *Weather and Climate*..........74
    Introduction................................74
    What Is Weather? 75
    What Is Climate? 75
    Weather and Climate in Our
        Daily Lives.............................75

🍁 CHAPTER 12: *Measuring the Atmosphere* ...77
    Introduction................................77
    Collecting Weather Data 78

Weather Instruments 80
Field Study 82

CHAPTER 13: *Battles in the Sky: Meteorology* 84
Introduction 84
Air Masses 85
Fronts 86
Clouds 87
Cyclones and Anticyclones 88
Forecasting the Weather 89
Career Profile: Meteorologist 91

CHAPTER 14: *Violent Weather: An Integrative Study* 94
Introduction 94
Hurricanes 95
Tornadoes 98
Blizzards 102

CHAPTER 15: *Climatology* 106
Introduction 106
Classifying Climate 107
Temperature Zones and Climate Classification 108
Climate Classification with Two Variables 108
The Köppen Climate Classification System 108
The Köppen Codes 109

CHAPTER 16: *Climate Regions of the World* 115
Introduction 115
The Tropical Rainforest Climate (Af) 117
The Savanna Climate (Aw) 118
The Monsoon Climate (Am) 121
The Desert (Bw) and Steppe (Bs) Climates 121
The Mediterranean Climate (Csa, Csb) 124
The Marine West Coast Climate (Cfb, Cfc) 125
The Humid Subtropical Climate (Cfa) 126
The Continental Climate (Dfa, Dfb, Dwa, Dwb) 127
The Subarctic Climate (Dfc, Dfd, Dwc, Dwd) 129
Polar Climates (ET, EF) 130

CHAPTER 17: *Microclimates* 132
Introduction 132
Factors Affecting Microclimates ... 133
Career Profile: Climatologist 133
Field Study 133

CHAPTER 18: *Global Warming: An Integrative Study* 134
Introduction 134
The Greenhouse Effect 135
Greenhouse Gases 135
Scientific Uncertainty 137
Solving the Problem of Global Warming 142

PART 4 The Lithosphere: Building Up the Land 145

CHAPTER 19: *Geological Time* 146
Introduction 146
Geological Change 147
The Geological Record 147
The Geological Time Scale 148
Absolute Time 151

CHAPTER 20: *The Structure of the Lithosphere* 153
Introduction 153
Why There Are Layers 154
The Core 154
The Mantle 154
The Asthenosphere 155
The Crust 155

CHAPTER 21: *Elements, Minerals, and Rocks* 157
Introduction 157
Three Classes of Rock 159
Field Study 162

CHAPTER 22: *Minerals in Our Lives* 163
Introduction 163
Career Profile: Geologist 164
Igneous Ore Bodies 165
Sedimentary Ore Bodies 165
Other Economic Minerals 167

CHAPTER 23: *Tectonic Processes: Building Continents* 170
Introduction 170
Plate Tectonics 171

How Continents Move 172

CHAPTER 24: *Tectonic Processes: Building Mountains* 175
Introduction..............................175
Fold Mountains 176
Features Caused by Faulting 179
Features Associated with Volcanism 183

CHAPTER 25: *Mapping Techniques: Contours, Profiles, and Models* 188
Introduction..............................188
Contours....................................189
Profiles......................................189
Three-dimensional Diagrams 189
Models......................................189
Field Study................................189

CHAPTER 26: *Earthquakes: An Integrative Study* 192
Introduction..............................192
Earthquake Damage 193
Measuring and Predicting Earthquakes..........................196

PART 5 Gradational Processes: Wearing Down the Earth 199

CHAPTER 27: *Forces that Shape the Earth* 200
Introduction..............................200
Base Level 201
Principles of Gradation 201

CHAPTER 28: *The Work of Gravity* 206
Introduction..............................206
Slope..207
Unconsolidated Material..............207
Natural Vegetation 207
Tectonic Stability 208
Moisture Content 209

CHAPTER 29: *The Work of Rivers* 210
Introduction..............................210
River Systems............................211
Drainage Patterns 211
River Systems and Flow Dynamics ..213
The Geomorphic Cycle 215
Karst Topography: Rivers Underground..................221

CHAPTER 30: *Measuring Rivers* 223
Introduction..............................223
Measuring Width and Depth 224
Measuring Flow Velocity 225
Field Study................................225

CHAPTER 31: *The Work of Wind* 227
Introduction..............................227
The Aeolian Landscape................228
Running Water in Arid Landscapes 228
Wind-blown Sand 228
Stages of Aeolian Development.....231

CHAPTER 32: *The Work of Ice* 233
Introduction..............................233
Types of Glaciers 234
Glacial Movement 234
Continental Glaciation................236
Alpine Glaciation........................241
Alpine Glaciation and Land Use Patterns..........................242

CHAPTER 33: *The Work of Waves*..........247
Introduction..............................247
Wave Size..................................248
Erosion and Wave Action 248
Wave Action and Deposition 249
Emergent Coastlines 249
Submergent Coastlines................253
Coral Formations........................253

PART 6 The Ecosphere: All Things Living Under the Sun 257

CHAPTER 34: *The Ecosphere and Our Place in It* 258
Introduction..............................258
Ecosystems................................259
Adaptation................................259
The Web of Life 259
The Effect of People on the Ecosphere 260
Modifying the Environment..........260

CHAPTER 35: *Natural Vegetation* 262
Introduction..............................262
Plant Needs................................263
Changes in Natural Vegetation 264
Vegetation Succession 264
Change Caused by Human Activity..265

Classifying Natural Vegetation 268
The Forest Community 268
The Grassland Community 269
The Scrub Community 270
The Desert Community 270
Local Variations 270

📖 CHAPTER 36: *Soils* 273
Introduction 273
The Properties of Soils 274
The Effect of Climate on Soils 277
⚙ Soil Management: A Fine
   Balancing Act 278
🍁 Field Study 280
Soil Classification 282

PART 7 People and Their Ecosystems:
Integrative Studies 287

⚙ CHAPTER 37: *You, Your Lifestyle,
and Nature* 288
Introduction 288
Sustainable Development 289
Studying Different Ecosystems:
   An Independent Study 290

⚙ CHAPTER 38: *Living in the
Tropical Rainforest* 292
Introduction 292
Climate 293
Plants and Animals 293
Soils and Landforms 295
Indigenous Peoples 296
Resource Development 297

⚙ CHAPTER 39: *Living in the
Boreal Forest* 302
Introduction 302

Climate 303
Plants and Animals 303
Soils and Landforms 303
Indigenous Peoples 305
Resource Development 305
Career Profile: Forester 308

⚙ CHAPTER 40: *Living in the
Tropical Grassland* 314
Introduction 314
Climate 315
Plants and Animals 315
Soils and Landforms 317
Indigenous Peoples 319
Resource Development 319

⚙ CHAPTER 41: *Living in the
Temperate Grassland* 325
Introduction 325
Climate 326
Plants and Animals 326
Soils and Landforms 326
Indigenous Peoples 327
Resource Development 327

⚙ CHAPTER 42: *Living in the
Maritime Ecosystem* 332
Introduction 332
Climate 333
Plants and Animals 333
Soils and Landforms 333
Indigenous Peoples 334
Resource Development 334

Glossary 338
Credits 347
Index 348

INTRODUCTION

The world in which we live is incredibly complex. Natural systems constantly change, sometimes rhythmically, sometimes chaotically, as they interact among themselves and with people. Today more than ever, people are concerned about the environment. And yet the human population continues to expand geometrically, drawing more and more resources from our finite world. It is important that we understand how the world operates so that our planet can continue to provide for our needs while remaining the delightful place it is.

There are many ways to study physical geography. This textbook accommodates four approaches:
- the traditional approach
- the thematic (disasters) approach
- the integrative studies approach
- the field studies approach

In the table of contents, icons indicate sections of the book that relate to each approach. By picking the sections carefully, teachers can design their own program of study.

Many people contributed to the production of this book. I would especially like to thank the following people for their support and advice. Jim Tilsley, maritimer, economic geologist, and vice-president of the Canadian Association of Environmental Geologists, spent many hours explaining current practices in resource development, not only in Canada but in Brazil and Africa. Ron Bonham reviewed the text. His integrity, years of teaching experience, and good sense helped me greatly throughout the project. The meteorologists and climatologists at Environment Canada provided up-to-date information on the state of the world's atmosphere. Loralee Case edited manuscript and provided many excellent suggestions. Finally, I would like to thank my family who were always supportive and understanding. I sincerely hope this book helps people across Canada to understand the beauty and power of our magnificent planet.

Ron Chasmer

Part 1

Physical Geography:
An Overview

CHAPTER 1

The Physical Environment: An Overview

INTRODUCTION

Geography and science are similar subjects: they deal with essentially the same concepts. The difference between the two lies in approach. Scientists look at the earth through microscopes. They study in minute detail the things that make up our planet. Geographers look at the world differently. It is as if they study the world through telescopes from some distant place. They are interested in how the various aspects of nature interact with one another rather than in how an individual element operates in isolation.

How would scientists and geographers view acid rain, for example? A team of scientists specializing in many fields would study it. Chemists would measure the pH level of lakes to determine acidity and develop a profile to show how acidity has changed over time. Biologists would study plants and animals to determine the effects of acidification. Geologists would study soils and rocks to determine their chemical composition. Often each specialist would work in isolation.

Geographers, on the other hand, may not even visit the site. They would use maps and data provided by scientists, as well as information from other sources, to study the interrelationships among the systems. The scientific data would be evaluated together with such geographic factors as prevailing winds, urban development, industrial processes, and environmental law. By synthesizing this information, the geographer gains an understanding of the overall problem and can offer solutions.

Physical geography is often confused with earth science. Like the study of earth science, physical geography deals with natural systems and how they interact. But physical geography puts people into the study as well. Rather

FIGURE 1.1   The Earth as Seen from Space

Ecosytem	Atmosphere	Lithosphere	Hydrosphere
Polar mosses, lichens, marine mammals	• below freezing • cold all year	• permanently frozen • poor soils	• ice, glaciers, snow • little snowfall
Desert cacti, few plants, nocturnal animals	• hot or cold • sunny most of the time	• sandy or rocky • poorly developed soils	• little surface water • little precipitation
Wetlands aquatic plants and animals	• varies • layers of organic matter	• water-logged soils • plentiful rainfall	• flooded land

FIGURE 1.2 *Characteristics of Three Ecosystems*

than just studying climate or rocks, physical geographers study how these elements affect people and how people affect these elements. It is important to remember the role of people whenever we study earth systems, for it is people that create many of the environmental problems that plague the planet today.

Elements of Physical Geography

Our planet is unique in our solar system. Oxygen in the **atmosphere**, minerals and fuels in the **lithosphere**, and water in the **hydrosphere** interact to create a dynamic planet that provides all of the elements needed to support the life zone called the **biosphere**.

Within the biosphere there are many **ecosystems**. These are regions unique in the ways in which the atmosphere, the hydrosphere, and the lithosphere interact. Figure 1.2 shows three ecosystems and their characteristics.

Often, elements of the four spheres overlap. Is precipitation part of the atmosphere or the hydrosphere? Is groundwater part of the hydrosphere or the lithosphere? Obviously, some elements are found in more than one sphere.

All things in nature can be found in one of the earth's four spheres. The atmosphere includes air and its composition, air temperature, and air pressure. The hydrosphere includes surface water in oceans, rivers, and lakes, as well as groundwater, water vapour, and icecaps. The lithosphere includes rocks, minerals, landforms, and soils. The biosphere includes all plants and animals, living or dead.

We might think of our planet as a spaceship. Spaceships provide the essentials for life, like oxygen. Water is stored in containers aboard ship just as it is in oceans, lakes, and aquifers on earth. Astronauts eat food stored on board; the biosphere provides food for the inhabitants of spaceship earth. Fuels propel the spaceship just as minerals and fuels provide for our needs. But although a spaceship and the planet are similar, there are significant differences. The spaceship is a **closed system**; nothing leaves or enters the craft. Planet earth, on the other hand, is an **open system**; it is constantly being renewed by its own forces. On the spaceship, life-support systems are stored materials that do not perpetuate themselves. Here on earth, our life-supporting elements are continually being created and destroyed by solar energy. Earth is a constantly changing environment. This will enable it to survive long after the spaceship can no longer support its occupants.

Things To Do

1 Define the following terms: *geography, physical geography, earth science, atmosphere, hydrosphere, lithosphere, biosphere.*

2 a) Prepare an organizer comparing earth with a spaceship.
  b) Design a spaceship that could function indefinitely as planet earth does.

Chapter 2

Maps: Important Tools for the Geographer

INTRODUCTION

Geographers use many tools to study the earth, but perhaps the most important of these are maps. Maps have incredible variety. They can be a sketch on the back of an envelope telling your friend how to get to your house. Or they can be complex digital images from outer space displayed on a computer screen.

Maps provide a bird's-eye view of a given area. To do so with any degree of accuracy, however, most maps include the following characteristics: scale, direction, a grid, a legend, and a projection. This chapter will examine each of these map elements.

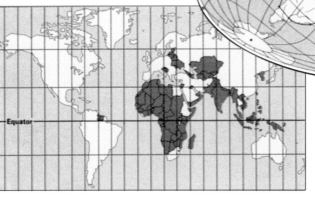

FIGURE 2.1   Some Types of Map Projections. Mercator's Projection (upper right), Oblique Aitoff Projection (oval), Gall's Projection (lower left), and Peter's Projection (lower right).

Scale

Just as model automobiles and building blueprints are made to scale, so too are maps. Every accurate map clearly displays its scale. This indicates the size of the map in relation to the real world. Scales are useful not only to get an idea of relative size, but to determine distances between places. Pilots, truckers, and other travellers need this vital information.

There are three types of scales: a **scale statement**, a **scale ratio**, and a **linear scale**. A scale statement is, quite simply, a sentence comparing distance on the map to the real world, for example, "1 cm represents 120 km." To calculate the actual distance between two places, use a ruler to measure between points and multiply the number of centimetres by its real-world equivalent. In this case, you would multiply by 120 because 1 cm represents 120 km.

The scale ratio expresses the same information numerically, for example, 1:12 000 000. In this example, 1 unit on the map represents 12 000 000 units in the real world. It doesn't matter what the units are; they could be centimetres or millimetres. However, the same units must be applied to both elements of the ratio. How do you measure distance using a scale ratio? Assume that the distance between two points on the map is 3.2 cm. By multiplying this measurement by 12 000 000, you can find the real-world distance. In this case, 3.2 × 12 000 000 = 38 400 000 cm. Usually it is convenient to convert this figure into a larger unit, such as kilometres. Since there are 100 000 cm in 1 km, the actual distance is 384 km (38 400 000 ÷ 100 000).

A linear scale is a line on the map showing the equivalent actual distance (see Figure 2.2). Finding distances with a linear scale requires no arithmetic. The distance between two places is marked on a piece of paper. The paper is then aligned along the linear scale and the distance is read off. If the distance is longer than the scale line, it is necessary to move the paper along the linear scale several times.

Maps are either large scale or small scale. Large-scale maps show small areas and have scales of 1:10 000 to 1:1 000 000. They are used in topographic maps, local road maps, and other small area maps that require detail. Small-scale maps have scales of 1:1 000 000 to 1:10 000 000 and illustrate large areas such as continents or oceans in a general way. Most maps found in atlases are small scale.

Direction

Most maps are aligned with north at the top. A **compass rose** is often used to indicate north. (See Figure 2.3 on page 8.) This clearly shows the **cardinal points** north, south, east, and west and the **intercardinal points** northeast, southeast, southwest, and northwest. These points provide a general indication of direction. For greater accuracy, direction may also be indicated using degrees. With north as 0°, we can use a protractor to see the number of degrees for each of the cardinal points—east is 90°, south is 180°, and west is 270°.

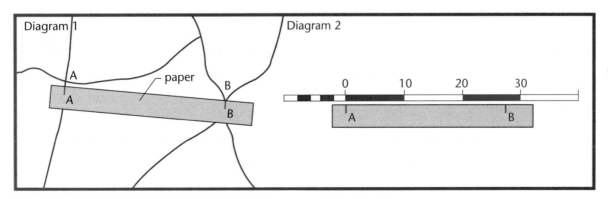

FIGURE 2.2 *How to Use a Linear Scale. First mark the distance between two places on a piece of paper (Diagram 1), then compare the distance to the linear scale to get the "real" distance (Diagram 2).*

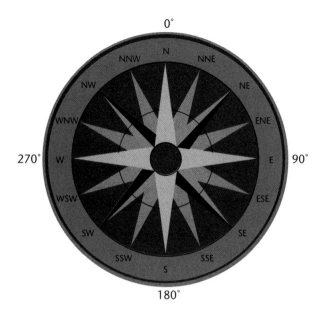

FIGURE 2.3 *Compass Rose Showing Cardinal and Intercardinal Directions*

Bearings are directional measurements given in degrees. If you are travelling by boat from one island to another, you would use a protractor to measure the bearings on your map. You would plot your course by drawing a line between the two islands and using a protractor to find the direction in degrees that you need to travel. If you aligned your compass with north and followed the same bearing as you read on the map, you would be sure to reach your destination. You can see how important maps are in navigation.

Grids

Grids are the horizontal and vertical lines that help you find places on maps. Each line is labelled with a number or letter. By stating the **co-ordinates** of the two lines, it is easy to find the square on the grid in which a place is located. There are three basic types of grids: **latitude** and **longitude, algebraic grids,** and **military grids**.

Most of us are familiar with latitude and longitude. Lines of latitude run east to west. They are measured in degrees north to south from the Equator. These parallels of latitude represent the angle formed by the Equator and the earth's axis. The North and South poles are 90° from the Equator. The half of the earth north of the Equator is called the Northern Hemisphere, while the half south of the Equator is the Southern Hemisphere. Since there are corresponding latitudes on either side of the Equator, the designation N for north or S for south is used. For example, 45°S runs through Argentina but 45°N runs through Canada.

Longitude is the name given to vertical lines that run north and south. They are measured in degrees from east to west. The **Prime Meridian**, which runs through the Greenwich Observatory near London, England, is 0°. Continuing east and west, these meridians all meet at the North and South poles. Unlike latitude, they do not run parallel to each other. Places to the west of the Prime Meridian are in the Western Hemisphere, while places to the east are in the Eastern Hemisphere. The designations W for west and E for east are used to indicate the hemispheres. For example, Tokyo, Japan, and Wainwright, Alaska, both have a longitude of 160°. Since Tokyo is in the Eastern Hemisphere it is 160°E, while Wainwright is 160°W.

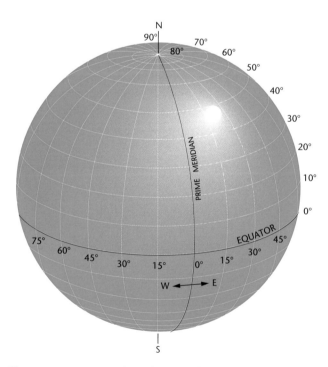

FIGURE 2.4 *Latitude and Longitude*

146   Gazetteer of Canada

A

Abbotsford British Columbia 42 H4 49 02N 122 18W
Aberdeen Saskatchewan 40 B2 52 20N 106 16W
Aberdeen Lake Northwest Territories 45 L3 64 30N 99 00W
Abitibi, Lake Ontario Quebec 35 E2 48 42N 79 45W
Abitibi River Ontario 35 D2 49 40N 81 20W
Abloviak Fiord Quebec 32 C4 59 30N 65 25W
Acadia Alberta 41 A1 50 56N 114 03W
Acton Ontario 38 B2 43 38N 80 04W
Actonvale Quebec 37 O6 45 39N 72 34W
Adair, Cape Northwest Territories 45 R5 71 32N 71 24W
Adams Lake British Columbia 43 E2 51 10N 119 30W
Adams River British Columbia 43 E2 51 23N 119 23W
Adelaide Peninsula Northwest Territories 45 M4 68 15N 97 30W
Adlatok River Newfoundland 30 B4 55 40N 62 50W
Adlavik Islands Newfoundland 30 C3 55 00N 58 40W
Admiralty Inlet Northwest Territories 45 O5 72 30N 86 00W
Admiralty Island Northwest Territories 45 M4 69 25N 101 10W
Advocate Harbour in. Nova Scotia 31 F7 45 20N 64 45W
Agassiz British Columbia 42 H4 49 14N 121 52W
Agassiz Ice Cap Northwest Territories 45 Q7 80 15N 76 00W
Agassiz Provincial Forest Manitoba 39 B1 49 50N 96 20W
Aguasaban River Ontario 35 C2 48 50N 87 00W
Akimiski Island Northwest Territories 45 P1 52 30N 81 00W
Akpotak Island Northwest Territories 32 C5 60 30N 68 00W
Ailsa Craig Ontario 38 A2 43 08N 81 34W
Ainslie, Lake Nova Scotia 31 F7 46 10N 61 10W
Airdrie Alberta 41 C2 51 20N 114 00W
Air Force Island Northwest Territories 45 R4 67 58N 74 05W

FIGURE 2.5   *A Sample from a Gazetteer*

To find the latitude and longitude of a place, you must look for the nearest intersection of these two lines. Then you follow each line to the margin of the map where the co-ordinates are named. This tells you the general latitude and longitude of a place. Sometimes, however, precise measurements are needed. Each degree of latitude or longitude is broken down into sixty **minutes** ('). Minutes are further broken down into **seconds** ("). So a place could be 44°32'10" N and 95°12'54" W. This degree of accuracy is seldom needed for map work in high school.

Often you want to find the location of a place on a map. The **gazetteer**, a special type of index in an atlas, lists urban centres, bodies of water, countries, and other geographic features alphabetically. Beside each listing is the page number of the best map for this feature along with the latitude and longitude. Using this information, it is easy to locate places.

Grids are also used to find places on road maps. Often an algebraic grid is used. Each grid line is given a number or letter. Those running north to south are numbered along the top and bottom margins of the map; those running east to west are lettered in the left and right margins. Thus each square on the map has a number and a letter reference, for example, B6. In the map index, each urban centre is listed alphabetically, along with its co-ordinates. By using the algebraic grid reference, it is easy to find a place on a road map.

The most sophisticated grid system is the military grid, so named because the armed forces developed it to determine precise locations on large-scale topographic maps. Grid reference numbers are listed along the margins of the map. Those along the top and bottom are called **eastings** because they measure from west to east; they name the spaces to the right of each line. Those on the left and right margins are called **northings**, so called because they measure from north to south; the numbers correspond to the space above each line. The four digit numbers found on the grids are called **four-figure grid references**. For greater accuracy, there are **six-figure grid references**. (See Figure 2.6 on page 10.)

Legends

Symbols and colours are used on maps to indicate different features. A legend that indicates the meaning of each symbol is shown somewhere on the map. It is not necessary to memorize these symbols, but some of them are so obvious that we know what they are instinctively. (See Figure 2.7 on page 11.)

Projections

Mapmakers, or **cartographers**, are faced with the challenge of making flat maps from the curved surface of the earth. In the process, distortions occur. You can see for yourself why this is so by carefully peeling an orange so that most of the skin remains intact. Now try to flatten the skin. The only way you can do it is to rip the skin at the top and bottom. But this creates distortion.

To solve the problem, early cartographers developed **map projections**. A light was placed in a glass globe with the continents and grid lines drawn on it. The globe projected the shadows of the lines onto a piece of paper. The lines were then traced on the paper, thereby

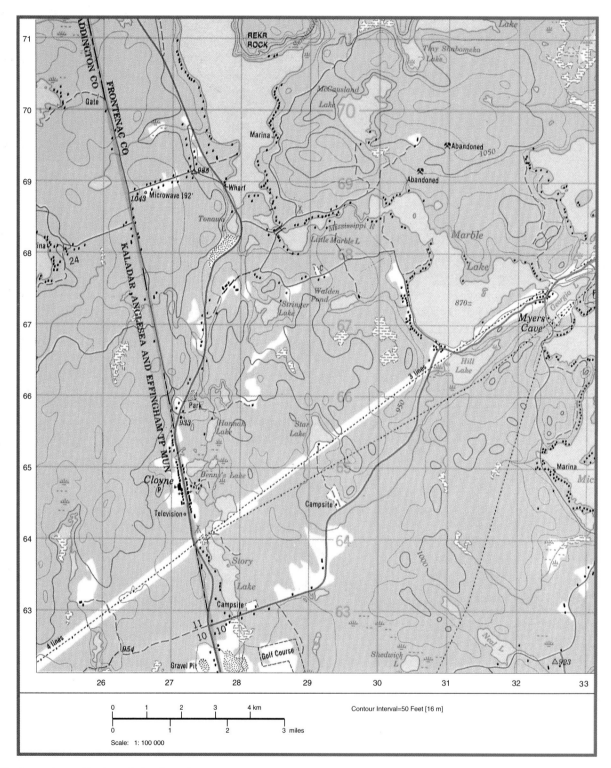

FIGURE 2.6 *Mazinaw Lake, Ontario*
  a) Locate the village of Cloyne in grid square 2764. The "27" refers to the easting, while the "64" refers to the northing. What recreational centre is found in square 2869?
  b) The campsite in square 2964 has a six-digit grid reference of 294645. What is the grid reference for the microwave tower approximately 8 cm north of Cloyne?

creating a flat map. Unfortunately, such a map is accurate only where the globe touches the paper. All other points are increasingly distorted the further away they are from this **standard point**. (See A in Figure 2.9 on page 13.) This **planar projection** is useful, however, especially when you are studying polar regions. Other map projections seldom show these high latitudes accurately.

Cartographers found that the accuracy of map projections increases when more of the paper touches the globe. For equatorial regions, **cylindrical projections** are appropriate because the paper is in contact with the globe along the Equator. As you move away from this **standard line**, the scale constantly changes and the shape becomes increasingly distorted. The **Mercator Projection** is the most famous cylindrical projection. (See Figure 2.1 on page 6.) It was widely used for centuries even though polar regions were monstrously distorted. It shows Greenland as being larger than South America, though it is in fact only one-eighth the size of the southern continent. Likewise, Antarctica appears as an enormous land mass at the bottom of the map. The main reason why the Mercator Projection was so popular despite these distortions was because it showed directions accurately. Navigators could determine what bearings they needed. For this reason this projection is still widely used by airline pilots, sailors, and other navigators.

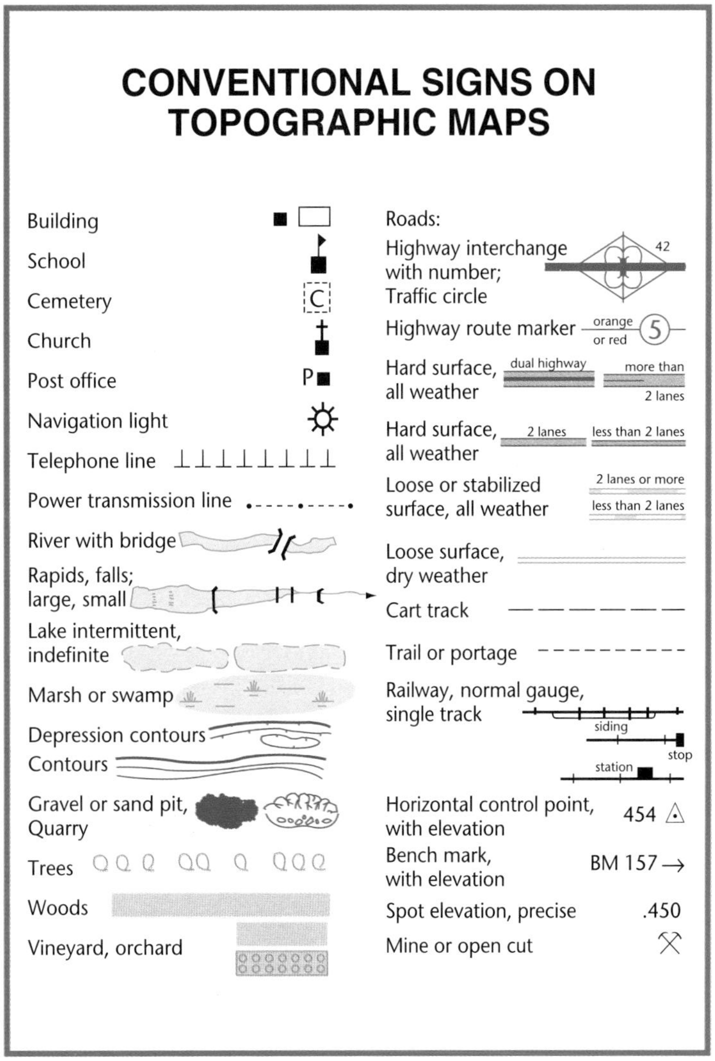

Conical projections are often used to map the middle latitudes. Instead of wrapping the paper in a cylinder around the globe, the paper is made into a cone with the globe touching it along one line of latitude. The distortion continues as you move away from the standard line, but the degree of distortion is less than with cylindrical projections because the distance from the standard line is reduced. Unfortunately, only one hemisphere at a time

CONVENTIONAL SIGNS ON TOPOGRAPHIC MAPS

Building	Roads:
School	Highway interchange with number; Traffic circle — 42
Cemetery	
Church	Highway route marker — orange or red — 5
Post office	Hard surface, all weather — dual highway, more than 2 lanes
Navigation light	Hard surface, all weather — 2 lanes, less than 2 lanes
Telephone line	
Power transmission line	Loose or stabilized surface, all weather — 2 lanes or more, less than 2 lanes
River with bridge	
Rapids, falls; large, small	Loose surface, dry weather
Lake intermittent, indefinite	Cart track
	Trail or portage
Marsh or swamp	Railway, normal gauge, single track — siding, station, stop
Depression contours	
Contours	
Gravel or sand pit, Quarry	Horizontal control point, with elevation — 454 △
Trees	Bench mark, with elevation — BM 157 →
Woods	Spot elevation, precise — .450
Vineyard, orchard	Mine or open cut — ✕

FIGURE 2.7 *Conventional Signs on Topographic Maps Using the topographic maps found in this book, locate as many of these signs as you can.*

can be mapped with this projection. The other half of the world is missing.

Maps today are seldom made using projections. Sophisticated computer programs and mathematical formulas now generate maps with as little distortion as possible. Often two or more different maps are used to create a composite map. The important thing to remember about map projections is that each one has its purpose. It is up to the geographer to choose the best projection for the task at hand.

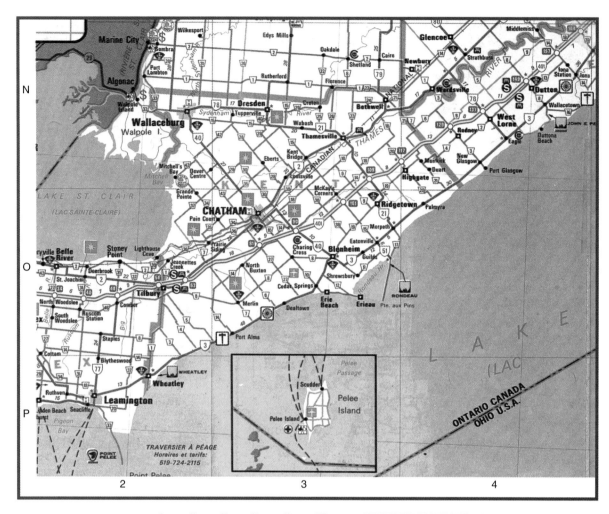

FIGURE 2.8 *A Car Rally*

Complete this car rally using this map or the Official Road Map of Ontario. ➡ *Start in the town of Glencoe (N4). Travel southeast to the first picnic site.* ➡ *Turn right and travel 33 km to the crossroads.* ➡ *Turn so that you cross the railway tracks until you get to the divided highway. You are at Checkpoint 1. Name this interchange.* ➡ *Travel westbound four interchanges and exit left just past the two service centres.* ➡ *Proceed on this road to the police station.* ➡ *You are at Checkpoint 2. What is the name of the recreation area at this site?* ➡ *Turn east and proceed to the county road that passes beside the airport and has access to Highway 401.* ➡ *Turn right and follow this road past the railway tracks to the river. Follow the river until you reach the town named after the river.* ➡ *You are at Checkpoint 3. What is the name of this town?*

Things To Do

1. a) Use a variety of atlas maps to determine the distances between the places in the chart shown here, then complete the chart in your notebooks. (Use the gazetteer to locate the cities.)
   b) Compare your answers with those of other students. Why could your answers be different and yet still be mathematically correct?
   c) Determine the direction from City 1 to City 2 for each pair of cities. Use the compass rose to determine cardinal directions or a protractor to determine bearings.

2. a) Use a globe to determine the shortest route from Edmonton to Moscow. Is it east or west?
   b) Approximately how many degrees apart are the two cities?

City 1	City 2	Scale	Distance
Vancouver, BC	Los Angeles, USA		
Ottawa, ON	New York, USA		
Halifax, NS	London, UK		
Dublin, Ireland	Moscow, Russia		
Paris, France	Cairo, Egypt		
Nairobi, Kenya	Chicago, USA		
Montreal, PQ	Cornwall, ON		

   c) Using the scale on a globe, calculate the distance between each city.
   d) Calculate the distance using the scale on an atlas world map.
   e) Explain why there may be a difference between the two distances.
   f) Is the globe or the atlas map more accurate?

3. a) Explain what a map projection is.
   b) Outline the advantages and disadvantages of the Mercator Projection.
   c) Study Figure 2.9.

Determine which projection would be the most appropriate to use under each of the following circumstances:
- studying polar regions
- studying Canada
- studying tropical regions
- measuring distances
- routing a flight plan
- studying migratory birds
- studying world climates
- determining global patterns

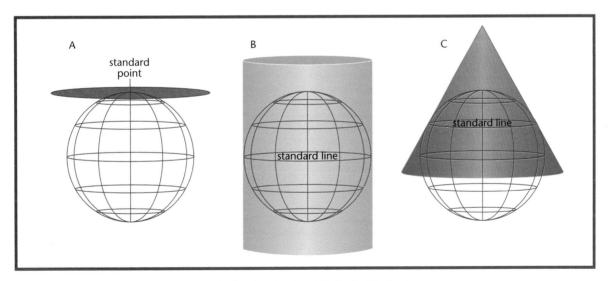

FIGURE 2.9 *(A) Planar Projection; (B) Cylindrical Projection; (C) Conical Projection*

FIELD STUDY

Selecting a Site

Physical geography comes alive when you go out into the field and study nature first hand. Throughout this course you will have a chance to do field studies. Selecting a field study site is important. Choose an area that is easy to get to. Find out who owns or manages the site and obtain permission to conduct your field study there. You might consider a farmer's woodlot, a portion of the school property, a section of land in an outdoor education centre, a site in a local conservation area, an urban ravine, or even your own backyard. No matter which site you choose, however, *make sure you leave the area the way you found it!*

1. Obtain a topographic map of the site. Trace a sketch map of the site, including scale, grid, and prominent features. Include a legend.

2. Take photographs of interesting features at the site. Describe each photo, then mount them with the map on a display board. Use a legend to indicate the location of features.

3. Include a map of the world showing the location of your study area. Indicate the location using latitude and longitude.

PART 2

Energy Systems:
What Makes the World Go Round?

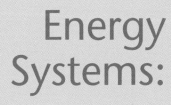

Chapter 3

Energy from the Sun

Introduction

The earth is a dynamic place. It is constantly changing. A raging river washes away sediment. A hurricane rips a path of destruction through a subtropical forest. A glacier retreats unnoticed up a mountainside. Some of these changes occur rapidly. Others are so slow that we don't even notice them. No matter how quick or how slow, or how large or how small, all of these changes require **energy**.

Figure 3.1 *An Aztec Temple*
The Aztecs, like many ancient civilizations, built temples to ensure the sun would keep shining.

The Sun as an Energy Source

It was no accident that ancient peoples revered the sun as a god. The mighty Ra was credited with growing the grain that made Egypt great. The ancient Greeks believed that their sun god, Apollo, protected their flocks and herds. The Aztecs offered human sacrifices every **solstice** to make sure the sun would return to produce the golden corn on which their empire depended. These ancient peoples understood the importance of the sun to their very existence.

The sun, either directly or indirectly, provides most of the energy found on earth. Minimal amounts of energy come from other sources. The tides are an example of kinetic energy that is caused by the pull of gravity from the moon and the sun. Some thermal energy comes from the decay of radioactive elements in the lithosphere. Light from the moon and the stars provides other sources of energy but the amounts are minuscule. Thus if the sun were to suddenly stop shining, the earth would cease to be a living planet.

Solar radiation travels through space at the speed of light. When it reaches the earth's atmosphere, subtle changes begin to occur. Some of this radiant energy is converted to thermal energy as the sunlight heats dust particles, carbon dioxide, and other greenhouse gases in the atmosphere. Some of the light is reflected back into space as radiant energy by clouds and ice crystals in the upper atmosphere. Most of the ultraviolet radiation is reflected back by the **ozone layer**. Most visible sunlight, however, makes it to the earth's surface. Once it strikes the earth, it is converted to different forms. Some of the energy is reflected back into the atmosphere as light. Some is absorbed by the earth's surface as thermal energy. And some is converted to chemical energy by plants.

Solar Energy and Natural Processes

The nature of the earth's surface determines how much solar energy is converted to heat or remains as light. **Albedo** refers to the percentage of solar radiation that is reflected back

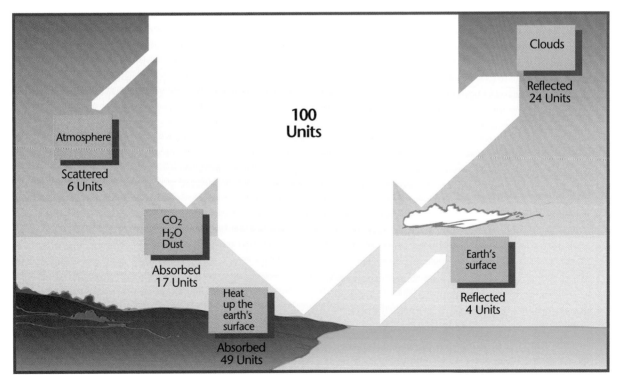

FIGURE 3.2 *Solar Inputs and Outputs*
For every 100 units of solar energy that enters the atmosphere, only 49 per cent actually heats the earth's surface.

Technology Update: Albedo and Design

Colour affects albedo, therefore it is important for designers and engineers to have a knowledge of colour principles. Here are a few examples.

- Black-shingled roofs are common in Canada. The heat absorbed by the dark colours melts snow on sunny winter days. In tropical countries, where excessive heat is a problem, bright, shiny ceramic tiles are often used. Why do you think this is so?

- Some solar heating systems rely on water to transmit heat. The tubes the water runs through are black so that the water will be heated by the sun.

- People in tropical countries often wear light-coloured clothing. Light colours have a high albedo, which helps people to stay cool and comfortable.

- Sand is often scattered on the snowy and icy roads that make winter driving treacherous. The sand not only gives cars more traction, but it lowers the albedo, which helps to melt the snow or ice.

into the atmosphere as light. The solar energy that remains is absorbed by the earth's surface and becomes thermal energy. If the albedo is high, this means that most of the radiation is being reflected as light. Therefore the air over these surfaces tends to be much cooler than the air over surfaces where the albedo is low.

From space, areas with high albedo appear as the brightest parts of the planet during daylight hours. Icecaps, snow fields, and bare ground all have high albedos. Generally speaking, light, shiny, smooth surfaces reflect a high percentage of the solar radiation they receive. Dark, dull, rough surfaces convert more solar radiation into heat and thus have lower albedos.

Balancing Energy Flows

Nighttime temperatures drop drastically on the moon and the sun's other planets, but not on earth. Why? The fact is, little solar energy comes directly to earth as heat. The sun provides us with radiant energy that the earth's surface converts to heat. This heat is then radiated back into the atmosphere. The atmosphere acts like a blanket, trapping the heat close to the earth's surface. It is not surprising that the **daily minimum temperature** usually occurs just before dawn. This is because the earth has been radiating heat without converting any solar radiation for several hours.

There is a balance between the amount of solar radiation that enters the earth's atmosphere and the amount of energy that leaves it. If more energy entered than left, the earth's temperature would increase. This is what many people believe is happening as a result of increased carbon dioxide in the atmosphere. This naturally occurring gas is being produced in great quantities by human activities. Carbon dioxide reradiates thermal energy back to the earth's surface. If the amount increases, more energy will be retained by the atmosphere and the temperature will rise. This so-called **greenhouse effect** could upset climate systems, raise sea levels, and disrupt human activities all over the world.

There could also be catastrophic results if the earth radiated more energy than it received. Temperatures would drop. Icecaps would get larger. The growing season would be shortened, which would mean we might not be able to grow the crops needed to support a global population of over 6 billion people.

Other Forms of Solar Energy

Sunlight and thermal energy are not the only forms of solar energy. Other natural processes are also caused by the sun. Winds and ocean currents, for example, are forms of kinetic energy created from differences in the heating and cooling of the earth's surface. When air or water is heated, it rises, and cooler air or water rushes in to replace it. This movement, called **convection**, is set in motion by solar energy.

Precipitation is another form of solar energy, although it is an indirect one. Thermal energy is required to evaporate water. The energy consumed in this process results in less energy for heating, so the temperature drops. The opposite happens when water vapour condenses. Heat is generated. You may have noticed that the temperature often rises during a snowstorm. This is because the condensation of water vapour around dust particles, which forms the snowflakes, generates heat; it is the same energy that evaporated the water in the first place. In this way, precipitation is an indirect form of solar energy.

The **fossil fuels** we burn to heat our homes and run our machines were created from solar energy. Take natural gas, for example. The gas (called **methane**) was created millions of years ago from decomposing swamp vegetation. While these plants were alive, they converted solar energy into organic matter through photosynthesis. After they decomposed, the energy was stored until it was extracted and released when the fuel was burned. Other fossil fuels, such as oil and coal, were formed in similar ways.

Even the food we eat contains solar energy. It is obvious that vegetables contain solar energy because it is necessary for plant growth. But what about meat? It, too, contains solar energy. Animals feed on grains and other vegetation. They convert the solar energy from these plants into meat, which we then consume. In turn our bodies convert this solar energy into chemical energy through the process of digestion.

Non-solar Energy Sources

The earth does create some energy of its own. Its hot interior, with temperatures up to 2600°C, originated billions of years ago when the earth and sun were first created. However, little of this **geothermal energy** reaches the earth's surface, although massive amounts are released locally when volcanoes erupt. Compared with solar energy, geothermal energy is negligible. All sources of geothermal energy combined amount to less than 0.0001 per cent of the earth's total energy supply.

Things To Do

1. Study Figure 3.2. Prepare an alternative organizer to show the energy balance in the earth's atmosphere. Use a balance sheet, a flow chart, or some other device.

2. Study five or six different surfaces exposed to direct sunlight. Using a thermometer, rank these areas from highest to lowest albedo. Explain what you observed.

3. List all the energy exchanges that have happened to you in the past twenty-four hours. Prepare an organizer like the one shown here to illustrate how each was derived from solar energy.

ENERGY EXCHANGE	FORM OF ENERGY
• got out of bed	• kinetic energy
• had a shower	• thermal energy
• ate breakfast	• chemical energy

4. Develop a mini thesis on the theme of energy consumption. Prepare a thesis statement and a one-page argument to support your position. Present your thesis to your group or class. Suggested topics include:

- Energy: Its Importance in Our Society
- Energy: Resources We Take for Granted
- Energy: What Happens When the Fuel Is Gone?
- Energy: Sustainable Alternatives

Chapter 4

Energy Flows in the Geosphere

Introduction

The earth, or **geosphere**, is made up of four separate spheres: the lithosphere, the atmosphere, the hydrosphere, and the biosphere. Because the earth is incredibly complex, it is easier to study energy flows within these individual spheres rather than study flows for the earth as a whole. Once we understand energy patterns *within* each sphere, we will be better able to understand patterns *among* spheres.

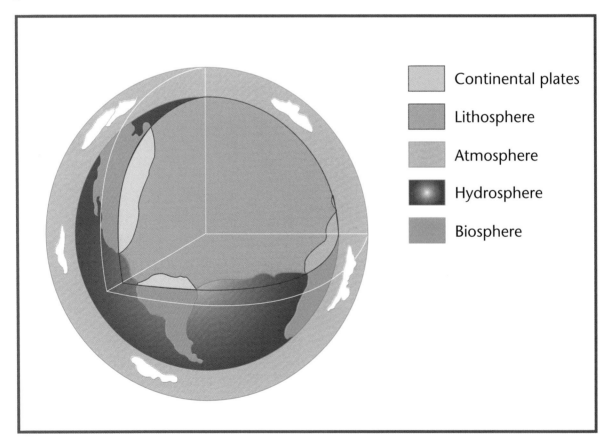

FIGURE 4.1 *The Geosphere*

Energy in the Lithosphere

The lithosphere was the first part of the geosphere to evolve. This is the solid part of the earth. It includes the soil, rocks, mountains, continents, **crust**, **mantle**, **core**, and all the other solid parts of the planet on the surface and beneath it. The lithosphere is ancient, dating back at least 4.6 billion years. To say that it was *created* billions of years ago is incorrect, however. While its origins date back billions of years, the lithosphere continues to change and develop. This process will go on as long as energy flows through the lithosphere. If the energy stops flowing, the earth will no longer be a living planet.

Energy in the Atmosphere

The atmosphere is made up of all the gases that encircle the planet. These are primarily nitrogen, oxygen, and carbon dioxide, but other gases are found in small quantities. When the earth was first formed, the atmosphere was practically non-existent. It started to form from gases given off by volcanic eruptions. Since these ancient beginnings, the atmosphere has evolved over time and continues to change today.

Energy in the Hydrosphere

The hydrosphere comprises all of the earth's water elements. It includes the oceans, seas, lakes, glaciers, rivers, and streams on the earth's surface; groundwater contained in aquifers and in the upper layers of the soil; and water vapour and precipitation in the atmosphere. The hydrosphere was first created from the lithosphere billions of years ago. But like the atmosphere, it is constantly evolving.

Energy in the Biosphere

The biosphere consists of all the planet's living elements. It includes plants, animals, and people. Life began in the hydrosphere (water) but later moved onto the lithosphere (land) as the atmosphere (air) became better suited to sustaining life. The biosphere is the most recent of the spheres to evolve. Although life first appeared on the planet about 4 billion years ago, as with the other spheres the biosphere continues to change today.

Things To Do

1. a) Define each sphere.
   b) What characteristics do all four spheres have in common?
   c) Explain how the sun is central to energy flows among the spheres.

2. What implications does the fact that the geosphere is constantly changing have for people today?

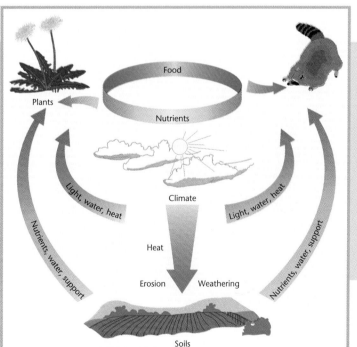

FIGURE 4.2 *Interrelationships in Physical Geography*

CHAPTER 5

Energy Systems in the Lithosphere

INTRODUCTION

Unlike the other spheres, energy flows within the lithosphere do not originate directly from the sun. They are derived from energy sources deep within the earth that have been there since the planet was created 4.6 billion years ago. Because the lithosphere originated when the solar system was created, it is made of the same cosmic material as the sun. Therefore, even the energy deep within the earth is, in a way, solar energy.

FIGURE 5.1 *Volcanoes*
Volcanoes send energy from deep in the lithosphere to the earth's surface.

Nuclear Energy

We know that nuclear energy powers the sun. But did you know that nuclear energy within the planet resulted in the greatest miracle in the universe? That miracle, of course, is life. Our planet is the only place in the universe known to have a biosphere. If it was not for nuclear energy deep within the planet, the conditions for life would never have evolved. To understand the role nuclear energy plays in this process it is necessary to look at the beginnings of the solar system.

The Origins of the Solar System: The Big Bang Theory

While there are still many mysteries surrounding the creation of the solar system, most astronomers accept the **big bang theory**. This states that the solar system began forming 15 to 20 billion years ago following an explosion of galactic proportions. This "big bang" spewed matter in all directions and a huge cloud of dust and gas began to rotate slowly in space. For reasons that are unclear, the cloud began to shrink and become denser. The increased density at the centre was so great that nuclear reactions produced huge amounts of energy. From this energy a new star we call the sun was created. It consumed 90-95 per cent of the original dust cloud.

The remaining material from this cosmic whirlpool formed into a giant disc 10 billion kilometres across. It continued to spin around the sun. Over time, small circular movements called **eddies** began to form within this cloud. As each eddy spun faster and faster, its denser matter clustered together in the centre. Finally, enough of the dense matter in each eddy gathered together to form a **protoplanet.** The lighter matter, gases like helium and hydrogen, spun to the outside. Some of this formed the great gas planets of the outer solar system. Much of the gas eventually disappeared into outer space.

Why Earth Has a Biosphere

Mercury, Venus, Earth, and Mars made up the four inner planets, but our planet was the only one that could support a biosphere. Why? Mercury and Venus were too close to the sun. Extreme heat and poisonous gases made life impossible there. Mars was too far from the sun. The cold temperatures, an absence of water vapour, and a thin atmosphere made the evolution of life unlikely. By contrast, Earth occupied a position that was perfect for the creation of life. Its size and orbit provided

FIGURE 5.2   *The Cooling Planet as It Might Have Looked 3 Billion Years Ago*

enough gravity to hold an atmosphere. Its distance from the sun ensured that temperatures were cool enough for life, but warm enough for water to exist as a liquid and a gas. These essential elements led to the development of life on our planet.

Our early planet was very different from what it is today. At first there was only the lithosphere. There was no atmosphere, no hydrosphere, and no biosphere. As the planet spun in its orbit around the sun, denser material accumulated towards the centre of the earth, while lighter material rose to the surface. The crust of the planet eventually cooled, but **radioactive decay** kept the inner rock hot.

Lighter gases like carbon dioxide and water vapour were trapped in this molten core. They eventually rose to the surface, where they escaped the lithosphere in violent volcanic eruptions. Gravity held these life-giving gases around the planet. As the amount of carbon dioxide and water vapour increased, the temperature of the planet stabilized. Water vapour that condensed from the air formed the early hydrosphere. Atmospheric oxygen was created as the sunlight broke down the water molecules into hydrogen and oxygen. The essentials of a living planet were in place as a result of the nuclear energy at the core of the still-cooling planet.

Conditions were now perfect for the biosphere to develop. Bacteria in the oceans converted carbon dioxide into living matter. Eventually, marine plants and animals evolved from these primitive beginnings. But the land was still too hostile an environment for life. It would be 3 billion years before plants and animals would venture out of the water onto the land.

Potential Energy

For billions of years, bacteria, algae, and later more advanced plants lived and died in the world's shallow primeval seas. While they were living, these plants used photosynthesis to produce oxygen and **organic matter** containing carbon. The oxygen became a part of the atmosphere. When the plants died and decomposed, the carbon-based compounds that were left drifted down to the ocean floor. Eventually some of these carbon remains formed into coal. Others developed into methane, or natural gas, and oil, the liquid form of carbon. These deposits remained trapped within the crust that formed under the primeval seas. Many of these seas have disappeared today. Forces deep within the lithosphere thrust the seabeds high into the air, often well above today's sea level.

It took tremendous amounts of solar radiation over billions of years to create these carbon deposits. All this solar energy was stored in the carbon. The key to releasing this energy was fire. Today these reserves of fossil fuels are extracted from the earth's storehouse. They provide the potential energy needed to fuel the machines of modern society.

Kinetic Energy

Although the ground beneath our feet seems solid enough, the lithosphere is constantly moving. We can observe this movement in such things as rock slides and **earthquakes**. Less obvious are gradual movements that are ongoing every day. **Plate tectonics**, for example, cause the continents to shift slightly each year. But the movement is usually undetectable to people. It occurs at a rate about as fast as your fingernails grow!

Another form of movement in the lithosphere is **isostasy**. Layers of the earth's rock move up and down depending on the mass upon them. This concept is easy to understand if you think of a mattress. When you lie on it, it sinks down. When you get up off the mattress, it rebounds to its original height. The same thing happens to the lithosphere. When there is a large mass like an ocean pressing down, the rock layers are depressed. If the ocean dries up, the pressure is removed and the rock layers rebound. This is one reason why the fossilized remains of ocean creatures can sometimes be found in the mountains.

Isostasy is often the reason why **hydrocarbons** may be located near the earth's surface and far away from any oceans. As we have discovered, the primeval oceans dried up. The layers of fossil fuels they left behind were thrust slowly towards the earth's surface by isostatic forces. Erosion wore away the overlying layers of accumulated rock, leaving these valuable fossil fuels close to the surface.

Chemical Energy

For the past 200 years, people have been using the earth's fossil fuels to produce chemical energy for the great industrial age. The carbon that was removed from the atmosphere by plants billions of years ago is now being returned to the atmosphere when fossil fuels are burned. Burning is a chemical process. Carbon is being combined with oxygen to create a new compound. This process is called **oxidation**. It is represented by the following equation: $C + O_2 = CO_2$.

In a relatively short period of time, billions of years worth of stored solar energy is being released into the atmosphere. This increase in the greenhouse effect could result in higher global temperatures. (For more information on global warming and the greenhouse effect see Chapter 18.)

Things To Do

1. Explain how the atmosphere and the hydrosphere were created from the lithosphere.

2. a) Critically assess the big bang theory. What are its strengths and weaknesses?
   b) Find out about other theories of the earth's origins.

3. a) Look through old geographic magazines and collect pictures that show energy in the lithosphere.
   b) Categorize your pictures by the type of energy they represent.
   c) Present your pictures in a series of four collages under the following headings: Nuclear Energy, Kinetic Energy, Potential Energy, and Chemical Energy.
   d) Select one picture that represents many different forms of energy. Write an explanation of the energy flows that are shown.

Chapter 6

Energy Systems in the Atmosphere

INTRODUCTION

Solar energy must pass through the atmosphere to reach the earth's surface. As it does, tremendous amounts of radiant, thermal, and kinetic energy are exchanged.

There are many complex interrelationships as energy flows through the layers of the atmosphere to the earth's surface.

FIGURE 6.1   *The Atmosphere*
The thin luminescent band along the edge of the earth shows the lower atmosphere.

Radiant Energy

People once thought the earth was the centre of the universe and the sun travelled across the sky to create day and night. Today we know this is not true. Day and night are caused by the rotation of the earth every twenty-four hours. The earth rotates as it travels around the sun. The part of the earth facing the sun experiences day, while the part of the earth facing away from the sun experiences night. It appears that the sun is moving, but it is really the earth.

Night does not come or go suddenly. Twilight occurs at sunrise and sunset each day. During this time there is some light, but it is not as bright as it is when our part of the earth is directly facing the sun. From space, however, this twilight zone is not visible. Instead the line between day and night is as sharp as the edge of a knife. This **circle of illumination** moves across the face of the earth each day.

Sunlight is reflected off dust particles and ice crystals in the atmosphere. These often create beautiful sunrises and sunsets of pink, gold, and orange. In the early 1990s, this illumination was particularly beautiful. There was more dust than usual as a result of the eruption of Mt. Pinatubo in the Philippines in June 1991. The volcano ejected 20 million tonnes of sulphur dioxide into the upper atmosphere. Within a month this gas had encircled the planet!

Earth-Sun Relationships

The earth is tilted on an angle of 23.5° from the vertical. (See Figure 6.3 on page 29.) This results in the lengths of day and night changing with the seasons. The exception is at the Equator, where there is exactly twelve hours of daylight and twelve hours of darkness each day. Moving away from the Equator, the lengths of day and night become more extreme until there are six months of daylight and six months of darkness at the North and South poles.

Twice a year, the entire planet receives twelve hours of daylight and twelve hours of darkness. Called the **equinox**, this occurs around March 20th and September 20th when the sun is directly over the Equator. The earth is neither tilted away from nor towards the sun.

Along the Equator, the equinox is just like any other day. The sun sets daily at 6:00 p.m. This is almost the opposite of conditions in the Canadian Arctic. Here people experience the extremes of daylight and darkness. During winter the nights seem almost endless. But in summer daylight lingers into the early morning hours. Most Canadians, however, live further south. Here the summer days are long. This enables people to spend several hours after supper enjoying outdoor activities. The opposite occurs in winter. It is dark after supper and most people spend their evening leisure time indoors.

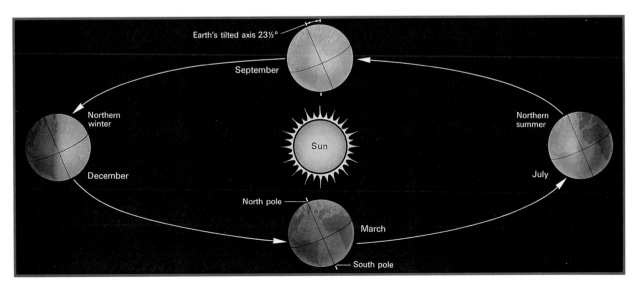

FIGURE 6.2 *The Effect of Revolution and Tilt on Seasons*

Things To Do

1. Using a small globe, an elastic band, and a tape measure, complete the following simulation.
   a) Place the elastic band around the globe so that it crosses over the North and South poles. Align the band so that it crosses the Equator at a 90° angle. One side of the globe represents day while the other side represents night. Use a stick-on label to remind yourself which side is which. Measure the number of centimetres in daylight and the number of centimetres in darkness along the Equator. The two distances should be the same. In other words, they should have a ratio of 1:1—twelve hours of darkness and twelve hours of light. Continue measuring for other latitudes as indicated in the organizer above. Complete the organizer in your notebook.
   b) Move the position of the elastic band to the west (left), so that it lies between the North Pole and the Arctic Circle. Adjust the elastic by sliding it to the east (right) in the Southern Hemisphere so that it is straight. Measure the distances of the dark and light halves of the globe along the Equator. These should always be equal. This simulation shows how the circle of illumination moves across the earth. Repeat the process in part a) in an organizer.
      i) How did the lengths of day and night change in the Northern Hemisphere as you moved the elastic?
      ii) How did the lengths of day and night change in the Southern Hemisphere as you moved the elastic?
   c) Follow the procedure in part b), but this time move the elastic to the east (right) in the Northern Hemisphere and to the west (left) in the Southern Hemisphere. Complete this simulation in an organizer.
      i) How did the lengths of day and night change in the Northern Hemisphere as you moved the elastic?
      ii) How did the lengths of day and night change in the Southern Hemisphere as you moved the elastic?
      iii) What part of Figure 6.2 does this simulation represent?

2. Compare Figures 6.2 and 6.5.
   a) Use these figures to name the earth's position in A.
   b) What is the name of the earth's position in B?
   c) What is the name of the earth's position in C?

Latitude	Equator 0°	Tropic of Cancer 23.5°N	40°N	Arctic Circle 66.5°N	40°S	Antarctic Circle 66.5°S
Distance						
• Day	____ cm	____ cm	____ cm	____ cm	____ cm	____ cm
• Night	____ cm	____ cm	____ cm	____ cm	____ cm	____ cm
Ratio day:night						
Hours						
• Day						
• Night						

Things To Do (continued)

3. Using a copy of Figure 6.3, a ruler, a protractor, a calculator, and a pencil, complete the following exercise.
   a) Refer to Diagram A in Figure 6.3. Draw a line separating the part of the earth in darkness from the part in light. The line should form a 90° angle with the sun's rays shown in the diagram. Label the line the *circle of illumination*. Shade in the side of the earth facing away from the sun.
      i) At what degree of latitude does the circle of illumination intersect with the earth's surface in the Northern Hemisphere? in the Southern Hemisphere?
      ii) What is this parallel of latitude called in each hemisphere?
      iii) How much day and night does the Arctic receive at this time?
      iv) How much day and night does the Antarctic receive at this time?
      v) Using a ruler, measure along the Equator and express, as a ratio, the length of the line in daylight to the length of the line in darkness. How does this translate into day and night?
      vi) Measure along each of the lines of latitude on the diagram. State the ratio of the length of the line that is in daylight to the length of the line that is in darkness. Translate this into the length of day and night.
   b) Follow the procedure in part a) to show the length of day and night in Diagram B of Figure 6.3.
   c) In Diagram C of Figure 6.3, assume that the sun is perpendicular to the paper.
      i) At what latitude are the sun's rays perpendicular (90°) to the earth?
      ii) Why is there nothing to shade in as darkness in this diagram?
      iii) How many hours of sunlight are there at the Equator?
      iv) How many hours of sunlight are there at the poles?
      v) What position of the earth-sun relationship is Diagram C illustrating? (See Figure 6.5.)

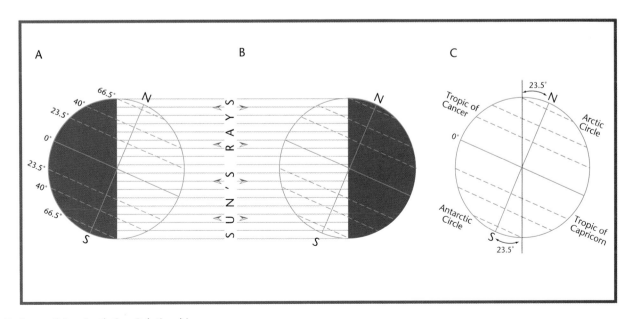

FIGURE 6.3 *Earth-Sun Relationships*

Things To Do

1. a) Explain how tilt and rotation cause day and night.
   b) Explain how tilt and revolution result in variations in the lengths of day and night.

2. Discuss how a change in the lengths of day and night affects people's lifestyles in the tropics, in the middle latitudes, and in the polar regions.

3. In what way are the Arctic and Antarctic circles significant lines of latitude with respect to day and night?

4. Explain the relationship between the earth's rotation and time.

5. In the movie *Superman* the super hero flew around the earth in a clockwise direction faster than the earth was moving in order to go back in time. In theory, do you think this could be possible? Explain your answer.

Thermal Energy

Latitude

Every winter thousands of Canadians travel south. The reason for this migration is obvious. Canada is a northern country. It lies in the high latitudes and has cold winter temperatures. Generally, temperatures decrease moving away from the Equator. Why do you think this is so?

You might consider the fact that the distance between the Equator and the sun is less than the distance between the poles and the sun. It is true that the curvature of the earth results in slight differences in distance between the sun and different parts of the earth. But these differences are too small to have any significant impact on temperature. You might think that the reason for the temperature difference could be because the sun's rays are stronger at the Equator. But this theory, too, is incorrect.

The reason it is colder at the poles than at the Equator has to do with the shape of the planet. The earth's curvature affects the **intensity** of sunlight. At the equinox, the sun is directly overhead at the Equator. This means that the angle of the noonday sun is perpendicular (90°) to the horizon. Moving away from the Equator, the angle decreases steadily until it reaches 0° at the North and South poles. At these latitudes, the sun sits right along the horizon.

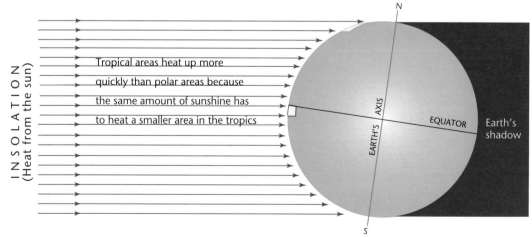

FIGURE 6.4 *Solar Intensity and Latitude*

As the sun's angle decreases, the area of the earth's surface that must be heated by a given amount of solar radiation increases. In other words, a larger area must be heated by the same amount of energy. In addition, the sun's rays have more atmosphere to pass through at high latitudes. Entering the atmosphere at an acute angle, they lose increasingly more energy as they pass through layer after layer of atmosphere. This increased volume of air reduces the amount of radiant energy that reaches the earth's surface. The less radiant energy that reaches the ground per given unit of area, the lower the temperature. Figure 6.4 shows this relationship.

Now that you understand why it is hotter at the Equator, our discussion becomes more complicated! If you were to stand on the Equator and plot the sun's path every day of the year, you would notice the following pattern. Each day the sun rises in the east and sets in the west. But the sun's angle above the horizon changes from day to day. Around March 20th the sun sits directly overhead at the Equator. The angle from the northern horizon to the sun at noon is 90°. Each day this angle decreases until June 20th. On this date the sun appears to stop moving closer to the horizon. This is called the **solstice**, which means *sun stand* in Latin. Of course, if you were to measure the angle from the southern horizon to the sun during this period it would be greater than 90°.

> **GEO-Fact**
>
> Equinox and solstice dates vary each year from the 20th to the 23rd. In the twentieth century, the most common date was the 21st. Astronomers tell us that the 20th will become the usual date for solstices and equinoxes in the twenty-first century.

After June 20th the sun appears to reverse its course. By September 20th, it is once again directly overhead at the Equator. Now the sun starts to dip towards the southern horizon. The angle from the northern horizon is greater than 90°. This continues until the winter solstice around December 20th. The sun then begins its journey north until it is directly overhead again on March 20th. And so this pattern of the seasons continues year after year.

Of course, we know that it is the earth and not the sun that is moving. As our planet revolves around the sun, the Northern Hemisphere is either tilted towards the sun or away from it. The opposite occurs in the Southern Hemisphere. Thus when the Northern Hemisphere is tilted towards the sun, it is summer here in Canada and winter down under in Australia.

The intensity of sunlight at any given latitude changes with the seasons. As the sun drops lower in the sky, the amount of radiant energy is reduced. Combined with the fewer daylight hours, this makes for colder temperatures in winter. Generally, as latitude increases, the severity of winter also increases. In summer, on the other hand, as the sun gets higher in the sky, the intensity of the sunlight and duration of daylight hours increases. Summer returns as the temperatures rise.

Date	March 20	June 20	September 20	December 20
Sun is directly over:	Equator	Tropic of Cancer	Equator	Tropic of Capricorn
Season (N. Hem.):	Spring	Summer	Autumn	Winter
Season (S. Hem.):	Autumn	Winter	Spring	Summer
Name:	Equinox	Solstice	Equinox	Solstice

FIGURE 6.5 *Seasons Organizer*

Place	Latitude	Mean Temp.	Place	Latitude	Mean Temp.
Aklavik, NWT	68°	-9°C	Alert, NWT	82°	-18°C
Atlanta, Georgia	33°	16°C	Baker Lake, NWT	64°	-12°C
Boston, Massachusetts	42°	10°C	Calgary, Alberta	51°	3°C
Chicago, Illinois	42°	10°C	Dallas, Texas	33°	18°C
Honolulu, Hawaii	21°	24°C	Key West, Florida	24°	26°C
Ottawa, Ontario	45°	5°C	Veracruz, Mexico	19°	25°C
Belém, Brazil	1°	27°C	Galapagos Islands	0°	25°C
Santa Cruz, Argentina	50°	9°C	Valparaiso, Chile	33°	14°C
Deception Island, UK	63°	-3°C	McMurdo Sound, Antarctica	77°	-18°C

FIGURE 6.6 *Latitude and Temperature—A Correlational Study*

Things To Do

1. a) Prepare a **scattergraph** to show the relationship between mean annual temperature and latitude using the statistics in Figure 6.6.
   b) Is there a positive or negative correlation between temperature and latitude?
   c) What conclusions can you draw from this scattergraph?
   d) What could account for any anomalies?

2. Speculate on the angle of the sun where you live. Now see how close your guess is by completing this exercise.
   At 12:00 noon on a sunny day, position yourself facing due south. Using a large chalkboard protractor, sight along the base until you see the horizon. Be sure to keep the base horizontal. Have a friend read off the sun's angle as you hold the protractor steady. How close was your guess?

3. How does the fact that temperature increases as you go towards the Equator affect the way people live?

4. Explain how the earth's tilt and revolution cause the seasons.

5. Locate each of the following latitudes on an atlas map. In a group, brainstorm how earth-sun relationships would affect lifestyle at one of the following locations as directed by your teacher. Report your ideas to the class.
   a) 45°S
   b) 20°N
   c) 0°
   d) 80°N
   e) 30°S.

Altitude

Have you ever noticed that there is often ice on top of a mountain even in summer? Why does the temperature drop as you go up a mountain? To somebody with little knowledge of physical geography, this may not make sense. Wouldn't the top of a mountain be warmer than the valley below since it is closer to the sun? In reality, the opposite is true. This is because heat does not come directly from the sun. Solar energy is converted to heat only after it reaches the earth's surface. So if you were to climb a mountain, you would be moving *away* from the source of heat—the earth.

Another reason why temperatures decrease as **altitude** increases is because lower altitudes have greater air pressure. The mass of air pressing down from above compresses the air

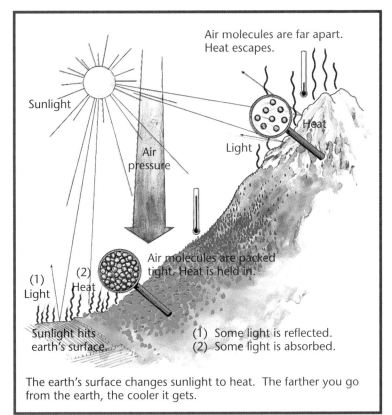

FIGURE 6.7 *Why It Is Cold on Top of a Mountain*

There is a negative correlation between altitude and air temperature. Climatologists have found that temperatures drop about 6.4°C for every 1000 m of altitude. This is called the **environmental lapse rate**. It may vary according to the amount of pollution and water vapour in the air. If there are lots of particles in the air, more sunlight is converted to thermal energy and the air is heated. Regions where there is considerable dust or high humidity often have shallower lapse rates. This is true in coastal British Columbia because of the high humidity of the maritime air.

Exposure

Exposure to direct sunlight also influences temperatures at different altitudes. A slope that receives a lot of direct sunlight will be warmer than a slope that lies in the shade. In Switzerland, for example, southern slopes get more sunlight than northern slopes. As a result, trees grow at higher elevations on the southern slopes. Strong winds also reduce temperatures. A mountain slope that faces the prevailing winds is often much colder than one that is sheltered.

molecules. At low altitudes air has a greater concentration of dust particles and higher humidity. These conditions allow the air in a valley to hold the heat radiating from the earth better than the air high in the mountains.

The environmental lapse rate has a great influence on people living in mountainous areas. In Canada's Western Cordillera, for example, the colder temperatures mean that snow arrives early and leaves late. This makes it possible for ski resorts such as BC's Whistler Mountain to thrive from November to April. Most people don't live at ski resorts year round, however. In British Columbia, much of the population is found along the coastal plains or in the valleys where the climate is warmer. In the tropics, the opposite is often true. The cooler mountain air is more invigorating than the steamy heat of the lowlands. Consequently, many people live at higher elevations—the reverse of the population pattern in the middle latitudes. Compare a relief map of Central America with a population map of the same region. Where do most of the people live?

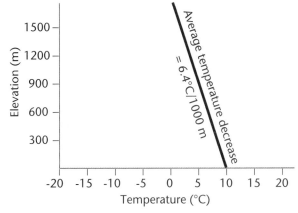

FIGURE 6.8 *Environmental Lapse Rate*
This graph shows the correlation between temperature and elevation.

Elevation and Population of the Ten Largest Tropical South American Cities

	Population (000 000)	Elevation (metres amsl*)
Mexico City, Mexico	13.0	2239
São Paulo, Brazil	12.6	762
Rio de Janeiro, Brazil	5.6	10
Lima, Peru	5.0	152
Bogotá, Colombia	4.0	2610
Caracas, Venezuela	3.0	914
Belo Horizonte, Brazil	2.5	762
Recife, Brazil	2.3	10
Guayaquil, Ecuador	1.5	6
Medellín, Colombia	1.4	1470

Elevation and Population of the Ten Largest Canadian Cities

	Population (000 000)	Elevation (metres amsl*)
Toronto, Ontario	3.0	116
Montreal, Quebec	2.8	57
Vancouver, BC	1.3	14
Ottawa-Hull	0.7	103
Edmonton, Alberta	0.7	676
Calgary, Alberta	0.6	1078
Winnipeg, Manitoba	0.6	240
Quebec City, Quebec	0.6	90
Hamilton, Ontario	0.5	91
St. Catharines, Ontario	0.3	92

*amsl-above mean sea level

FIGURE 6.9 *Elevation and Population—A Comparison*

Kinetic Energy

Wind

Heat always moves from a hot substance to a cold one. This is the main principle in cooking. Thermal energy moves from the stove element into the saucepan and then into the food. The same principle is found in the atmosphere. When there are differences in temperature between two places, air moves the heat from the hot area to the cold one. This is the basis for all winds.

As you probably know, when you heat something it expands. This is why you can open a tight screw cap on a bottle after you run it under hot water. Similarly, heat causes air to expand. The molecules rotate, vibrate, and translate rapidly, causing the space between them to increase. This causes the molecules to spread out and become less dense. When this happens, the air rises. We say the air has **low pressure** because there is less air pressing down. When the air is cold, the opposite occurs. The molecules remain close together, creating much denser air. This is called **high pressure**. Air always moves from an area of high pressure to an area of low pressure. This movement is commonly known as wind.

Things To Do

1. a) Refer to Figure 6.8. Does the environmental lapse rate have a positive or a negative correlation?
   b) Why does the environmental lapse rate occur?
   c) If you were to climb a mountain, describe what would usually happen to each of the following: temperature, air pressure, pollution, humidity.

2. How are altitude and latitude similar in the ways in which they influence solar radiation?

3. Explain how the environmental lapse rate affects people's lifestyles.

4. Study Figure 6.9.
   a) Calculate the mean (average) elevation and mean population for the ten largest South American cities.
   b) Calculate the mean elevation and mean population for the ten largest Canadian cities.
   c) What patterns do you notice in comparing a) and b)? Why do these patterns occur?

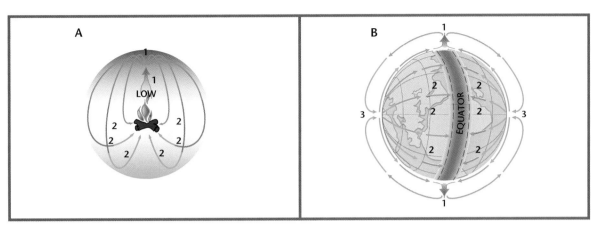

FIGURE 6.10 *How Convection Currents Occur*
(A) When you sit facing a campfire, you feel cold air on your back. Wind blows into the vacuum that occurs when the air over the fire is heated. (B) Where on the earth does the same action occur?

Convection Currents

Winds are a type of kinetic energy. They form when **convection currents** attempt to equalize energy in the atmosphere, as shown in Figure 6.10. Buoyant warm air rises (1). Heavier cool air moves in to take its place (2). At the same time, the rising warm air cools because of the environmental lapse rate. It becomes heavy and sinks to the ground (3). Convection currents are found in all matter that is under pressure, including air, water, and even the lithosphere.

Maritime Air Flows

Whenever two fluids have different temperatures, a **convectional flow** occurs. Consider a coastal area. The air over land is different from the air over water. Water takes longer to heat up and cool down. In the summer, the air over water is often colder than the air over land. This is because it has not yet warmed up from the cold winter temperatures. This cool maritime air typically has a high pressure. Over land the air is warmer and so a low pressure cell develops. The air in the high pressure cell over the ocean is drawn into the low pressure cell. This warmer air rises and the cooler maritime air blows in to take its place. This is why coastal cities often have cooler summer temperatures than cities inland. In the winter this pattern reverses. The winds blow from land out to sea.

Convection currents cause maritime winds to change direction each day and night. During the day, the air over land is often warmer than the air over the sea, so the winds blow onto shore. For tourists on the Carolina beaches of the US Atlantic coast, for example, these are welcome winds. Mosquitoes breed in the inland swamps. The cool ocean breeze keeps these insects inland and people on the beach are not bothered by them. At night, the flow often reverses and the winds blow from the land to the sea. Why do you think this happens?

Katabatic Winds

In polar regions and high mountain elevations, **katabatic winds** often occur at night. These winds result when the air over glaciers becomes colder after the sun goes down. The air pressure increases rapidly. The bitterly cold, dense air flows down from the glacier into the lowlands. It is so dense that it actually sinks through gravity. The warmer valley air is pushed upwards, forming a **temperature inversion**. The low-lying air is cold while the air aloft is warmer. Katabatic winds are common in Antarctica.

> **GEO-Fact**
>
> Balloons illustrate the principles of air pressure well. If you hold a balloon at the neck so that no air can escape, the air inside the balloon has a higher pressure than the air outside. When you release the neck of the balloon, the air moves from the high pressure to the low pressure—that is, from the inside out.

Global Wind Systems

The principles that apply to local winds also apply to global wind systems. English meteorologist George Hadley first theorized in 1735 that convection currents move hot air from the Equator to higher latitudes. When the air over the Equator is heated, it expands and rises, creating a low pressure cell. Cooler air then rushes in to replace the rising air. Once the equatorial air has risen, it cools and starts to fall back to earth. The rising air keeps pushing the cooling air further and further away from the Equator in the upper atmosphere. Eventually this air sinks back to earth in the high latitudes far from where it originated. As it drops, the air pressure increases and the air becomes warmer. In this way equatorial heat is distributed around the planet.

Back in the eighteenth century, Hadley could not prove his theory. It was not until much later, in the middle of the twentieth century, that scientists studying prevailing winds found evidence to support it. There was one difference, however. Instead of a single convection cell in each hemisphere, they discovered three separate cells in the Northern Hemisphere and three in the Southern Hemisphere. Figure 6.11 is a simplified model of these cells.

The Coriolis Effect

The cellular model of prevailing winds is complicated by other factors. One of these is the earth's rotation. This causes winds to be deflected to the right in the Northern Hemisphere and to the left in the Southern Hemisphere. This phenomenon is called the **coriolis effect**. A simple example shows how it operates. Suppose you are playing catch with a friend while you are both on a merry-go-round. Your friend is on the outer edge of the merry-go-round but you are standing at the stationary centre. If you throw the ball to your friend, she will not be able to catch it. It will pass behind her because she is moving. To compensate, you must throw the ball in front of your friend. Viewed from above, the ball seems to curve. But in fact it is travelling in a straight line.

> **GEO-Fact**
>
> Matter exists in three different states: solid, liquid, and gas. Liquids and gases can flow from one place to another. Because of this they are sometimes called *fluids*. Solids are not usually thought of as being able to flow, but they can be fluid if they are compressed under great pressure.

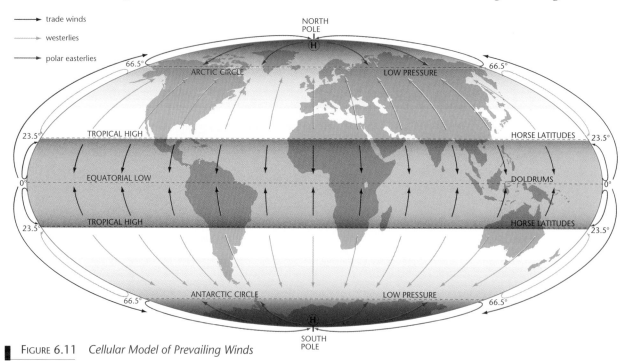

FIGURE 6.11 *Cellular Model of Prevailing Winds*

(A) If the pitcher throws the ball to the catcher at position #1 (pitch a), the ball will go behind him because he has moved to position 2 on the merry-go-round by the time the ball gets to the edge of the revolving platform. Pitch **b** reaches him in time because the pitcher has compensated for the rotation of the platform.

(B) Looking from above in diagram A, pitch **b** seems to go straight, but to the catcher on the ground it seems to curve to the right relative to where he is.

(C) To fly to Montreal from the North Pole, you have to compensate for the rotation of the earth.

FIGURE 6.12   *The Coriolis Effect*

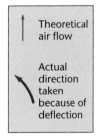

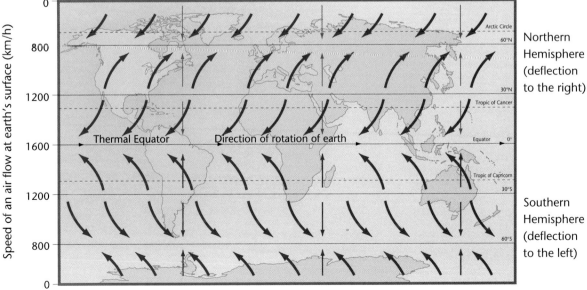

FIGURE 6.13   *The Cellular Model of Prevailing Winds and the Coriolis Effect During the Equinox*

The same thing happens on earth. If you were travelling by plane from the North Pole to Montreal you might end up in Winnipeg if the navigator did not allow for the earth's rotation. Instead of flying directly to Montreal, the pilot must fly to the east of the city in order to arrive there. While this route appears to be curved, it is really a straight line. The earth's surface moved relative to the airplane. Figure 6.12 illustrates the coriolis effect.

Prevailing Winds

Winds that blow towards the Equator do not come *directly* from the north or south. They come from the northeast in the Northern Hemisphere and from the southeast in the Southern Hemisphere. This is because of the coriolis effect. In the days when sails propelled trading ships from one port to another, these winds provided the energy. Hence the name the **trade winds**. In tropical climates these

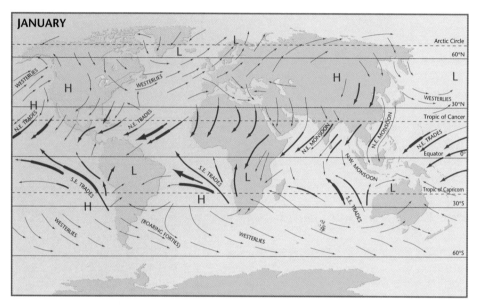

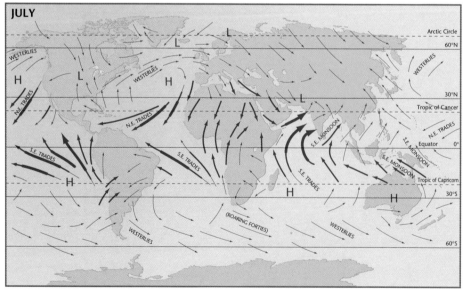

LEGEND

H High pressure cell

L Low pressure cell

Prevailing winds
The heavier the arrow, the more constant the wind direction

FIGURE 6.14 *Prevailing Winds for January and July*

refreshing winds help to relieve the oppressive heat of the endless summer. At the Equator, of course, there is little wind because the air is rising in convectional updrafts. The hot, muggy, still air of these **doldrums** creates extremely uncomfortable conditions.

Winds in the middle latitudes are also deflected by the coriolis effect so that they blow from the west. These are aptly known as the **westerlies**. In polar regions, the air is descending over the poles. The prevailing winds, called the **polar easterlies,** deflect once again to blow from the northeast in the Northern Hemisphere and the southeast in the Southern Hemisphere. Thus we can see how the earth's rotation and the coriolis effect greatly modify the cellular model of prevailing winds.

Seasonal Variations in Prevailing Winds

Global wind patterns are modified by the seasons. The sun is only directly over the Equator twice a year. At other times it is directly over a latitude between the Tropic of Cancer and the Tropic of Capricorn. Geographers refer to the latitude directly under the noonday sun as the **thermal equator.** Its movement affects wind systems. The trade winds, for example, move north or south with the thermal equator. The westerlies and the polar easterlies also march back and forth in response to the apparent movement of the sun.

Things To Do

1. Explain how convection currents cause local winds due to differences in temperature over land and water.

2. In point form, explain prevailing winds. Include the following concepts: convection currents, Hadley Cell, the coriolis effect, trade winds, westerlies, doldrums, polar easterlies, low pressure, high pressure, thermal equator.

3. Figure 6.13 shows the prevailing winds during the equinox. The sun is shining directly down on the Equator. To get an idea of how wind belts move with the thermal equator, try this simulation.
   a) Trace the prevailing winds on a piece of tracing paper or acetate using Figure 6.13 as a guide. Use the Equator, the tropics, and the polar circles as guidelines. Gradually slide the overlay north so that the thermal equator is over the Tropic of Cancer. Now slide the overlay south so that the thermal equator is over the Tropic of Capricorn.
   b) List latitudes where the winds reverse from season to season as you move the tracing paper. What effect could this have on the local climate?

4. a) Place the overlay you made in activity 3 over Figure 6.14 at either January or July. Be careful to centre the thermal equator directly over the Tropic of Capricorn for January or the Tropic of Cancer for July.
   b) Find at least three places where the prevailing winds differ between the overlay and Figure 6.14. Mark the locations on the overlay.
   c) The winds shown on the overlay are theoretical. No allowance has been made for differences in the earth's surface. The winds shown in Figure 6.14 are what actually happen. Surface variations affect how the winds blow. Explain why the winds are blowing differently in the places you identified in b).

Chapter 7

Energy Systems in the Hydrosphere

INTRODUCTION

Water is found virtually everywhere on our planet. Yet it is one of the rarest minerals in the universe. Without water the earth as we know it would not exist.

Over 70 per cent of the earth is covered by sea water. In addition, water is found in lakes, streams, ponds, and other surface features. In its solid state, water forms snow, icecaps, and glaciers. In the lithosphere, it lies trapped between layers of impermeable rock in aquifers. It is also in the atmosphere as water vapour.

Water is unique among the earth's minerals in that it exists in three forms: as a liquid, a solid, and a vapour. In its liquid form, water is **translucent**, stores solar energy, and flows readily from one place to another. Each of these properties plays an important part in the earth's energy systems.

FIGURE 7.2 *Water: A Powerful Source of Energy*

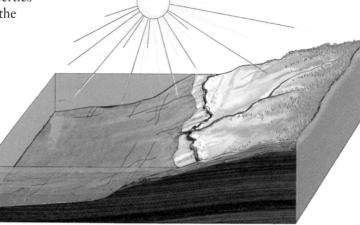

FIGURE 7.1 *The Moderating Influence of Large Bodies of Water*

A Storehouse of Solar Energy

As we discovered in the last chapter, water takes longer to heat up and cool down than land. This is because it is translucent, and so it absorbs sunlight to greater depths than land. Therefore, it takes more solar energy to heat sea water than to heat land. In addition, convection currents similar to those in the atmosphere are present in large bodies of water. These currents mix the warmer surface water with colder water from below. This turbulence further slows down the heating process. The ocean is therefore a tremendous storehouse of thermal energy.

Have you ever wondered why Vancouver is so much warmer than Winnipeg in the winter? The two cities are approximately the same distance from the Equator and have similar elevations. The reason Vancouver is warmer is because of the moderating influence of the Pacific Ocean. All summer long the coastal waters of British Columbia store solar energy. When winter arrives, this heat is returned to the atmosphere. Winnipeg, however, is situated in the middle of North America where there are no large bodies of water to moderate temperatures.

Of course, the opposite is also true. Summer temperatures on the coast are usually cooler than they are inland. The ocean is relatively cooler than the land in summer. It takes longer to heat up, resulting in cool coastal temperatures. This is one reason why many people like to spend their summer vacations on the coast. The cooler temperatures and coastal breezes make the summertime heat a little more bearable.

		Mean Temperature (°C)		Latitude
		January	July	
1	Alice Springs, Australia	28	12	23°S
	Noumea, New Caledonia	26	20	22°S
2	Bordeaux, France	6	21	44°N
	Bucuresti, Romania	-3	23	44°N
3	Kraków, Poland	-3	19	50°N
	Isles of Scilly, England	8	16	50°N
4	Cabo de Finisterre, Spain	10	18	43°N
	Toronto, Canada	-5	19	44°N
5	Reykjavik, Iceland	0	12	64°N
	Baker Lake, Canada	-32	10	64°N
6	St. Louis, United States	0	26	39°N
	San Francisco, United States	10	15	39°N

FIGURE 7.3 *Temperature Statistics for Selected Pairs of Centres*

Things To Do

1. Study the climate statistics for each pair of locations in Figure 7.3. Each pair is on approximately the same parallel of latitude.
   a) Subtract the January and July temperatures to find the annual temperature range for each location.
   b) Speculate which location is coastal and which is inland. Check your answers in an atlas.
   c) What generalization can you make about temperature range and coastal locations?

2. Explain why water heats and cools more slowly than land.

3. Explain how the moderating influence of large bodies of water could influence each of the following: a) leisure activities; b) agriculture; c) transportation; d) housing.

Kinetic Energy

The ocean is a vibrant, powerful force. The water seems to be in perpetual motion as breakers crash on the shore and froth spews into the air. This kinetic energy is released in four ways: through convection currents, waves, ocean currents, and tides. All but one of these—tides—originates with the radiant energy of the sun.

Convection Currents

Convection currents operate in the oceans as well as in the atmosphere. Warm water is lighter than cold water, so you would expect that the warmer water would stay on the surface. In sea water, however, there is vertical movement. The reason for this is salt. You may have noticed that you float better in salt water than in fresh water. This is because salt water is denser than fresh water, so your body is more buoyant. The greater the salinity of the water, the denser and heavier it is. Thus sea water that contains a lot of salt tends to sink, even if it is warmer than the less salty water beneath it. The amount of salinity varies with depth. Surface water is usually saltier than water at lower depths. This is because surface water is exposed to the air; evaporation then reduces the amount of water relative to the salt content. This creates warmer, saltier water at the surface.

Rates of evaporation and precipitation are not uniform across the earth's surface. Desert climates exist even over the oceans on either side of the Tropics of Capricorn and Cancer. (See pages 121-124 for an explanation of desert locations.) Equatorial oceans receive heavy precipitation. Consequently, ocean waters along the Equator

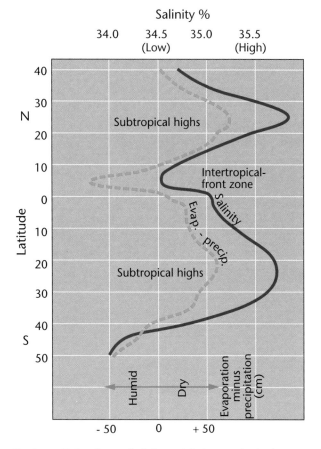

FIGURE 7.4 *Ocean Salinity as it Relates to Latitude*

and middle latitudes where rainfall is heavier are less salty than waters at the Tropics of Capricorn and Cancer. The salt is diluted by all the extra water. This less salty water flows towards the Tropics of Cancer and Capricorn, creating a convectional flow. (See Figure 7.4.)

Waves

People frequently confuse waves with ocean currents. Although you may think that waves move water forward, this is not actually the case. Waves move water around and around in a circle as crests and troughs flow across the water's surface. Energy moves through the water without actually carrying the water anywhere. Currents, on the other hand, actually move the water from one place to another.

Waves are created by winds. If you gently blow over a basin of water, little ripples will form on the water's surface. If you blow harder, the ripples will become bigger. Waves form in much the same way. Thus wave size depends on the speed of the wind, how long the wind blows in one direction, and the distance it blows across the ocean. Because oceans are larger than lakes, waves are usually bigger in oceans.

Ocean Currents

Currents are like rivers running through the ocean. Water actually flows from one place to another. Four dynamics are at work in creating ocean currents: prevailing winds, convection currents, the coriolis effect, and the shape of the ocean basin. If you compare a map of ocean currents with one of prevailing winds, you'll notice that the two maps are similar. Where the trade winds blow from the east, the equatorial current also flows from the east. Similarly, the North Atlantic Drift coincides with the Westerlies. The constant drag of the prevailing winds on the surface of the ocean causes the currents to flow in a similar path. This redistributes the energy generated by the heating and cooling of the atmosphere.

The convection currents that result from differences in water temperature and salinity accentuate the currents created by the prevail-

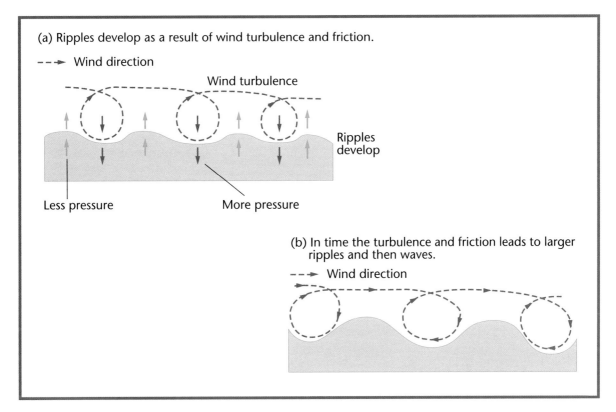

FIGURE 7.5 *Wave Formation*

ing winds. Frigid polar waters sink as the hot salty waters from the tropics flow towards the poles to replace them. Thus the heat of the tropics is distributed to colder waters, while the polar waters flow towards the Equator at great depths. It has been estimated that it takes as long as a thousand years for polar waters to re-emerge on the ocean's surface in the tropics.

As with winds, the coriolis effect constantly acts upon ocean currents. Currents are deflected to the right in the Northern Hemisphere and to the left in the Southern Hemisphere. This factor, combined with the shape of the North Atlantic Basin, results in the Gulf Stream. This incredible ocean current carries warm tropical water from the Caribbean Sea across the North Atlantic to the southern shores of England. The Isles of Scilly off England's southwest coast are noted for their tropical palm trees and Mediterranean-like plants. Yet these islands lie at the same latitude as northern Ontario, a place that is far from tropical!

The shape of the Atlantic Basin directs the North Atlantic Drift, an extension of the Gulf Stream. Originating in warm tropical seas, the Gulf Stream is directed northeast by the North American coastline. Frigid arctic currents deflect the warm current further east until it reaches the northeast coast of Europe. Other ocean currents are similarly affected by the shape of ocean basins.

When ocean currents move towards the poles they are warm currents. Like the Gulf Stream, they bring thermal energy from the tropics to colder waters. Currents that flow from the poles to the Equator are colder than the surrounding waters. They are usually found on the west coasts of continents. The Galapagos Islands, off the coast of Peru near the Equator, lie in the cold Peru Current. Even though these islands have many tropical plants and animals, the waters are so cold that penguins and seals migrate there from Antarctica to spend part of the year.

Tides

Tides are different from other movements of ocean water. The daily rise and fall of the oceans is not caused by solar energy. It is the gravitational pull of the moon that causes tides. The side of the earth that is in line with the moon has high tide, while the side that is out of the line of pull has low tide. Tides are an interesting phenomenon. But they have nothing to do with the global distribution of solar energy and so are not dealt with in any detail here.

Things To Do

1. Prepare an organizer to compare convection currents, waves, and ocean currents. Include the following criteria: characteristics, causes, influence on human geography, and energy distribution.

2. Explain how salinity affects oceanic convection currents.

3. Study Figure 7.6. What generalizations can you make about: a) equatorial currents, b) mid-latitude currents, c) polar currents, d) Southern Hemisphere currents, e) Northern Hemisphere currents, f) west coast currents, and g) east coast currents?

4. Compare the Ocean Currents map (Figure 7.6) with the Prevailing Winds map (Figure 6.15). How are the patterns on these maps similar? Where do differences exist? What could explain these differences?

5. Explain how kinetic energy in the oceans redistributes heat energy from the tropics to the polar regions.

6. Explain how kinetic energy in the oceans affects the way people live. Include such topics as leisure, fishing, shipping, agriculture, clothing, and lifestyle.

7. Study the technology update "Measuring the Seas" on page 46. Outline how methods for studying the oceans have changed.

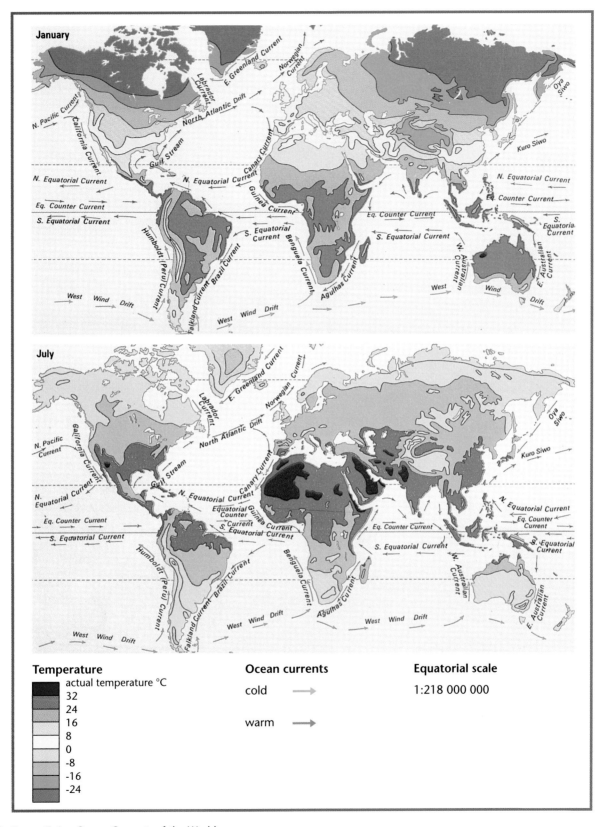

FIGURE 7.6 *Ocean Currents of the World*

TECHNOLOGY UPDATE: MEASURING THE SEAS

The importance of sea water as a modifying influence on climate and the ability of these waters to hold vast amounts of carbon dioxide have made the study of the seas a priority for climatologists studying climate change. Until recently, oceanographers used primitive methods to gather data about the sea. The direction and speed of ocean currents, for example, were determined by drift bottles. These were sealed bottles that were let afloat from various locations. They contained notes asking those who found the bottles to mail the notes back to the scientists, with the locations where they found the bottles recorded on them!

Today new technology enables oceanographers to study the seas much more accurately. Data are collected by sophisticated sensors adrift in the oceans. These sensors are designed to float at different depths to study oceans three dimensionally. The sensors transmit radio messages to satellites, which then relay the data back to earth. This enables scientists to provide the kind of detailed analysis that was impossible in the past.

One study currently under way is the World Ocean Circulation Experiment. The main purpose of the study is to measure **mesoscale eddies**. These are ocean currents similar to storms that occur in the atmosphere. It is believed that these eddies determine, to a large extent, how ocean currents operate. By studying the eddies, scientists hope to gain a better understanding of how ocean currents distribute heat from the Equator to higher latitudes. It is speculated that anywhere from 25 to 75 per cent of thermal energy transfer from the tropics to the poles is caused by ocean currents.

A joint France-US study is using satellite technology to determine exact sea level. This will enable scientists to determine whether sea levels are rising as a result of global warming. Ocean temperatures are also being accurately measured for the first time. Sound travels through water at measurably different speeds, depending on water temperature. In 1993, a worldwide network of broadcasting stations began sending sound waves through the oceans. Receiving stations determine the speed they travel and thereby establish the water temperature. Over time, this process will enable scientists to determine if the seas are warming up or cooling down.

The Orbital Sciences Corporation, based in Virginia, is launching another satellite to study how the colour of the oceans changes over time and from place to place. Colour is determined in part by plankton, the microorganisms that are essential to marine food chains and the production of atmospheric oxygen. By studying the distribution of plankton, scientists can determine if increases in ultraviolet radiation are affecting plankton growth.

All this scientific activity should tell us more about the oceans than we have ever learned in the past. Super computers will analyse the data and build a better understanding of how the oceans work, how they affect global climates, and how important they are to our daily lives.

Adapted from "Drift-nets for Data," *The Economist*, 16 May 1991.

Thermal Energy

The Hydrologic Cycle

The oceans are not the only part of the hydrosphere in which energy flows are found. Water is also found in the atmosphere and on land masses.

The hydrologic cycle is an important part of global energy systems. Figure 7.7 illustrates how water moves through nature. Although this flow chart shows the hydrologic cycle as a simple circular pattern, in reality water movements are much more complex.

Water is constantly evaporating. This is true whether it is salt water in the ocean, fresh water in a mountain stream, or melt water on a glacier. In all cases, however, only surface water exposed to the air evaporates.

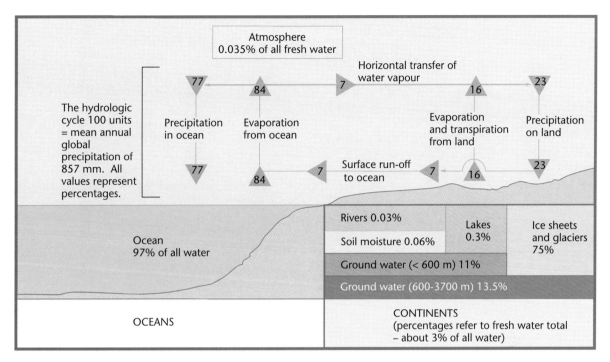

FIGURE 7.7 *The Flows and Storage of the Hydrologic Cycle*

The rate of evaporation depends on a variety of factors. One of these is air temperature. Evaporation rates are much lower in frigid climates like the high arctic than in climates with warmer temperatures. In arctic climates water is locked in ice for much of the year. Moreover, cold temperatures reduce the movement of water molecules. Since the molecules are relatively still, they do not escape into the air.

Another factor that affects evaporation is the amount of water vapour in the air. Evaporation rates in humid climates are lower than those in desert regions because humid air has little room to hold additional water vapour. Winds also affect evaporation. Dry winds increase the rate of evaporation because they have more room to hold water vapour than humid tropical winds. In Africa, for example, dry desert winds called **harmattans** often blow from the interior into coastal regions south of the Sahara Desert. These winds bring welcome relief from the humid hot air that usually dominates equatorial Africa. Similar local winds occur in other regions where heat and humidity are high.

Once water becomes vapour it is pure. Any impurities, including pollution, are left behind in the original water source. In desert regions, high rates of evaporation draw water to the surface from deep within the soil. This water carries with it many of the soil's minerals. When the water evaporates, the minerals remain on the ground. These form the **saltpans** and **alkaline flats** that are common in desert regions.

Humidity

The moisture content of air is called **humidity**. The amount of moisture the air can hold varies. Warm air is able to hold more water vapour than cold air. When a warm air mass containing a lot of moisture is cooled, the amount of water vapour it can hold is reduced. The excess water vapour changes back to a liquid through the process of **condensation**. The temperature at which condensation occurs is called the **dew point**. Dew is the moisture you often see on a cool summer night following a warm, humid day.

Temperature rises as air pressure increases. When air is compressed, the molecules are forced together. The temperature goes up because the energy is concentrated in a smaller space. The opposite happens when air pressure

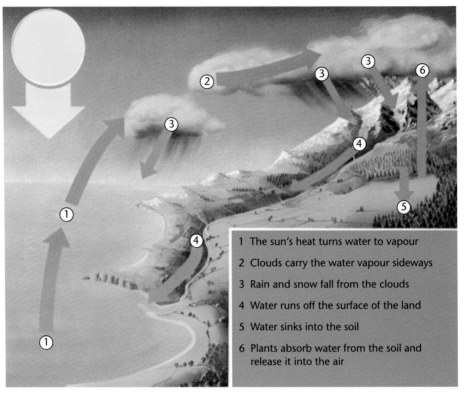

FIGURE 7.8 *The Water Cycle*

is lowered. The air is spread out, so the temperature decreases. In either case, the amount of potential energy remains the same. It is just distributed differently. The rate at which the temperature changes, called the **dry adiabatic lapse rate**, is about 10°C for every 1000 m.

When air rises, it cools and its ability to hold water vapour decreases. Once moist air has reached the dew point, clouds occur. The dry adiabatic lapse rate has reduced the temperature of the air so much that it can no longer hold as much water vapour. When this happens, condensation takes place. But the water vapour has to have a surface on which to condense. Just as the water vapour created when you shower condenses on your bathroom mirror, the water vapour in the air condenses on dust particles to form clouds. Now the pure water vapour may become polluted again, depending on the nature of the dust particles.

In these clouds the air continues to rise, still cooling all the while. But the rate of cooling is lower than the dry adiabatic lapse rate. The energy that was used to evaporate the water in the first place has now returned to the air as the water vapour becomes liquid once again. So the air does not cool down as fast above the dew point. This rate of cooling is called the **wet adiabatic lapse rate**. It varies between 3°C and 6°C for every 1000 m the moist air rises in the clouds.

As more and more water vapour condenses on the dust particles, water droplets are formed. These particles become heavier and eventually join together to form precipitation. Depending on the air temperature, this precipitation can take many forms—rain, sleet, snow, hail, and anything in between. Once the precipitation reaches the earth, it eventually flows into bodies of water and the water cycle continues.

Precipitation

Precipitation occurs when moist air in the atmosphere rises above the dew point. This happens in one of three ways: through **convection precipitation, orographic precipitation,** and **frontal precipitation**. Although these are three distinct processes, the dynamics in each are the same. As air rises, its ability to hold water decreases. The dew point is reached, and condensation forms on dust particles. Precipitation then starts to fall. Let's consider how these dynamics operate in each of the three precipitation types.

Convection Precipitation

Convection currents in the atmosphere carry water vapour aloft in a steady vertical flow of air. Because these currents occur only when there is hot air, convectional precipitation is often experienced during the Canadian summer when hot, humid air develops. The sun heats the ground all day until, by late afternoon, the air pressure has dropped and convection currents are in operation. Strong gusts often accompany the storm as the cooler air rushes in to replace the air that has risen. Friction between air molecules rising in the current often creates static electricity, which manifests itself as lightning and thunder. Thunderstorms can be violent. They are one way in which nature discharges huge concentrations of thermal energy.

An interesting weather phenomenon that often accompanies thunderstorms is hail. Have you ever wondered why these balls of ice fall from the sky even though the temperature may be quite warm? Here's what happens. Water vapour condenses on dust particles but the raindrops do not fall. Convection currents keep pushing them higher and higher in the atmosphere until they reach heights where the temperature is below freezing. The raindrops freeze and become heavy. They start falling below the dew point. More water vapour condenses on the ice balls as convection currents carry them aloft once again. This up and down movement gradually increases the size of these frozen raindrops until eventually they are too heavy to be supported by the rising column of air. At this point the hailstones fall to the ground. Smaller stones often melt on the way

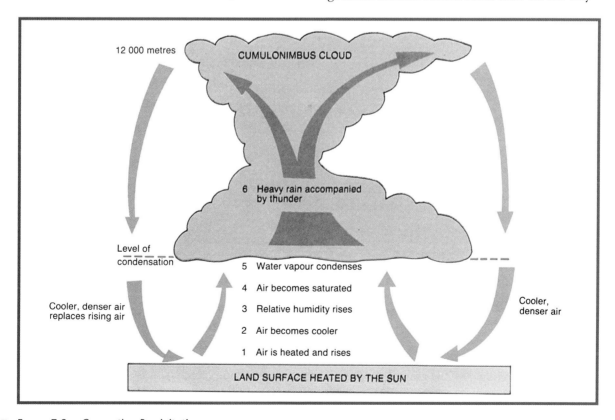

FIGURE 7.9 *Convection Precipitation*

down, creating huge drops of rain. Others are so big that they remain frozen. You may have heard the expression "hail the size of golf balls." These giant hailstones can cause considerable damage.

Convectional rain does not just occur in Canada. It happens all over the world wherever heat and convection currents form. In equatorial regions, thunderstorms are common most afternoons as the rain reduces the energy built up during the hot daylight hours.

Orographic Precipitation

Orographic or relief precipitation is similar to convection precipitation in that air is forced to rise above the dew point. In this case, however, it is not hot air that causes this movement. Instead it is a change in surface elevation. Orographic precipitation is often found in mountainous regions, such as the west coast of North America. Here the Westerlies carry warm, humid air formed over the Pacific Ocean inland over the coastal mountains. Once this air reaches the mountain barrier, it has nowhere to go but up and over the mountains. Eventually the air is so high that the dew point is reached and rain or snow starts to fall.

As the air continues up the mountains more and more moisture is released. By the time the air reaches the highest point, it has lost much of the water vapour it had when it was over the ocean. The air now descends the eastern slopes of the mountain range. Air molecules begin to spread apart. The relative humidity drops even more as the air grows warmer because of the dry adiabatic lapse rate. Between mountain ranges near-desert conditions can prevail because of these dry winds. Called the **rainshadow**, these areas are found between the mountain ranges of the Western Cordillera as well as in many other mountainous parts of the world.

Dry adiabatic winds can cause freakish weather conditions on the leeward side of

> **GEO-Fact**
>
> January 27, 1962, was a day they'll never forget in Pincher Creek, Alberta. At midnight the temperature was -19°C and the relative humidity was 83 per cent. Within one hour the temperature rose 22°C, the humidity dropped to 56 per cent, and the snow simply disappeared! Can you imagine the amount of energy it took to cause this sudden weather change?

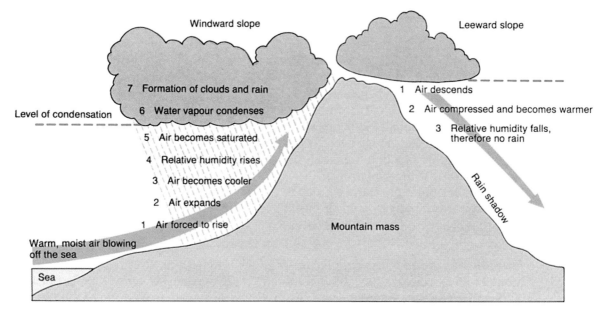

FIGURE 7.10 *Orographic Precipitation*

mountain ranges. Around Calgary, Albertans enjoy a winter wind called the **chinook**. This is a warm, dry wind that occasionally descends the eastern slope of the Rockies. It causes temperatures to rise as much as 20°C in a matter of minutes! Snow virtually disappears. This is a welcome relief from the frigid air that usually dominates the Alberta winter.

Temperature (°C)	Water Vapour (g/m^3)
-15	1.4
-10	1.9
-5	3.0
0	4.9
5	6.9
10	9.4
15	12.6
20	17.3
25	23.0
30	30.4
35	39.8

FIGURE 7.11 *Maximum Capacity for Holding Water Vapour at Given Temperatures*

Frontal Precipitation

Frontal precipitation occurs when two air masses with different characteristics meet. Warm, moist air is forced to rise when it comes in contact with colder, heavier air. Moisture in the lighter air mass condenses when the temperature drops below the dew point. This happens because the elevation of the lighter air mass has increased. Frontal precipitation is much like orographic precipitation. Instead of landforms forcing the air to rise, it is a denser air mass that pushes the air higher until the dew point is reached.

In the Canadian prairies, cyclonic storms march across the region throughout most of the year. These storms occur when a cold, dry, arctic air mass comes in contact with a warm, moist air mass from the Gulf of Mexico or the Atlantic Ocean. The cold air is heavier and forces the less stable warm air to rise. As it rises, the dry adiabatic lapse rate causes the air to cool until the dew point is reached and precipitation starts to fall. Frontal precipitation can drag on for days as a front slowly moves out of a region. Dull, wet weather or cold winter storms resulting from frontal systems are common in many mid-latitude locations, including much of southern Canada, Europe, and Russia.

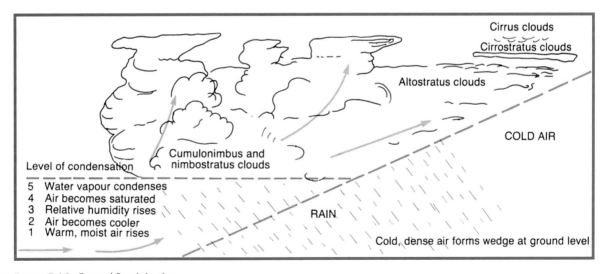

FIGURE 7.12 *Frontal Precipitation*

Things To Do

1. Use the information in Figure 7.7 to prepare circle graphs to show the distribution of continental water in rivers, lakes, glaciers, etc.

2. a) Prepare an organizer listing all the ways in which your family uses water over a one-week period. Devise a way to determine the quantity of water consumed for each use. Include this in your organizer.
   b) Suggest ways in which your family could reduce water consumption without reducing your standard of living.
   c) Provide creative solutions to solving the municipal water shortage for a region or city of your choice. Present your ideas to your class or group.

3. According to Figure 7.7, the world's total water supply is 1.356×10^9 km³. Use this figure to calculate the value of each percentage in the diagram. For example, water in oceans is $97\% \times 1.356 \times 10^9 = 1.315 \times 10^9$.

4. a) In your notebook, complete an organizer similar to the one at the top of this page comparing different winds of the world.
   b) From the winds listed in the organizer, create two groups based on similarities in their formation.
   c) Explain how these winds can affect the daily lives of people in these areas.

NAME	LOCATION	DESCRIPTION	EXPLANATION
Harmattan			
Chinook			
Sirocco			
Bise			
Föhn			
Simoom			

5. a) Prepare a scattergraph of the information in Figure 7.11.
   b) What correlation do you notice? Why does this relationship exist?

6. The relative humidity measures the percentage of water vapour in the air compared to the absolute humidity (the maximum capacity for holding water vapour at a given temperature). For example, assume that the amount of moisture in the air was 25 g/m³ and the air temperature was 35°C. Figure 7.11 shows that the water vapour holding capacity of air at 35°C is 39.8 g/m³. Therefore the relative humidity would be

$$\frac{25}{39.8} \times 100 = 62.8\%.$$

   a) Use the graph you created in activity 5 or Figure 7.11 to calculate the relative humidity for each situation B through H in the chart at the top of page 53.
   b) Which of these would be experiencing precipitation? Explain.
   c) Use the scattergraph in activity 5 to read the temperatures at which each dew point would be reached. For example, in A, 25 g/m³ is reached at about 27°C.
   d) Use a dry adiabatic lapse rate of 1°C per 100 m to determine how much each body of air would have to rise before humidity reached 100 per cent (the dew point). For example, in A, the temperature has to drop 8°C. Therefore the air would have to rise 800 m before the humidity reached 100 per cent and the dew point was reached.
   e) Take one example from the organizer in part a). How much would the humidity drop if the air flowed 2000 m down the side of a mountain?

7. Explain the differences in the environmental lapse rate, the wet adiabatic lapse rate, and the dry adiabatic lapse rate. Include criteria such as the rate of cooling, why it happens, where it happens, and its effect on water vapour.

Things To Do (continued)

	A	B	C	D	E	F	G	H
Temperature	35°C	25°C	10°C	5°C	2°C	0°C	-5°C	-10°C
Absolute humidity	25 g/m³	10 g/m³	7 g/m³	3 g/m³	4 g/m³	2 g/m³	4 g/m³	2 g/m³
Relative humidity	62.8%							
Dew point temperature	27°C							
Altitude to reach dew point	800 m							

8 Prepare an organizer comparing convection, orographic, and frontal precipitation.

9 Study the photographs in Figure 7.13. Decide which type of precipitation each photo is illustrating. Be prepared to explain your answers.

FIGURE 7.13 *Types of Precipitation*

Potential Energy

The hydrosphere is set in motion by energy from the sun. Waves, ocean currents, rain, and evaporation all result from the heat energy formed as the earth converts sunlight to thermal energy. Another form of kinetic energy is the energy of rivers and streams. As water flows over the continents and back to the oceans, tremendous energy is generated.

Water under the pull of gravity always flows downhill until eventually it reaches the ocean. When the route to the ocean is cut off by high land, the water is detained temporarily in ponds, lakes, inland seas, or underground aquifers. The water stored in these natural reservoirs has tremendous potential energy. It is only when the water starts flowing again, however, that kinetic energy is produced.

Any water above sea level has potential energy as it flows towards the sea. In Canada, Ontario and Quebec have a lot of opportunities to utilize this energy. Many hydroelectric plants use running water to generate electricity. This is a clean and natural way to provide energy for the large urban population of the Windsor-Quebec City Corridor.

Things To Do

1. Study Figure 7.14.
   a) Write a point-form summary or prepare a sketch map outlining the route taken by water on its way from Quetico Provincial Park to the Atlantic Ocean.
   b) Prepare a flow chart showing the route in a). Include cycles in the lithosphere (caused by absorption), the atmosphere (caused by evaporation), and the biosphere (caused by photosynthesis) throughout the journey.

2. Select a place anywhere in the world, but preferably far from the ocean. Prepare a sketch map or flow chart showing how water could get to the sea from this location.

3. Explain how fresh water has potential energy.

4. a) List at least ten ways in which water is used.
   b) Categorize each use under one of the following headings: Sanitation, Human Consumption, Industrial, Irrigation, Cooling, Recreational.
   c) Rank each category from most important to least important. Give reasons for your ranking.
   d) Use your rankings in c) to decide which situation should have priority in each of the following scenarios:
   - irrigating pasture or providing municipal water
   - irrigating a golf course or supplying water to a chicken farm
   - serving industrial plants or maintaining a swimming pool

5. Where else in the hydrosphere are there tremendous amounts of stored potential energy? How could this be harnessed to provide electricity?

6. In what ways could the harnessing of hydroelectricity affect wildlife habitats?

Let's trace the movement of water as it travels from northwestern Ontario to the Atlantic Ocean. It starts as snow in Quetico Provincial Park. In the spring the meltwater runs along the land in rivulets, eroding little channels in the soil as it goes. Some of the water evaporates. Some is absorbed by the soil. Some becomes part of the spring vegetation. The water that remains enters the *Pigeon River* along the US–Canada border. From here it moves into *Lake Superior*, where it may remain for years. In time it enters *Lake Huron* at the *Straits of Mackinac*. At this point the water is relatively free of pollution, although it does contain natural minerals from the soil.

From *Lake Huron* the water flows into the *Detroit River* and then into the shallowest of the *Great Lakes*, *Lake Erie*. Here pollution begins to invade the water. It continues flowing along until it tumbles over *Niagara Falls*. Rapids in the *Niagara River* carry the water to the lowest of the *Great Lakes*, *Lake Ontario*. From here the water is carried into the *St. Lawrence*. More rapids and lower elevations speed the water through the *Gulf of St. Lawrence* to its final destination, the *Atlantic Ocean*. Had this water started its journey a little to the north of Quetico, it may have ended up in Hudson Bay; a little to the south and it could have flowed down to the Gulf of Mexico.

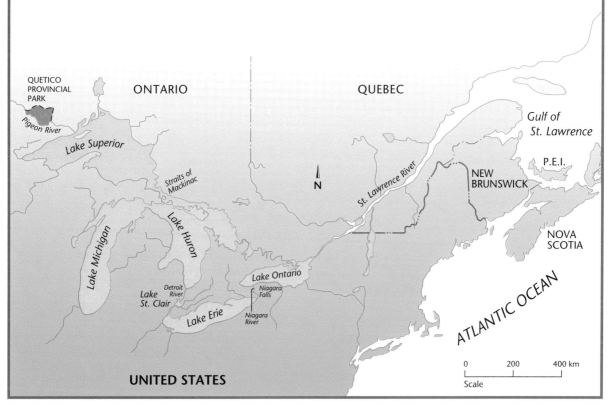

FIGURE 7.14 *The Route Water Takes to the Atlantic*

Chapter 8

Energy Systems in the Biosphere

INTRODUCTION

Energy is required for all movement in the biosphere. When a tree grows, a minnow swims, or a person walks to school, energy is expended. Kinetic energy created by living things results from chemical energy as plants and animals grow. Potential energy is stored in living tissue, waiting to be used by the organism that created it. **Metabolic processes** convert energy from the sun into kinetic, chemical, and potential energy. Let's examine how these processes work.

Kinetic Energy

We know that energy flows through the biosphere because of the constant change exhibited by plants and animals. Every movement is an example of energy being used. Some movements, such as the hopping of a frog, are easy to detect, but others are more gradual. For instance, you cannot actually see a geranium growing, but you know that it is because it becomes larger. Some changes happen so slowly that we don't notice them even in a lifetime. Trees in the New Forest of southern England are 900 years old. Local residents say the forest is unchanged. But is it really?

Plants and animals exhibit energy as they alter their structures to adapt to their environment. For example, where the winters are so cold that water freezes and prevents photosynthesis, broadleaf trees lose their leaves before winter arrives. Polar bears have adapted to the white wilderness that is their habitat by growing heavy white coats.

Kinetic energy can be illustrated over short and long periods of time. The daily struggle

FIGURE 8.1 *The New Forest in Southern England*

for survival is really the story of how an organism gets energy. Over the long term, plants and animals evolve so as to utilize energy as efficiently as possible. But when environmental change is sudden, they may not have time to adapt.

When changes in the environment are slow, it is possible for life forms to adapt by evolving physically or behaviourally. But when change is rapid, plants and animals can survive only by locating someplace else. Much wildlife migrates seasonally or permanently. If some caribou did not migrate south in search of food and water during the hard winters of the Canadian Arctic, they would not survive. Similarly, when their natural habitats are destroyed by urban development, foxes and deer move away in search of ravines and woodlots.

Even plants migrate. Their seeds are carried by wind and water and even on the coats of animals to new locations. If they end up in suitable habitats, they produce new plants. This "seed migration" is important, particularly if the environment in which the parent plant originated is destroyed. Thus the range of plant species in any area can change over time. Figure 8.2 shows how plants in Great Britain have changed since the last ice age.

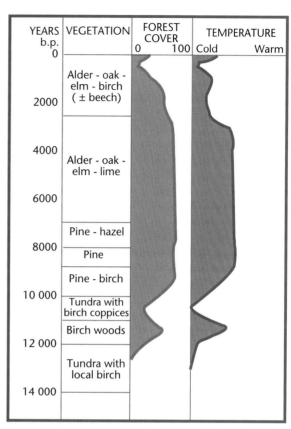

FIGURE 8.2 *Changes in the Vegetation of Britain During the Late Glacial and Postglacial Periods.*

Things To Do

1. List the different ways in which plants and animals show kinetic energy. How do plants differ from animals in the way they move?

2. How do plants and animals respond to environmental change?

3. Study Figure 8.2. What correlation is there between temperature and forest cover? Why does this relationship exist?

4. Select a natural area in your local community that is experiencing change.
   a) Prepare an inventory of plants and animals in the area.
   b) Find out what provisions have been made to protect indigenous species.
   c) Evaluate these provisions to determine if they meet the ecological needs of the region. Indicate any changes that you would implement to make the plan more ecologically sound.
   d) If the plan is a good one, write a letter to the editor of a local paper praising it. Give reasons to explain why you support the plan and make any suggestions you feel would make the plan even better. If the plan is a poor one, explain why you cannot support it. Inform friends and neighbours about your concerns. Write letters to municipal officials and politicians. Attend council meetings. Do what you can to prevent your community from losing its valuable natural lands.

Chemical Energy

The biosphere is made up of all living things. These can be divided into two distinct categories based on energy use: **producers** and **consumers**.

Producers are plants. They are called producers because they are the only living things that can transform solar energy into chemical energy. This energy is used to grow leaves, stems, roots, and all the other essential plant organs.

Consumers are animals. They are unable to convert solar energy directly into food. Consumers get their energy in one of two ways: from eating producers with their stored chemical energy and/or from eating other consumers. Primary consumers, or **herbivores**, are those animals that eat plants directly. Cows, goats, and sheep are all examples of herbivores. Secondary consumers, or **carnivores**, are those animals that eat consumers. Lions, sharks, and eagles are examples of carnivores. A third category of consumers, **omnivores**, eats both plants and animals. This is the category to which the human species belongs, along with other primates, dogs, and pigs. But no matter what mix of plants and animals consumers eat, they all depend on plants either directly or indirectly to obtain their energy from the sun.

Photosynthesis

The way in which plants convert solar energy is not fully understood. However, we do know that it is a chemical reaction brought about by sunlight acting on carbon dioxide and water in specialized plant cells. This process is called **photosynthesis**, a term meaning "putting together with light." Carbohydrates (plant food) are formed when atoms of carbon, hydrogen, and oxygen are combined. When carbohydrates are produced, oxygen is released into the atmosphere. The energy that went into making the carbohydrates is then stored in them. It is only when these carbohydrates are used by the plant or are burned or eaten that this energy is released. This simple chemical equation summarizes the process of photosynthesis:

$$H_2O \text{ (water)} + CO_2 \text{ (carbon dioxide)} + \text{sunlight} \longrightarrow CHOH \text{ (carbohydrates)} + O_2 \text{ (oxygen)}$$

The carbohydrates produced through photosynthesis are significant to the study of energy

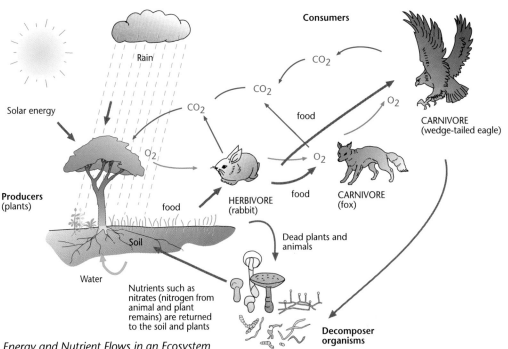

FIGURE 8.3 *Energy and Nutrient Flows in an Ecosystem*

flows for several reasons. Not only do they provide stored solar energy for consumers, but solar energy is often stored in fossilized remains. Oxygen is produced as a by-product of photosynthesis. This gas is vital to the survival of consumers. In addition, carbon is held in vegetation. When the plant rots or is burned, the carbon reacts with oxygen to form carbon dioxide. This greenhouse gas affects the amount of heat energy trapped in the atmosphere. The biosphere is intimately linked to the lithosphere, the atmosphere, and the hydrosphere through the process of photosynthesis.

Studies of air bubbles locked in ancient sediments or in continental glaciers indicate that the atmosphere is much different today than it was when it was first formed. Then it contained almost no oxygen. Today about 20 per cent of the atmosphere is oxygen. Almost all oxygen came from plants, most of which grew in ancient oceans. We know that plants depend on the atmosphere for life. But the atmosphere is equally dependent on plants. What would happen if people destroyed all the forests and polluted the seas so that the **phytoplankton** died? The oxygen vital to our survival could be lost and the planet could revert to the way it was before animals colonized land.

It is also necessary for plants to convert the energy in tissues in order to grow and reproduce. To use energy stored in carbohydrates, plants take back oxygen, usually through the roots, to reverse the process of photosynthesis. Carbon dioxide and water are combined with nutrients in the soil to produce plant structures. Figure 8.4 shows the nutrients plants need. If even one of these chemical elements is lacking, the plant will not grow. Only two elements, carbon and oxygen, come from the atmosphere. The rest originate in soils and rocks. Water is also vital to plant life. Rainwater dissolves nutrients. The plant roots then absorb the water, and with it the nutrients plants need.

Nutrient	Amount Needed (mg/L)	Function	Sources
Oxygen	–	respiration	atmosphere
Carbon	–	photosynthesis	atmosphere
Nitrogen	15	protein and vitamin creation	soil bacteria
Sulphur	1	protein and vitamin creation	minerals from soil (pyrite, gypsum)
Calcium	3	cell membrane formation	minerals from soil (limestone, feldspar)
Potassium	5	protein creation, phosphorus synthesis	minerals from soil (clay, mica, feldspar)
Magnesium	1	chlorophyl, enzymes	minerals from soil (dolomite, clay)
Phosphorous	2	organic molecules, energy source	iron, aluminum, calcium phosphates in soil
Iron	0.1	respiration	iron oxides in soil
Trace minerals*	0.2	enzymes, respiration, cell division, nitrogen synthesis, cell function	igneous rock, sea spray (boron, sodium)

*includes manganese, copper, zinc, boron, molybdenum, cobalt, sodium, silica, chloride

Adapted from *Fundamentals of Physical Geography*, Briggs *et al*, Copp Clark Pitman, 1989. Reprinted by permission of Copp Clark Longman.

FIGURE 8.4 *Essential Plant Nutrients*

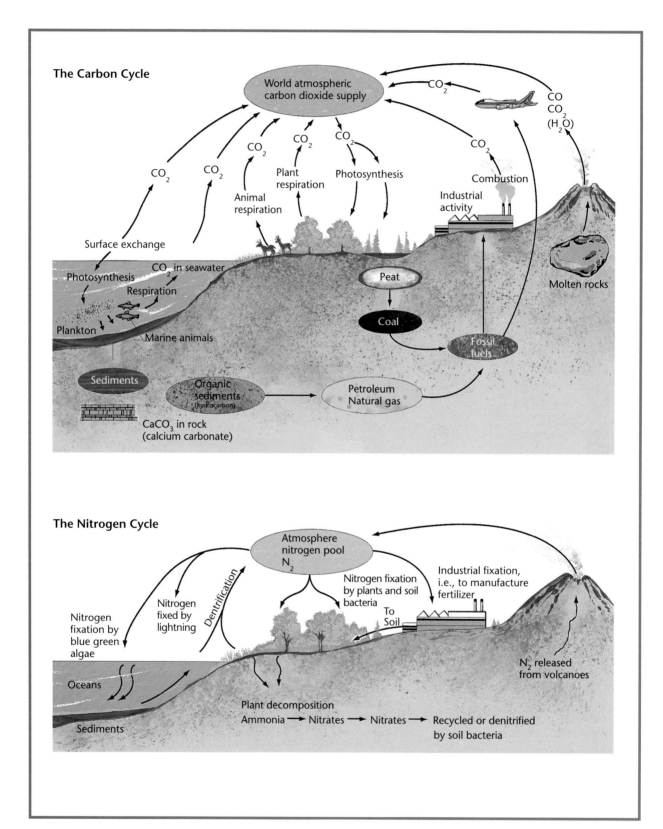

FIGURE 8.5 *The Four Cycles*
Describe the flows shown in each diagram.

The Phosphorus Cycle

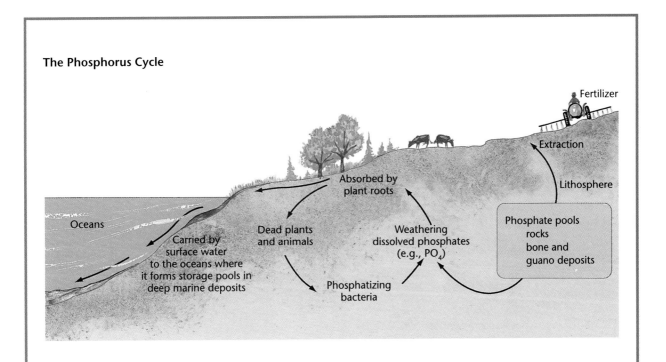

The Oxygen Cycle

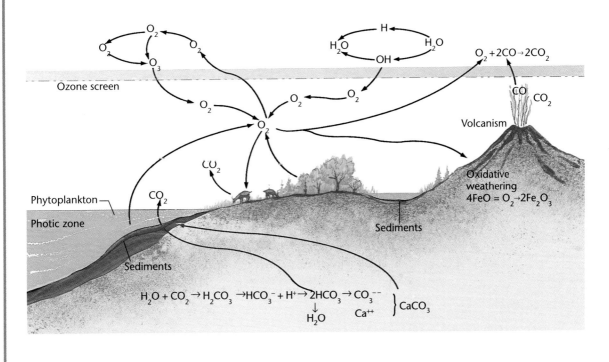

Transpiration

Plants also serve a vital function in the hydrologic cycle. Once water has been used to transport nutrients to plant cells, the plant expels the water through a process called **transpiration**. You may have noticed that some leaves are moist on the surface. This is transpiration in action. The moisture evaporates from the leaf and re-enters the atmosphere. This process ensures the continuation of the hydrologic cycle.

What happens when there are no plants and, therefore, no transpiration? Studies in Africa and Brazil have shown that local precipitation patterns change when forests are removed. Deforestation is especially devastating in tropical countries. Unlike air masses in the middle latitudes, tropical air masses do not travel far from their source. Convection currents move the air up and down in a relatively narrow band, so incursions of humid air from outside the cell are unlikely. Thus if the vegetation in these areas is removed, so is the moisture supply. As a result, today, in places like Sudan in Saharan Africa, the desert has encroached across a land that was once covered with grass and trees.

Soil

Soil is the main nutrient source for most plants. But how do nutrients get in the soil?

Many are contained in the lithosphere. They get from the rocks into the soil through weathering. **Mechanical weathering** breaks rocks into fragments through the actions of water, wind, and ice or as the rocks rub and grind against one another. **Chemical weathering** occurs when chemical changes alter the rock's form and dissolve vital minerals. Nutrients also enter the soil from plant and animal waste. Specialized consumers, called **decomposers**, break down these wastes into simple compounds that can be absorbed into the plants. Actually the term waste, in this sense, is an inaccurate one. Nothing in nature is wasted. Bacteria, worms, termites, and other decomposers eat away at decaying material until it is transformed into usable nutrients. Nitrogen and potassium are the main nutrients provided by these natural fertilizers.

FIGURE 8.6 *Termite Mound in Australia*
Termites provide the largest amount of animal biomass of any creature on earth.

Things To Do

1. How is solar energy transferred to all living things, either directly or indirectly?

2. Study Figure 8.4.
   a) Prepare a circle graph showing the values of nutrients needed for plant growth. Combine all fractional elements; title them Trace Minerals.
   b) Identify the different soil components that provide essential nutrients. Under each one list the nutrients.

3. Study Figure 8.5.
   a) Explain one of the natural cycles.
   b) Provide examples to show how people modify this cycle.
   c) How can human intervention be both positive and negative?
   d) Create a strategy whereby the negative impact of human intervention can be reduced.

Potential Energy

There is a tremendous amount of potential energy stored in the biosphere. The solar energy used to make carbohydrates in plants is stored in tree trunks, foliage, roots, and even the tiny plankton that inhabit the oceans. Animals convert the stored radiant energy of the plants they eat into animal tissue, thereby storing even more potential energy. When a consumer eats a producer, however, only a small amount of the potential energy is passed on. The rest is lost through body heat, respiration, and incomplete digestion. So a meadow mouse would need to eat 100 g of grass to produce 10 g of mouse tissue.

As the energy flows through the food chain, less and less of it is passed on. If a meadow mouse were eaten by a hawk, the hawk would receive only 10 per cent of the mouse's potential energy. You can see that it would take a lot of mice to provide all the food requirements of the hawk! You can look at it this way: 1 g of hawk fibre requires 10 g of mouse fibre, which requires 100 g of plant fibre. Each level requires more potential energy from the level below it. Because of this, the population of consumers decreases with each layer. So there might be several families of mice but only one family of hawks in the meadow.

The biosphere is not the same all over the earth. Some areas are able to store enormous amounts of potential energy in their **biomass**, while others cannot. There is obviously a strong correlation between the amount of biomass and the availability of inputs from the lithosphere and biosphere (nutrients), the hydrosphere (water), and the atmosphere (carbon). Where the inputs are high, the amount of biomass would be high; where they are low, the biomass would be low. Figure 8.8 shows the difference in biomass produced by the major ecosystems. It is interesting that rainforests provide the greatest biomass and are found where solar energy is greatest. Solar energy is not the only variable, however. Look at the category "Desert and Semi-desert." These areas have high energy inputs from the sun but are poor biomass producers. Obviously, one or more of the essential ingredients is lacking, in this case water. As you study this organizer, try to determine how inputs vary between ecosystems.

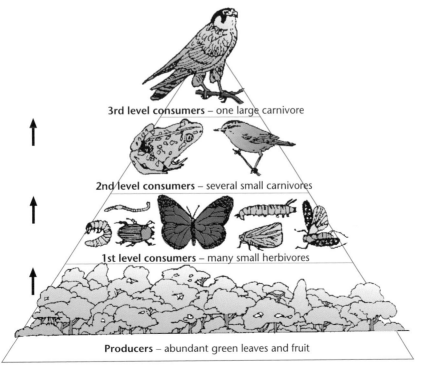

FIGURE 8.7 *A Food Pyramid*

Type of Ecosystem	Area (millions of km²)	Net Productivity per Unit Area (dry grams/m²/year) Normal Range	Mean	World Net Productivity (billions of dry tonnes/year)
Tropical Forest	24.5	1000-3500	2000	44.8
Temperate Forest	12.0	600-2500	1250	13.4
Boreal Forest	12.0	400-2000	800	8.7
Woodland and Scrubland	8.5	200-2000	700	5.4
Savanna	15.0	200-1500	900	12.3
Temperate Grassland	9.0	10-400	600	5.0
Tundra and Alpine	8.0	0-250	140	1.0
Desert and Semi-desert	42.0	100-3500	40	1.5
Cultivated Land	14.0	800-3500	650	8.3
Swamp and Marsh	2.0	100-1500	2000	3.6
Lake and Stream	2.0		250	0.5
Total Continent	149.0		773	104.5
Open Ocean	332.0	2-400	125	37.6
Continental Shelf, Upwelling	27.0	200-1000	360	8.9
Algal Beds, Reefs, Estuaries	2.0	500-4000	1800	3.4
Total Marine	361.0		152	49.9
World Total	510.0		333	154.4

Reprinted with the permission of Simon & Schuster from the Macmillan College text *Communities and Ecosystems* 2/E by Robert H. Whittaker. Copyright © 1975 by Robert H. Whittaker.

FIGURE 8.8 *Net Primary Productivity of Major Ecosystems*

Things To Do

1. a) Make a flow chart to illustrate a food chain in one of the following ecosystems: tundra, rainforest, ocean, temperate forest, prairie.
   b) In the food chain you created in part a), eliminate one of the levels. What effect would this have on the food chain?
   c) How could people interrupt this food chain?
   d) What repercussions could this have on the ecosystem?

2. Use Figure 8.8 and the map of natural vegetation on page 271 to prepare a graded shading map of the mean net biomass productivity of different ecosystems. Use a natural vegetation base map.
   a) Make a scale of four to six categories, with a legend allocating one shade to each category. For example:
   1500-2000 g/m²/year— very dark
   1000-1500 g/m²/year— dark
   500-1000 g/m²/year— light
   0-500 g/m²/year— very light
   b) Colour the map according to the legend.
   c) Rank the ecosystems based on the amount of biomass stored in each.
   d) Write an explanation of how biomass varies across the earth's surface.

CHAPTER 9

Landscape Systems

INTRODUCTION

There is a tremendous amount of vitality in the world around us. Movement is everywhere. Often you can see energy flowing through the environment. Clouds skim overhead. Birds dart through the sky. Streams gurgle and splash. Other energy movements are more gradual and, therefore, less obvious. Water evaporates from puddles. The ground shifts to maintain its isostatic equilibrium. Bacteria eat away at the leaves under our feet. The force behind all this vitality is energy.

Energy flows are not the same everywhere, although there are general energy patterns that can be applied to any place in the world. A **landscape system** is the way in which energy flows are organized in a particular place. When you study a landscape system, you are not just studying what is there; you are studying *why* it is there. The process is more important than the product. The study of systems is the same no matter what the subject. Take computer science, for example. A systems analyst studies how information flows through the computer hardware. It is not important what the equipment looks like. What is important is how the equipment processes information. Similarly, geographers are more interested in process—why things are the way they are.

An Alpine Landscape System

Have you ever visited the Swiss Alps? The region is famous for its incredible beauty. Figure 9.1 on page 66 shows one small part of this magnificent landscape. Look carefully at the photograph. Try to visualize the energy that is flowing through this landscape system.

If you study the picture from the foreground to the background, you can imagine movement in each of the four spheres. In the foreground the grass is growing. Photosynthesis is taking place, and carbon dioxide is flowing into the plants. Down the hill in the valley below, crops are growing in the fields and the trees are swaying. In the background trees are growing in the forest, rocks are breaking down as nature's forces act upon them, and patches of snow are melting. In the sky above the mountains, clouds have formed from the water vapour; precipitation may return some of the water to the river in the valley below. Energy is at work everywhere.

FIGURE 9.1 *The View from the Juras Mountains to the Alps across the Swiss Plains*

Techniques for Studying a Landscape System

In order to understand the processes that shape a landscape, geographers rely on these techniques.

1. **A Field Study:** The best way to begin studying a landscape system is with a visit to the site. Geographers cast a keen eye over the area and make notes on what they observe.

2. **Photographs:** Good photographs serve as a visual record of the site. Moreover, pictures reduce the scale of the scene so that it is easier to see broad patterns. Taking a picture from above provides an excellent perspective of the site. Geographers find a clear vantage point from the top of a hill or a tall building. Air photos can also be purchased from the government. For this type of study, **oblique air photos** taken from a low angle are better than **vertical air photos**.

3. **Diagrams:** As the cliché goes, "You can't see the forest for the trees." Usually, there is so much detail in a photograph that general patterns are hard to see. Tracing the photograph outlining only the components of the system enables researchers to "see the forest *despite* the trees."

4. **Maps:** Maps show the "bird's-eye view" of a site. They illustrate information through the use of lines, colours, and symbols. In addition, they are often drawn to scale. Maps are essential to any geographic study. They show spatial relationships and patterns better than any other technique.

5. **Flow Charts:** A systems analyst uses flow charts to show how data flows through the system. This is a useful exercise for geographers as well. They can make assumptions based on what they know about how each element of the system operates. This is probably the most difficult part of a field study.

6. **Quantitative Analysis:** Now the geographer can measure changes and confirm that assumptions made earlier were accurate. When changes are rapid, as with weather, measurements can be made easily. When movements are slow, as with forest growth, measurements are minute and more sophisticated methods are needed.

7. **Written Analysis:** This part of the study brings together all of the geographer's knowledge of the site to explain how the landscape system operates. The flow chart serves as an organizer.

Figure 9.2 *A Tracing of Figure 9.1*

Things To Do

Before beginning this activity, review the section "Techniques for Studying a Landscape System" on page 67.

1. Trace the photo on page 66 using onionskin, rice paper, or a transparency. Show the different elements of the landscape system. Figure 9.2 illustrates a tracing.

2. Label each element of the landscape system. Use a colour key for each different sphere. For example: hydrosphere—dark blue; lithosphere—brown; atmosphere—pale blue; biosphere—green. The more labels you have, the better. Try for fifteen to twenty. If you wish, lightly colour your scene, but do not cover up important details. It is a good idea if you go over your tracing with a fine-tipped black marker to improve presentation. Mount the tracing on a piece of stiff paper or cardboard.

3. Make a flow chart showing energy flows in the Swiss Alps. (Refer to Figure 8.5 for samples of flow charts.) Each of the elements you labelled in step 2 is a feature of the landscape. Print these labels on a piece of paper and draw boxes around each one. Draw arrows between each of the boxes to indicate where a process is occurring, and write the process on the arrow. For example, between leaves and clouds you could draw an arrow labelled *transpiration*. When you have completed drawing the arrows, redo the flow chart, rearranging boxes that go together. Use a ruler and fine-tipped marker to draw your finished flow chart.

4. Draw a sketch map showing what the valley would look like from directly overhead.

5. Use the flow chart to write an analysis of how energy flows operate in the alpine valley.

FIELD STUDY

In the previous exercise, field work was impossible because the landscape is not accessible to the class. However, you have access to natural landscapes in your own community that are every bit as interesting and vital as the Swiss Alps. In Chapter 1, you selected a site for extensive field study. Conduct a field study on this site using the same methodology as in the previous exercise. Before beginning, review the section "Techniques for Studying a Landscape System" on page 67.

1. Take several photographs, preferably from a hilltop or building. Choose the best photo and have it enlarged. Indicate on the site map you made in Chapter 1 where the photo was taken.

2. Trace the photo.

3. Revisit the site with the mounted tracing or a photocopy of it. Study various places on the tracing and make notes about details in the area. Use a legend or military grid to show the places where detailed observations were made. Don't worry about taking measurements at this point. The quantitative analysis will be dealt with later.

4. Prepare a flow chart using information from the tracing and your notes.

5. Finish up with a written report on the energy flows in your field study landscape.

PART 3

Studying the Atmosphere:
Weather and Climate

Chapter 10

The Atmosphere

INTRODUCTION

The thin outer shell of the earth is called the atmosphere. Because it is made up of gases and is usually transparent, we often forget that the air that surrounds the earth is part of the planet. From space, we can see the lithosphere, the biosphere, and the hydrosphere clearly. All we can see of the atmosphere, however, are the clouds that swirl across the earth's surface and the pollution that hangs over cities.

The atmosphere is essential to life on our planet. Without it, earth would be lifeless. There are several reasons why the atmosphere is necessary for life. It:
- provides oxygen for animals to breathe
- provides the carbon dioxide that is essential for photosynthesis
- helps to stabilize the differences in temperature between tropical and polar regions
- traps solar energy in the form of heat and prevents it from escaping into space
- recirculates water, which is essential for all life, and
- shields life forms from the sun's harmful ultraviolet rays.

Figure 10.1 *Cloud Formations in the Atmosphere*

Properties of the Atmosphere

Before we can study the atmosphere, there are several important properties or characteristics we must first understand.

The Atmosphere Is Fluid

Because the atmosphere is made up of gases, it is fluid and flows from one place to another. Winds and other atmospheric movements spread these gases throughout the atmosphere. Manufactured chemicals, such as pesticides and **chlorofluorocarbons**, as well as chemicals created as a result of volcanic eruptions and other natural processes, enter the atmosphere at a particular place and eventually flow around the planet. Many of the synthetic chemicals are unsafe once they enter the atmosphere. Some upset ecosystems and endanger both plant and animal life. Consider the example of DDT, an obsolete pesticide that is not **biodegradable**. While this poisonous chemical has not been used for over twenty years, traces of DDT are still found in the tissues of Antarctic penguins thousands of kilometres away from where DDT was originally emitted into the atmosphere!

Gases in the Atmosphere

The atmosphere is made up of many different gases, but nitrogen (approximately 78 per cent) and oxygen (approximately 21 per cent) are the main components. Nitrogen is an important nutrient used by both plants and animals. Oxygen is essential for life and combines with fuels when they are burned. While small in percentage breakdown, however, the remaining gases are not small in importance. Carbon dioxide (0.03 per cent) is essential for plant development. Water vapour (0-4 per cent) is vital to both plants and animals to transport nutrients through cell membranes. These gases are not evenly distributed, however. Carbon dioxide is often more concentrated over crowded cities where it is emitted by cars and factories in the burning of fossil fuels. Higher levels of water vapour, or humidity, are often found over bodies of water and forests.

Mass and the Atmosphere

Just because air is hard to see does not mean that nothing is there. Like all matter, air occupies space and has mass. This mass of air, or **air pressure**, varies from place to place on the earth's surface and according to elevation. Places that are hot have lower air pressure than places that are cold. As the atmosphere tries to equalize the air pressure, winds blow from areas of high pressure to areas of low pressure. This is the basis for weather and climate. Altitude also affects air pressure. This is because there is less air pressing down from above so the mass of air is less. Air pressure at higher elevations is so low that people have to breathe more to get the oxygen they need.

Layers in the Atmosphere

Figure 10.2 shows the four layers of the atmosphere. They are the troposphere, the stratosphere, the mesosphere, and the thermosphere. Not shown in the diagram is the magnetosphere, the force field that extends far into space. While each layer possesses different characteristics, the most important difference is temperature.

GEO-Fact

Sometimes clouds formed in the stratosphere from the dust of major explosions affect weather in the troposphere. Volcanic eruptions, meteor collisions, and even nuclear explosions can send enormous quantities of dust high into the atmosphere where it may remain for years.

Paleontologists speculate that dinosaurs became extinct 65 million years ago when one or more meteors smashed into the planet. The dust that resulted blocked the sun from providing the light and heat needed for photosynthesis. The earth was plunged into darkness for more than three months. Dinosaurs literally starved to death as the plants they needed to survive died off. In modern times, scientists have speculated that a nuclear winter several years long could result if there was a nuclear war because of the massive stratospheric clouds that would be formed.

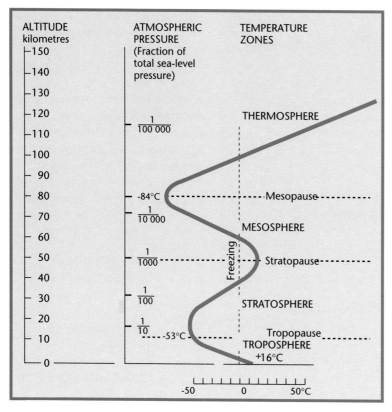

FIGURE 10.2 *Layers of the Atmosphere*

Imagine that you are in a hot-air balloon that is floating up to the top of the atmosphere. You would start your journey directly above the earth's surface in the **troposphere**. This is the most important layer because it is closest to the biosphere. Most life on earth depends on this layer of the atmosphere for survival. The troposphere is influenced most by differences in surface heating and the frictional drag of the planet's rotation. For these reasons, this is the most turbulent layer of the atmosphere. Vertical and horizontal mixing of air is common. Your balloon would be swept this way and that way as it was buffeted by storms and prevailing winds. Another significant characteristic of the troposphere is that the temperature decreases as altitude increases. (See pages 32-33.) By the time the balloon reached the top of this layer, it would be very cold. You would also have passed through 90 per cent of the earth's atmosphere by mass.

At the top of the troposphere is a transition zone called the **tropopause**. The thickness of the troposphere varies. At the Equator, the troposphere begins at about 16 km above the earth's surface. It gets thinner moving towards the poles, where it is only 9 km thick. The turbulent weather of the troposphere is left behind in the tropopause. The temperature stops falling and stabilizes at about -55°C. Your balloon would have entered a cold, still realm.

Once you pass through the tropopause, you enter the stratosphere. This layer contains the jet stream and the ozone layer. As in the troposphere, the air does not move up and down, but there are considerable horizontal layer-like movements of air. Often storms in the troposphere follow the paths of these rapidly moving upper altitude wind currents created by the coriolis effect.

Spread throughout the stratosphere are molecules of **ozone**. This special form of oxygen absorbs ultraviolet radiation. The energy transfer that occurs as the ozone is broken down increases the temperature in the stratosphere. Heat from the ozone layer radiates up and down, so that the lower layers of the stratosphere are colder than the upper layers where most of the ozone occurs. Thus as your balloon rises through the stratosphere, the temperature gradually increases until it exceeds the freezing point (0°C) 40 km above the earth. The **stratopause** is another transition zone where temperature characteristics change. At this level, the temperature stops increasing as stratospheric ozone thins out with altitude. You are entering the outer atmosphere.

After the stratopause, and for the next 30 km, your balloon would pass through the **mesosphere**. Now the temperature begins to drop steadily as you move further away from the earth. Eventually it reaches about -84°C. At this point you enter another transition zone called the **mesopause**. Here the temperature stabilizes, and then starts to increase once again. In the mesopause, the air is so thin that it seems as if you are in outer space.

The next layer, the **thermosphere** (also called the ionosphere), is the thickest one, ranging from 80 to 480 km. The gravitational pull of the earth is so slight that the atmosphere is now quite thin. In the thermosphere, the temperature starts to increase again. The sun's rays break down the molecules of the atmosphere electrically, somewhat like what happens with the ultraviolet rays in the ozone layer. As the **electrons** and **ions** are created in the thermosphere, radiant energy is converted to thermal energy and the temperature rises.

Outside these four layers of atmosphere is the **magnetosphere**. Here the atmosphere gradually gets thinner until there is no more air. It is impossible to tell exactly where the atmosphere ends and outer space begins. Solar dust sometimes gets trapped in the magnetosphere. The rotation of the earth makes the planet behave like a magnet. The charged particles from the sun align themselves with the North and South poles, forming streaks of light in the sky. These Northern Lights, or **aurora borealis**, often light up the dark winter nights in northern Canada.

Like all components of the earth, the atmosphere is a complex part of our existence. Without it, life as we know it could not exist.

Things To Do

1. Explain why the atmosphere is essential to life.

2. Describe the different properties of air.

3. Prepare an organizer comparing the different layers of the atmosphere. Include such criteria as width, air temperature, air movement, and distance from the earth's surface.

4. Explain why the temperature varies with each layer of the atmosphere.

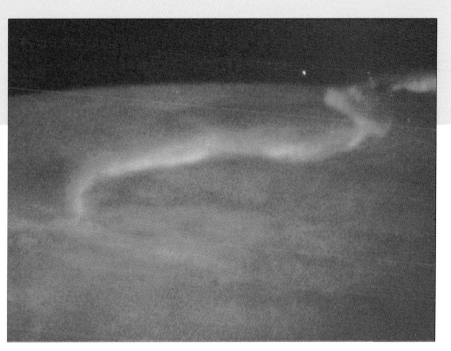

FIGURE 10.3 *The Northern Lights*

Chapter 11

Weather and Climate

INTRODUCTION

You've heard the expression "Everybody talks about the weather but nobody ever does anything about it." Why is it that we talk about the weather so much? Probably because it is the one thing we all have in common.

FIGURE 11.1 *A Tropical Storm*
What violent weather patterns exist in the region in which you live?

What Is Weather?

But what exactly *is* weather? A simple definition is "all the atmospheric activities that occur at a given place at a given time." These include air temperature, wind speed and direction, humidity, amount and type of precipitation, barometric pressure, and cloud cover. Weather consists of a variety of forces—none of which we are able to control!

People are weak and feeble creatures compared with the forces of nature. At best we can try to predict what the weather will be like and prepare for it. But even the science of predicting weather is an imprecise one. There are so many variables, forecasters can predict only what will happen a day or two in advance. And as we all know, even then the forecast may be wrong! Armed with a vast knowledge of weather patterns, powerful computers, sophisticated measuring devices, and geostationary satellites, we still have only a small grasp of what will happen in the weather in the immediate future.

What Is Climate?

Climate is like weather but on a different time scale. Climate can be defined as "the weather of a given place averaged over a long period of time." For example, the climate during the summer in the Prairie provinces is warm and dry. But in the summer of 1993, the weather was unusually cool and wet.

People compare the weather—what occurs at a particular time—with the climate—what is normally expected. For instance, in Florida, where the winter climate is warm and sunny, it is considered cold if the temperature dips below 10°C in February. In Winnipeg, however, where winter temperatures can be extremely cold, 10°C in February is considered incredibly warm.

Like weather, climate is continuously changing. A century ago, winters were considerably colder in Toronto than they are today. Old prints show people ice skating, playing hockey, and riding in horse-drawn sleighs on Lake Ontario. Yet these activities are impossible today because the lake no longer freezes over. Even in more recent memory the climate has changed. Torontonians now experience winter thaws when the temperature rises above freezing for a few days and the snow melts. This phenomenon seldom occurred thirty years ago. In those days, snow that fell in December stayed put until March!

Weather and Climate in Our Daily Lives

Atmospheric conditions have an incredible influence on our daily lives. Almost everything we do is affected by weather and climate. When you go to school, your choice of clothes is determined by the weather. If there's a heat wave, cool and comfortable is the only way to dress. The way you get to school may also be determined by the weather. On a nice day you probably walk, but if it's snowing you may get a ride. And on really bad "snow days," school may even be cancelled!

Buildings are designed to keep out the weather. In Canada, our schools are well insulated. Double-paned windows keep out the winter cold, while boilers provide warmth inside. In summer, some schools are air conditioned; in others screened windows provide ventilation. Schools are quite different in tropical countries, however. Heating is unnecessary but ventilation that allows cross breezes is extremely important. Because of the midday heat, many schools begin classes early, then break for an extended lunch period. Classes are completed in the late afternoon when temperatures begin to cool off. This relief from the midday heat is the basis for the *siesta* that is common in many hot climates.

Even our sports are affected by weather and climate. It is no accident that Canada, the most northerly of all countries, enjoys hockey as its national pastime. The Canadian Football League season begins and ends earlier in the year than its American counterpart because winter arrives earlier in our northern climate. Toronto's Skydome with its retractable roof allows the Blue Jays to play baseball in April without being concerned about freak snowstorms. And in 1988 when Calgary hosted the Winter Olympic Games, a chinook almost destroyed the skiing events by melting most of the snow overnight! In what other ways do weather and climate affect Canada's sports and recreational activities?

Weather and climate also affect the economy, particularly resource development industries. In the wet summer of 1993, prairie farmers could not get their grain dried out enough to store it. In southern Manitoba, fields and even entire farms were under water! Mining and lumbering activities often increase in winter because the freezing of **muskeg** allows heavy equipment to move across swamps that are impassable in summer. On the east coast the fishery is affected because the weather is often too severe for boats to venture into the North Atlantic. No matter what the primary industry, the weather plays an important part.

Weather has also played a part in shaping history. In 1281, for example, the Japanese were saved from invasion by typhoon winds. These "divine winds" allowed Japan to remain independent while the rest of Asia fell under the control of Genghis Khan. Three hundred years later, in 1588, a violent storm blew the 129 ships of the Spanish Armada off course, sparing England from conquest.

In their respective campaigns, Napoleon and Hitler were not defeated by the Russian army when they invaded that vast country; they were defeated by the severe Russian winter. The Russians retreated east, burning crops and villages as they went. When winter came, the enemy troops became bogged down in deep snow. Supplies could not reach the front and there was no food because the Russians had destroyed it. The French and German armies suffered as much from hunger, exposure, and fatigue as from Russian arms!

The only war fought on Canadian soil was also greatly affected by climate. The battles of the War of 1812 were not fought in the winter because of the difficulty in moving troops. (The only exception was the Battle of New Orleans, which was fought in January. Why would this battle have been the exception?) Canadian elections are seldom held in winter because bad weather would result in low voter turn-out. So you see, there is little doubt that weather and climate affect all aspects of our lives.

Things To Do

1. Prepare an organizer comparing weather and climate.

2. Study the section Weather and Climate in Our Daily Lives.
   a) Make two lists, one for weather and one for climate. Decide which examples in this section relate to weather and which relate to climate. Summarize each in the appropriate list.
   b) Add your own examples to illustrate how weather and climate influence your life.

3. a) Using clippings from newspapers and magazines, keep a scrapbook or prepare a bulletin board showing how weather and climate influence peoples' lives around the world.
   b) Prepare a map showing the locations of each of these news stories.

4. For a book or movie of your choice, explain how weather or climate affects the plot of the story. For example, Shakespeare's play *The Tempest* unfolds as it does because the characters are shipwrecked on a deserted island.

5. Interview an older Canadian to find out how the climate has changed in your area over the years.

CHAPTER 12
Measuring the Atmosphere

INTRODUCTION

FIGURE 12.1 *Red Sky at Night*
According to weather folklore, what does this night sky tell you about the weather tomorrow?

The weather is so important to our lives that a tremendous amount of time and effort are spent every day studying it. Long ago, superstition affected people's perception of weather. They tried to predict the weather by studying nature. They looked for patterns based on their memories of what happened under similar conditions. Sometimes their predictions were accurate. More often, however, they were wrong. Consider the myths described in the feature "Weather Myths: True or False?" Speculate on which ones are true and which ones are false. You'll find the answers on page 83.

Collecting Weather Data

Many of these myths are far from accurate. In modern society, a more accurate way of forecasting the weather is needed. While observation is important, it needs to be backed up by carefully recorded measurements. Once the information is collected, it must be analysed. Then a forecast can be made, either by an experienced forecaster or by a computer simulator. Because the weather moves from place to place, it is not enough to record what is happening in one place only. Patterns across the country must be studied so that accurate weather forecasts can be made. But how do we collect this information?

Severe Weather Watchers

Volunteers across the country notify their local weather office if they see a violent storm approaching. On 31 July 1987, it was a severe weather watcher that first reported the Edmonton tornado to the weather office. Thousands of people were able to take shelter because of this report. Other volunteers also help the weather service. They take measurements of precipitation twice each day and send their observations to the weather office once a month. This does not help in daily forecasts, but it does provide Environment Canada with valuable information for use in the study of climate change.

Weather Observation Stations

Meteorologists and volunteers make observations twice daily, four times daily, or hourly at the stations. Some stations also measure the upper air using weather balloons. These hydrogen-filled balloons are sent aloft twice each day. Instruments attached to the balloon

WEATHER MYTHS: TRUE OR FALSE?

1. **Groundhog Day:** On February 2nd, Wiarton Willy and other groundhogs across the country come out of their burrows in the morning, or so the story goes! If it is clear and sunny, the groundhogs are frightened by their shadows and run back to their burrows to sleep for six more weeks of winter. If it is dull and cloudy, they do not see their shadows, which means spring must be just around the corner.
2. **Cricket chirps:** The number of cricket chirps reflects the temperature. Crickets chirp more frequently in hot weather than in cold weather.
3. **Sunsets and sunrises:** There are several sayings about the colour of twilight. Here is one:
   Red sky in morning, sailors take warning;
   Red sky at night, sailors delight.
   If the morning sky is red, it will be a stormy day. If the night sky is red, the weather will be clear the following day.
4. **Squirrels gathering nuts:** If squirrels are busier than usual gathering nuts in the fall, it is believed that the coming winter will be more severe.
5. **How March "comes in":** If March comes in like a lion, it is said to go out like a lamb. This means that if it is blustery on March 1st, the end of March will be pleasant. The opposite is also said to be true—a pleasant March 1st means a nasty blast of winter at the end of the month.

measure the atmosphere and radio the information back to the weather office. Within minutes of being received the information is radioed to the district and then to the central weather office for all of Canada. Increasingly, automatic stations are being used in remote areas. These remarkable devices use sensors to measure atmospheric conditions. These are radioed to district offices in digital form by a microcomputer. The whole apparatus is powered by a photovoltaic cell that utilizes solar energy and dry-cell batteries.

Weather Radar

Radar is used to measure the amount of precipitation contained in clouds. You often see weather radar in television weather reports. Microwaves are emitted from a slowly rotating scanner. These waves can travel through clouds or fog, but when they hit a raindrop or snowflake, some of the wave is reflected back to a receiver. The amount of this echo is proportional to the amount of precipitation. The location of the storm can be determined from the time it took the echo wave to go out and return.

Weather Satellites

Data collected by satellites are also used in weather reports on television. Meteorologists are able to see what the weather is like across whole continents using images from outer space. Fronts, storm systems, fog, and jet streams cannot only be observed, they can be plotted as they move across the land. Using this information, the meteorologist can figure out how long it will take for different weather conditions to enter a region.

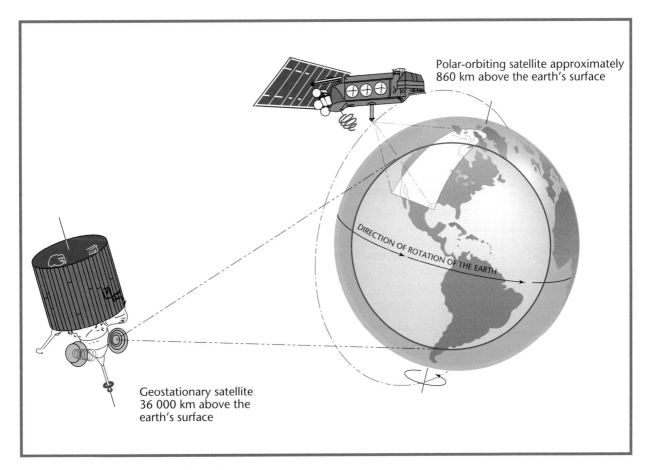

FIGURE 12.2 *How the Earth Is Viewed from a Polar-orbiting Satellite and a Geostationary Satellite*

TECHNOLOGY UPDATE: WEATHER SATELLITES

Weather satellites are our "eyes in the skies." They provide many of the images we see on our television weather forecasts. But they also do much more than show where the clouds are and where they're heading. Satellites today are sophisticated measuring devices. They contain sensors to measure the atmosphere, a computer to store and process the data collected, transmitters and receivers to send the information to the receiving station on earth, and solar panels to power all of this equipment.

Canada uses four polar-orbiting satellites and two geostationary satellites. The polar satellites travel around the earth every 105 min, taking readings all the time. The geostationary satellites are positioned over the Equator at 75°W and 136°W longitude. Their orbits are the same speed as the earth's rotation, so they remain over these same longitudes. Each satellite monitors the earth's surface and produces a new image every eighteen minutes. Cloud cover, snow cover, ice in waterways, and the heights of ocean waves are among the many things these satellites can determine. Wind speed and direction are established by the amount of movement of weather systems from one image to the next.

The satellites use microwaves to determine conditions at different levels of the atmosphere. Sophisticated equipment determines the temperature and water content at twenty different elevations reaching up to the stratosphere. The computer back on earth is able to create a three-dimensional grid showing atmospheric conditions for the troposphere over the entire continent. Other satellites continue the analysis so that complete data for the entire planet are obtained. This process provides meteorologists with an enormous amount of information that can be used to produce the most accurate forecasts possible.

A computer in Montreal combines data from satellites, air balloons, and ground stations to analyse the atmosphere. It produces maps showing atmospheric conditions at ground level and at different elevations in the troposphere. No longer does a meteorologist have to hand draw the fronts and weather symbols on a map. The computer does it automatically. But the capabilities of this enormous "number cruncher" do not end here. The computer uses physical-mathematical models to simulate the behaviour of the atmosphere in the future.

Weather Instruments

Lasers and microwaves have revolutionized measuring the elements of weather. They enable readings to be taken from great distances. One example of this new technology is the **ceilometer**. It is used to measure the height of the ceiling, or cloud cover. This instrument beams a laser pulse directly overhead. Whenever it meets an obstacle, the laser is reflected back to the instrument. By measuring the time it takes the laser to go up and come back down, the machine can determine how high the clouds are. The ozone layer can be measured using an ozonometer. Monitoring the ozone layer has become an important part of the weather service in recent years as the ozone layer thins and threatens to affect life on earth.

Even though tremendous advances have been made in the development of weather sensors, most equipment used today can be found in many homes and schools. The thermometer is still used to measure air temperature. As a liquid like mercury expands or contracts, it moves up or down a tube of glass. By comparing the position of the liquid to known temperatures, it is possible to calibrate the thermometer. Once the scale is known, the thermometer can be used to determine temperature.

Humidity is measured with a **psychrometer**. This instrument consists of a wet-bulb thermometer and a dry-bulb thermometer. The bulb of the wet-bulb thermometer is kept moist by a thin cloth bag connected by a wick to a bowl of water. As water evaporates from the wet bulb, energy is used up so the temperature is lower than it is for the dry bulb where there is no evaporation. The temperature difference between the two thermometers indicates the level of humidity. Tables are used to show the relative humidity for given wet-bulb and dry-bulb temperatures.

Barometric pressure is measured with an instrument called an **aneroid barometer**. As the air pressure or mass of the air changes, the needle on the barometer moves one way or the other. Changes in air pressure are used in weather forecasting. When barometric pressure drops, it is usually a sign that a storm is approaching; when barometric pressure rises, fair weather and clear skies are usually in the forecast.

Making Your Own Weather Instruments

It is possible to make simple devices to measure the atmosphere. Precipitation is easy to measure. All you need is a tin can. When rain falls, some of it is collected in the can. This precipitation can then be measured by placing a ruler in the can. If you are concerned that the measurement may be inaccurate on hot days because of evaporation, fit a funnel having the same circumference as the can into the top of the can. The water still accumulates, but evaporation is limited because the funnel reduces the water's exposure to the air.

Snow is a little harder to record accurately. You have to melt the snow first to take the measurement. As a general rule, snow of a given mass fills ten times the volume of rainwater of the same mass. If you are worried that blowing or drifting snow might influence the amount of snow in the can, place the can in a sheltered spot that is protected from the wind. Of course, whether measuring rain or snow, the water should be removed after each reading.

Wind direction is also easy to measure using a wind vane. These decorative instruments are often found on the tops of barns and garages. It is easy to make one using dowelling and scraps of wood. The important thing is to

Beaufort Number	General Description	Observations for Estimating Wind Speeds	Wind Speed in km/h at 10 m Above Ground
0	Calm	Smoke rises vertically	Less than 1
1	Light air	Smoke, but not wind vanes, shows direction of wind	1-5
2	Light breeze	Wind felt on face; leaves rustle; wind vanes moved	6-11
3	Gentle breeze	Leaves and small twigs moving constantly; small flags extended	12-19
4	Moderate breeze	Dust and loose paper raised; small branches moved	20-28
5	Fresh breeze	Small leafy trees swayed	29-38
6	Strong breeze	Large branches in motion; whistling heard in telegraph wires	39-49
7	Near gale	Whole trees in motion	50-61
8	Gale	Breaks twigs off trees	62-74
9	Strong gale	Slight structural damage occurs	75-88
10	Storm	Trees uprooted; considerable structural damage	89-102
11	Violent storm	Very rare; widespread damage	103-117
12	Hurricane	Very rare	118 and over

FIGURE 12.3 *The Beaufort Wind Scale*

position the wind vane so that "N" (north) is facing the proper direction!

Wind speed is harder to measure than direction. An **anemometer** is used to measure wind speed. Cups catch the wind and cause the spindle to rotate. The more rotations it makes in a given time, the greater the wind speed. While this instrument is hard to make, the **Beaufort Wind Scale** allows you to make remarkably accurate estimates of wind speed by observing the way things react in different wind velocities. (See Figure 12.3.)

FIELD STUDY

Focusing
Choose one of the following questions about the weather or create one of your own.
- How accurate are weather forecasts?
- How accurate is the weather compared with the *Farmer's Almanac*?
- How does the weather vary within a community?
- How is the weather different this year from what it is usually like (climate)?
- How do different weather variables affect each other?

Organizing
Prepare an organizer to collect the information needed to complete the study. Consider making a template for a spreadsheet program on your computer. If you do not have access to a computer, a paper and pencil organizer is another alternative.

Locating
Find the information needed to complete your study. Depending on the question you chose, it could be climate statistics for your city, weather forecasts from the newspaper, or long-range forecasts from an almanac.

Recording
Collect weather data over a period of time for the area you selected for in-depth study at the beginning of the course. The time period should be at least two weeks. Use weather instruments to record the following: temperature, precipitation, wind speed, wind direction, and barometric pressure. Remember, it is important to take all measurements at exactly the same time and place each day. Record the information in your organizer.

Evaluating and Assessing
Depending on the question you chose, compare your data with either published data or information provided by other students. If you chose to see how one set of weather measurements affects another, compare your data to see if there are any correlations. Possible correlations might be: wind direction and precipitation, wind direction and temperature, temperature and precipitation, barometric pressure and precipitation, etc.

Synthesizing
From your study, determine the patterns and trends. In other words, answer your focus question. Sometimes calculations are helpful in determining patterns. For example, you may want to determine the number of times the forecasted temperature was correct as a ratio to the number of times it was wrong. Another measure could be the percentage accuracy of rain forecasts. If you are comparing weather patterns in the community, it will be necessary to plot different readings on a map.

Concluding
Prepare a series of graphs to compare the information you synthesized. If you used a spreadsheet program, it is often easy to create a computer-generated graph. Be sure to choose the correct type of graph for the job. A time-series line graph is great for showing trends, but is probably not well suited to this study. Bar graphs and circle graphs are good for comparing

information; these would probably be most useful here. Scattergraphs show correlations and are useful if you want to determine how one variable affects another.

Applying
Make generalizations based on your findings and on your focus question. What recommendations would you make based on your focus question?

Communicating
Prepare a presentation to show others what you have discovered. Your presentation should include:

- the focus question
- your organizer showing all data collected from publications and your own measurements
- a map showing the location of the study area relative to the school and (where necessary) other study areas
- graphs comparing the information
- a conclusion written in paragraph form answering the focus question, with supporting evidence
- a written explanation accounting for differences that were either expected or unexpected.

The presentation could be in the form of a display, a bulletin board, a booklet, or even a video.

Weather Myths: Answers

1. **Groundhog Day:** *False.* Statistics show that groundhogs make terrible weather forecasters! The incidence of clouds on February 2nd has no influence on long-term weather. The legend probably evolved from a European proverb of the Middle Ages about Candlemas Day (February 2nd):

    If Candlemas Day be fair and bright,
    Winter will have another fight;
    If Candlemas Day brings cloud and rain,
    Winter is gone and won't come again.

2. **Cricket chirps:** *True.* Count the number of cricket chirps in eight seconds and add four. The answer should be within one or two degrees of the actual temperature.

3. **Sunsets and sunrises:** *True.* There is some truth to this ancient belief. In the middle latitudes, the weather usually moves from west to east. If there is a red sky at night, it is probably clear to the west because there are no clouds to block the sun. Therefore it is likely that it will be sunny the next day. When it is red in the morning (clear in the east), there is no indication whether it will be stormy or clear since the weather in Canada does not usually come from the east.

4. **Squirrels gathering nuts:** *False.* Squirrel activity is determined more by how good the crop was than by how bad the weather will be. Squirrels forecast the weather about as accurately as groundhogs!

5. **How March "comes in":** *False.* Environment Canada studied March weather in Halifax over a ten-year period. March came in like a lamb and left the same way seven times. Once it roared in and roared out. Twice it followed the pattern in the myth. Twenty per cent accuracy is not considered a reliable forecast!

Chapter 13

Battles in the Sky: Meteorology

INTRODUCTION

The primary reason for studying weather is so that accurate forecasts can be made. According to a recent national survey, 89 per cent of people listen to weather forecasts and use the information to help plan their activities. It has been estimated that weather forecasts are worth at least $1 billion dollars each year to the Canadian economy. Decisions made in industries such as agriculture, construction, fisheries, forestry, transportation, and recreation are all influenced by weather forecasts.

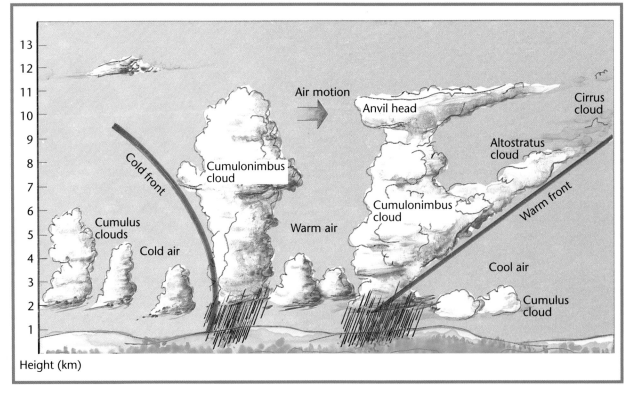

FIGURE 13.1 *Cloud Development in Cold and Warm Fronts*

Air Masses

The title of this section, "Battles in the Sky," is an appropriate one. Whenever we have a storm, a real battle is going on high above us in the troposphere. Different **air masses** are battling it out to see which one will dominate the weather in a given locale.

Air masses are bodies of air that develop over huge areas of the earth's surface. Two conditions are necessary for an air mass to form. First, the air must stay over the **source region** for a long period of time so that it can take on the characteristics of that region.

ular characteristics. For instance, if they develop over water, they have a high moisture content. These maritime air masses collect water vapour that has evaporated from the water below. The temperature of these air masses tends to be lower than that of the land in the summer but higher than the land temperature in the winter. This is because oceans moderate the temperature of the air mass. By contrast, continental air masses form over the continents. These air masses tend to be relatively dry compared to maritime air masses and more extreme in temperature. In winter, continental air masses are colder than maritime air

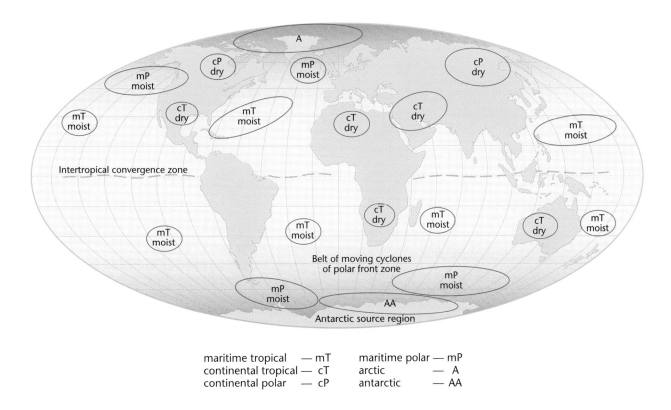

FIGURE 13.2 *Principal Air Masses of the World*

Second, the air must be stable. Air masses usually form where air is descending back to earth. These high pressure cells line up roughly over the polar circles and the tropics.

Depending on the nature of the land over which they develop, air masses take on partic-

masses. In summer, the opposite is true. These two different types of air masses affect the weather considerably in the middle latitudes where most Canadians live.

Variations in air mass characteristics are also determined by latitude. Air masses that form at high latitudes, such as over northern Canada, are much colder than those that form over

Arizona in the southwestern United States. Differences in **intensity** and **duration** of sunlight determine how warm the air is. Over the southern US, the amount of solar radiation received in winter is much greater than it is in northern Canada. Even in the summer, lower latitudes receive more solar energy than higher ones, although the difference is not as great. Because summer days are so much longer in the Canadian Arctic than they are further south, the amount of solar radiation is high compared with the winter when the number of daylight hours is less than six per day. Polar air masses are therefore quite warm in the summer compared with the extremely cold temperatures that characterize them in the winter.

Fronts

If air masses stayed where they were, the weather would be very boring! There would be few storms because the air would just sit over the land without changing. Of course, we know this is impossible. Prevailing winds and the rotation of the earth move air masses from one place to another. When air masses with different characteristics come into contact, a "battle" occurs. The line where the two air masses meet is called the **front**. Depending on the nature of the air masses, the temperature changes rapidly, winds pick up, and precipitation often occurs, especially if the "battle" is between a moist maritime air mass and a dry continental one.

The battleground where most of these conflicts occur is between polar and tropical source regions. These middle latitudes experience regular changes in weather as cold air masses meet warm ones and moist air masses meet dry ones. Once the battle between the two air masses is over, one air mass dominates. It is modified by the land beneath it. For example, in the winter, cP (continental Polar) air masses often move into the southern Prairie provinces and remain for several days. They bring crisp, cold days with sunshine and bitterly cold nights. Over two or three days the earth absorbs solar energy from the brilliant sunshine. This energy then modifies the air mass. Air temperatures rise, and air pressure begins to fall as the air becomes unstable.

Air masses are also formed over the Equator and the poles. The equatorial air mass seldom ventures far from equatorial oceans. Air flowing into the region from the tropics in the form of the trade winds keeps the air mass contained. Storms that occur in the region result from convection currents as the air is heated each day. Similarly, the arctic and antarctic air masses seldom venture into the middle latitudes. These air masses are responsible for severe blizzards in polar regions.

Meteorologists study and plot fronts as they move across the land. Observers can estimate when a storm will occur by studying clouds. Satellite images show the whole picture. Clouds and the fronts associated with them can be observed and their speed and direction determined.

There are two types of fronts. A warm front occurs when a warm air mass moves into a relatively cool region. This often happens in the spring as a southerly flow of air from the Gulf of Mexico flows north into southern parts of Quebec. The cold, dry air over Quebec is much heavier than the warm, moist air of the southern states. The lighter air rises above the heavier cold air, creating frontal precipitation. As the front moves north, the air temperature rises. Thus the precipitation, which often starts as snow, turns into freezing rain, and eventually becomes a rain storm. If the southerly flow of air is strong enough, it pushes the cold, dry air north, back to the polar continental source region.

Cold fronts occur when a relatively cold air mass moves into a warm region. The heavy, cold air pushes its way under the lighter warm air. As the warm, moist air is forced to rise, the temperature drops because of the adiabatic lapse rate. (See page 48.) The water vapour condenses and frontal precipitation occurs. As with a warm front, the type of precipitation can vary depending on the temperature. If the northerly flow is strong enough, the cold polar air mass will dominate the area until another frontal system moves in.

Clouds

Clouds are a telltale sign that a front is coming. When a warm front approaches, a layer of **cirrus clouds** develops, usually in the western or southwestern sky. These wispy, high altitude clouds indicate that warm air is flowing over a colder air mass; condensation has started to occur because of the differences in temperature. These clouds gradually thicken into **cirrostratus** and **altostratus clouds**. The increased cloud cover and loss of sunlight indicate that a storm is on the way. Light rain or snow starts to fall, sometimes accompanied by fog.

Fluffy white **cumulus clouds** found at lower altitudes usually indicate fair weather. These may gradually give way to **nimbostratus clouds**, which mark the beginning of a full-fledged storm. Clouds often build up, sometimes stretching as far as 10 km above the earth. **Anvil heads** form as **cumulonimbus clouds** build huge cloud fortresses in the sky. Heavy rain, thunder, lightning, and even hail may accompany these clouds, depending on the ferocity of the battle between the fronts. When the front finally passes through, the storm ends. The clouds dissipate, the temperature rises, and the sun comes out. Fluffy cumulus clouds may dot the sky once again. The wind usually drops, but it changes to a clockwise direction. Fog may linger until the sunlight burns it off. These types of storms are common in the summer, but they can occur in any season. The severity of the storm is determined by the difference between the two air masses. The greater the difference, the more severe the storm.

When a cold front passes into a region, the pattern is similar. Fair weather cumulus clouds develop into towering columns of cumulonimbus clouds where the two air masses meet. The length of the storm is short-lived. Thunder, lightning, and hail may accompany heavy rain, again depending on how great the difference between the two air masses. Temperatures drop and the wind shifts in a clockwise direction. Cold fronts also occur at any time of the year, although they are most common in the spring and fall. Sometimes extremely violent storms occur when a cold front moves through an area. If a moist, tropical air mass has dominated a region for some time in the late spring or early summer, a tornado may occur when the cold front moves through. These most violent of all storms kill many people each year and create awesome destruction of property. Meteorologists are particularly watchful when a cold front moves into an area of unstable maritime air.

Things To Do

1. Study Figure 13.2.
   a) Prepare a sketch map of North America. Indicate the location of source regions for each air mass. Shade the area between source regions and label it "Mid-latitude Storm Belt."
   b) Prepare a comparison organizer to describe each air mass. Include such criteria as: name (continental Polar, maritime Tropical, etc.); symbol (cP, mT, etc.); source region; relative temperature (hot, warm, cool, cold); and relative humidity (moist, dry).

2. Explain what happens to temperature, water vapour, and cloud cover when two air masses come into contact.

3. Describe the weather conditions associated with each of the following cloud types: a) cirrus; b) cumulus; and c) nimbostratus.

4. Keep a record of the clouds that pass overhead daily for one week. (Use a cloud indentification chart to help you.) Make weather predictions based on what you observe. Study the satellite images on TV weather broadcasts to verify the accuracy of your forecasts.

5. Why are storms more common in the middle latitudes than elsewhere?

Cyclones and Anticyclones

The fronts between polar and tropical air masses are stormy places. **Cyclones** are low pressure cells that often develop along these fronts. A ripple or wave forms in the front. Warm tropical air moves north along the front to the polar air mass. At the same time cold dry air from the north moves south along a cold front, forming a cyclonic storm or low pressure cell.

The cyclonic storm has three components: a warm front, a cold front, and the air in between them. As the warm front moves slowly north, its relatively warm, humid air overrides the cooler, drier air of the polar air mass and the air pressure drops. On the ground, the temperature is still low compared with the warm, humid tropical air high above. Water vapour condenses where the two air masses converge: heavy clouds form and precipitation begins to fall. Drizzle signals the approach of the warm front. The precipitation gradually becomes heavier until the warm front finally passes. The rain can last for two or three days as these continent-sized storms march across North America. When the warm front passes, the skies clear and the temperature rises. But this will not last long because the cold front is coming on fast behind the warm front.

The cold, dry air of the cP air mass bulldozes the warm, humid air high into the troposphere, creating precipitation. The spinning of the earth causes the cyclone to spin in a counterclockwise direction (to the left) in the Northern Hemisphere and

(a) Cold air moves westwards while warm air moves east. As a result of these opposing air flows and friction, a wave will develop in the front.

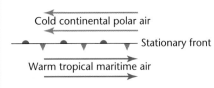

(b) Warm air advances polewards into the wave along the front. Thus, the air pressure falls towards the apex of the wave. The air circulates around the low pressure centre in an anticlockwise direction.

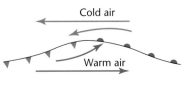

(c) The cyclonic circulation of air results in warm air advancing from the southwest. The warm air rises over the cold air flowing from the southeast. This is the warm front. At the rear of the depression, the cold air flowing from the northwest undercuts the southwesterly air. This is the cold front.

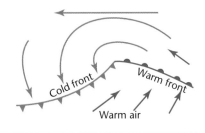

(d) The process described in (c) continues.

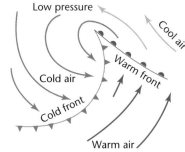

(e) The cold front generally advances faster than the warm front. It will finally displace the warm air at the earth's surface. The warm sector is thus gradually closed. The depression is at its maximum intensity.

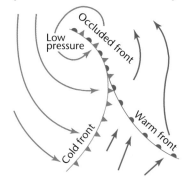

(f) Ultimately, the whole warm sector will be lifted upwards and replaced by cold air at the earth's surface. Once this happens, the depression ceases to exist.

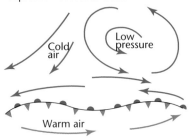

FIGURE 13.3 *Stages in the Development of a Middle Latitude Cyclonic Storm*

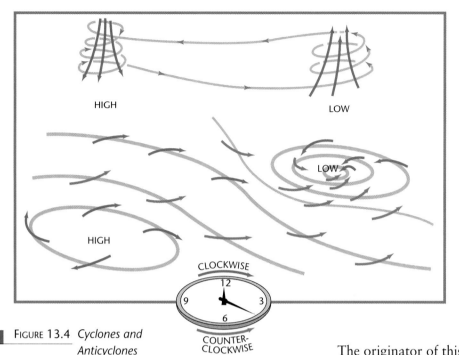

FIGURE 13.4 *Cyclones and Anticyclones*

in a clockwise direction in the Southern Hemisphere. This coriolis effect increases the strength of the storm. Wind speed picks up, the storm centre shrinks, and the storm begins to spin faster. (See Figure 13.4.)

Satellite images show cyclonic storms well. (See Figure 13.8 on page 93.) The clouds swirl into the low pressure cell along the cold and warm fronts. Because the cold front moves faster than the warm front, it often catches up to it. When it does, the warm front disappears from the earth's surface but remains high in the atmosphere above the earth. This is called an **occluded front**.

The storm is now at full intensity, but it will not last much longer. Once the temperature of the two air masses becomes similar and the humidity drops due to precipitation, the storm fades away. The air along the cold front becomes stable again and pleasant weather returns. These calm periods are called **anticyclones**, or high pressure cells. Dry, stable air descends to earth and dominates the weather.

Forecasting the Weather

Predicting the weather can be fun. But it can also be frustrating! So many variables affect the weather that a small change in any one condition can cause an abrupt and total change. Scientists are beginning to realize how incredibly complex natural systems are. A new scientific philosophy is evolving called the **chaos theory**. As important to modern scientific thought as the theory of relativity, scientists are beginning to measure random events and determine patterns even when none seem possible.

The originator of this theory was a meteorologist named Edward Lorenz. In the late 1960s, Lorenz was working on his weather model at the Massachusetts Institute of Technology when he had an incredible insight. Minute changes in data changed results radically. Applied to the real world, this suggested that a minute change in one region could affect weather the world over. For example, if a volcano erupts in Asia, temperature and rainfall patterns could be affected in Canada. Similarly, a change in ocean currents in the South Pacific could influence the weather in Europe. In the extreme, you could theorize that a butterfly flapping its wings in China could have an influence on the weather in Winnipeg! The **butterfly effect** sounds farfetched, but there is some truth to the view that atmospheric disturbances can affect weather thousands of kilometres away. You might understand this better if you think of a more familiar example. Let's say you stubbed your big toe. It is a minor injury but it hurts and you are forced to limp. The change in the way you move could affect your spine and might even result in a stiff neck. So although the toe was a minor problem, it affected your body in places far away from the original source. The same principle applies in the atmosphere.

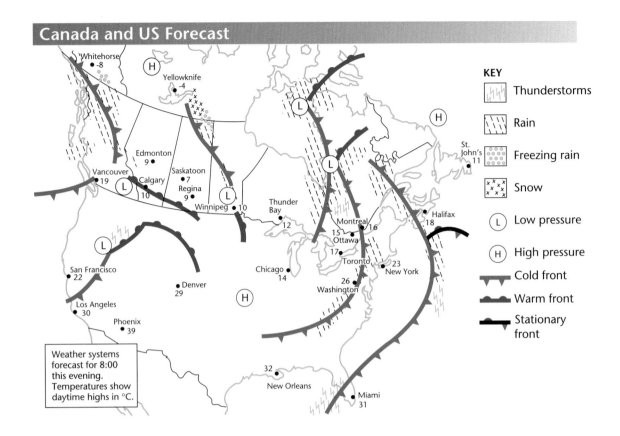

FIGURE 13.5 *Weather Map*

One event that can drastically affect the weather is a warm water current called **El Niño**. Every three to seven years, the waters of the Pacific Ocean off the coast of Peru become much warmer than usual. Patterns of air pressure change so dramatically that the trade winds blow from west to east instead of the other way around. This brings heavy rains to the west coast of South America. What seems to be a minor local event has a tremendous influence on weather all over the world. The El Niño effect upsets the world's patterns of prevailing winds. Deserts are flooded and fertile lands are made arid. In 1983, the El Niño cost hundreds of lives and billions of dollars in damage and destroyed ecosystems as droughts and floods struck towns and villages from India to Texas.

Sophisticated climate models can now predict an El Niño. Using infrared satellite images of the Pacific Ocean, the surface temperature of the water can be plotted by computer. When conditions match those of past El Niño years, researchers can accurately predict this potentially destructive ocean current. Two researchers at Columbia University predicted El Niños in 1986, 1991, and 1995. What used to be considered a random event has become predictable because of technology and an understanding that chaos does not necessarily mean "without order."

Storms seem to follow set pathways. If you study weather maps over several weeks, you will notice that storms flow across North America in the same pattern, usually from northwest to southeast, often dipping far to the south. Upper atmosphere studies have determined that storms follow **jet streams** that exist at altitudes between 9000 and 12 000 m. These winds often reach speeds in excess of 300 km/h.

One such jet stream exists over Canada. The powerful winds in this jet stream meander from side to side. As these waves become more exaggerated, they extend further south, often invading tropical regions. Cyclones are dragged along by the jet stream so that when it loops south, cold arctic air moves south with it. Cyclonic storms soon result as tropical and polar air masses collide.

Meteorologists are able to monitor jet stream activity using weather balloons and satellites. If they know where the jet stream is, they can easily predict where storms will move since they follow the same paths. Studying changes in jet streams could hold the key to long-range (three- to five-day) weather predictions.

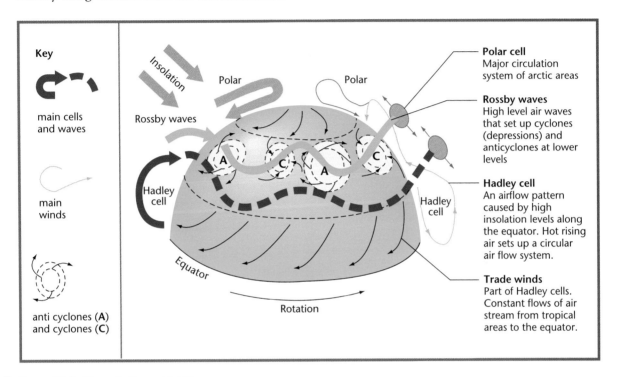

FIGURE 13.6 *Storm Paths and Jet Streams*

CAREER PROFILE: METEOROLOGIST

Are you fascinated by the weather? Are you good in geography, science, computers, and mathematics? Do you want a challenging career where every day is different? If you answered "yes" to these questions, then you may have what it takes to be a meteorologist.

The Nature of the Work
There are three different but interrelated fields for meteorologists:

Weather Forecasting: Forecasting the weather is a field everyone knows. Meteorologists use sophisticated mathematical models and computer/communications technology to measure and analyse the atmosphere and make weather forecasts. The person you see on television presenting the weather is not always a meteorologist. Most often this person is a broadcaster who simply delivers a report prepared by the meteorologists at the Atmospheric Environment Service.

Scientific Services: These scientists provide support services to governments and private organizations requiring meteorological input. Advice about such matters as air pollution, industry, transportation, recreation, and resource management are often required from meteorologists.

Research and Development: As with any scientific career, new methods and technologies are constantly being developed to improve our knowledge of the weather. Meteorologists are developing new equipment, computer programs, and methods of analysis so that our understanding of the weather can continue to grow.

Qualifications
The main employer of meteorologists in Canada is the Atmospheric Environment Service, located in Whitehorse, Vancouver, Edmonton, Winnipeg, Toronto, Montreal, Halifax, and Gander. Forecasters also work in armed forces bases around the world. Programs in meteorology are offered at the universities of Alberta, McGill, Toronto, Quebec at Montreal, Dalhousie, York, and British Columbia. Once hired, apprentice meteorologists undergo a rigorous training program at the AES training centre in Toronto.

Meteorological Technician
If you are more of a hands-on person who likes to work with high-tech equipment and would enjoy the prospect of working in a remote region, you might want to become a meteorological technician. These highly trained personnel work as weather observers, aerological technicians, ice observers, inspectors, data processors, and research assistants. A specialized degree is not necessary, but a high school diploma and high marks in physics, mathematics, and geography are essential. The Atmospheric Environment Service or your guidance counsellor can provide more information.

Things To Do

1. a) Examine Figure 13.5 and describe the weather patterns you observe.
   b) On a copy of Figure 13.5, colour the different types of precipitation.
   c) Label the air masses.
   d) Draw in the wind direction for cyclones (lows) and anticyclones (highs).
   e) In an organizer, describe the weather (temperature, precipitation, wind direction, and air pressure) for the following places: Yellowknife, Los Angeles, Phoenix, Toronto, Miami, St. John's, Vancouver.
   f) Predict the weather for the day *after* this map was made. Justify your forecast.

2. Clip the weather map from the newspaper each day for a week to predict the weather for the following day. Record the accuracy of your predictions.

3. Study the satellite image in Figure 13.7. Notice that there are "streamers" of clouds coming off Lake Superior, Georgian Bay, and Lake Erie.
   a) Which is warmer, the water or the air? Explain.
   b) Why do you think clouds are forming over the lake?
   c) What time of year is it? Give reasons for your answer.
   d) What is your forecast for areas down wind of the lakes (Collingwood, Buffalo, London)?
   e) How could this satellite image help in weather forecasting?

4. Make a photocopy of the satellite image in Figure 13.8.
   a) Label the cyclone in the image as a "low."
   b) Draw in the fronts and use arrows to indicate the wind direction around the cyclone.
   c) Label areas where "highs" might be.

5. Outline how studying jet streams could facilitate long-range forecasting.

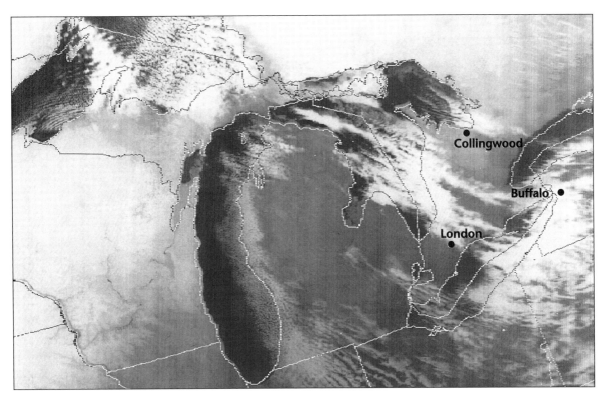

FIGURE 13.7 *A Satellite Image Showing the Great Lakes Region*
Notice the streamers of cloud coming off the lakes.

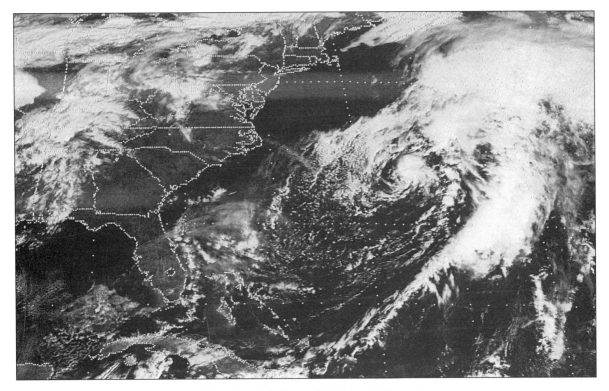

FIGURE 13.8 *A Satellite Image of a Well Developed Cyclone off the Island of Bermuda*

Chapter 14

Violent Weather: An Integrative Study

INTRODUCTION

There is nothing more awesome than a violent storm. Whether it be a severe thunderstorm, a winter blizzard, or a killer tornado, violent weather reminds us that we are tiny, vulnerable creatures compared with the forces of nature. Today we know much more about violent storms than we used to, but we are still powerless to control them. We can study storms, however, to understand them better and to prepare ourselves for the destruction they bring.

FIGURE 14.1 *A Hurricane as Seen Via Satellite*
Notice the distinctive eye and the heavy accumulation of cloud.

Hurricanes

Hurricanes, or tropical cyclones, are the most dangerous of all storms. About eighty hurricanes a year form in the Atlantic, the Pacific, and the Indian oceans. It has been estimated that in an average year up to 15 000 people are killed and over $5 billion worth of property is destroyed by hurricanes.

In Canada we are lucky in that we do not get many hurricanes. But when they do occur they cause tremendous damage. On the 14th and 15th of October 1954, Hurricane Hazel roared through southern Ontario. Rainfall totalling 182 mm poured down; winds reached speeds of 125 km/h. The entire Humber Valley floodplain became one huge river. A trailer park was swept away. Eighty people died. Farm animals drowned. Damage reached $100 million.

In the aftermath of the storm, the Metropolitan Toronto Conservation Authority was formed to prevent similar destruction in the event of another major storm. The organization took action to reduce the risk of flooding. Prior to the hurricane, it had been an unusually wet autumn in Toronto. The rivers were swollen and the ground was so sodden that it could not absorb any more rain. When Hurricane Hazel hit, the water from the torrential downpour could only overflow the river banks. After the disaster, the Conservation Authority built reservoirs to hold back water and planted thousands of trees to hold moisture in the soil, reduce erosion, and slow the rate at which water enters rivers. River channels were improved so that water could flow rapidly to the lake without backing up at bottlenecks. Many rivers and creeks were dredged, straightened, and contained in cement channels. Conservation areas were established north of the city in the Humber and Don river **watersheds**. They serve as wilderness and park areas where water can drain slowly into the rivers a little at a time. Though Hurricane Hazel was a tragedy, it resulted in Toronto becoming a greener and more attractive city.

Hurricanes are more common in tropical areas. In Japan they are known as typhoons and in Australia they are called willy-willies. No matter what the name, however, these storms share common characteristics. They are similar to the mid-latitude cyclones studied earlier (see pages 88-89), except that they grow to enormous proportions.

These storms are created when the hot, sunny weather heats the ocean in the tropics all summer long. By late summer or early fall, an enormous amount of thermal energy is stored in the tropical oceans. The air over the water is heated by the sun and the water. A

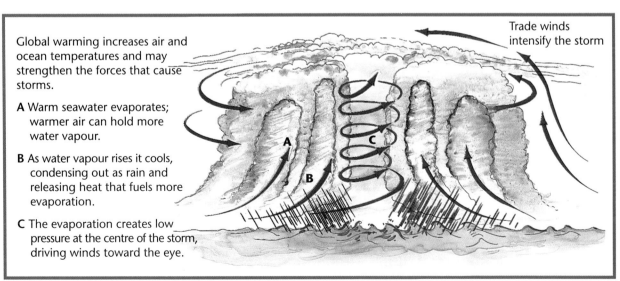

Global warming increases air and ocean temperatures and may strengthen the forces that cause storms.

A Warm seawater evaporates; warmer air can hold more water vapour.

B As water vapour rises it cools, condensing out as rain and releasing heat that fuels more evaporation.

C The evaporation creates low pressure at the centre of the storm, driving winds toward the eye.

Trade winds intensify the storm

From Newsweek, 7 September 1992, Blumrich.

FIGURE 14.2 *Anatomy of a Hurricane*

GEO-Fact

With apexes at Bermuda, Puerto Rico, and Fort Lauderdale, Florida, the Bermuda Triangle is a region where many ships and planes have mysteriously disappeared. A new theory has been proposed that may explain the mystery. Large pockets of methane, some as large as 5 km in diameter, lie under the sea floor. Occasionally this natural gas boils up through the floor and the water and out into the atmosphere. The water is full of methane bubbles which are less dense than the surrounding water. This causes ships entering the area to sink. Airplanes flying overhead may also hit a methane pocket. Since the amount of oxygen in the air is reduced, the plane's engines stall and the plane crashes. So why are the wrecks not found? Because they are buried under the sea floor, which has turned to mud as a result of the eruption. Does this sound farfetched? It is still a theory but scientists are investigating to see if there is some truth to it.

convection current starts. The warm, moist air rises rapidly. Winds blow into the partial vacuum that is created as this warm air rises. The coriolis effect starts the system spinning. As more and more air and water are sucked into the low pressure cell, the storm intensifies. The air pressure drops in the centre, winds pick up, and the rain starts to fall. At this point, it is just a thunderstorm. It would remain so if it were not for a "kicker" that causes the storm to become even stronger.

The trade winds make the thunderstorm into a full-fledged hurricane. Blowing steadily from the east, day after day, these strong prevailing winds intensify the counterclockwise spinning of the storm. As the spinning increases, a hurricane is born.

Meteorologists use satellite images to identify the classic cloud formation of a hurricane. Clouds swirl in spirals several hundred kilometres away from the storm centre. Close to the centre the clouds are so dense that individual swirls are no longer visible. Dead in the centre of the storm is the "eye." There are no clouds here and the sea is visible below.

The speed and direction of hurricanes can also be determined using satellite images. The position of the eye is plotted on a map each hour. The distance the hurricane has moved is calculated using the scale on the map. The direction is determined by projecting the path of the storm along the line it seems to be taking. Direction is not always easy to predict, however, because storms frequently take unexpected turns.

Meteorologists also study hurricanes using airplanes that actually fly at low altitudes right through the storm! They measure wind speed, atmospheric pressure, rainfall, and, if possible, wave height. This is a dangerous assignment, but one that provides vital information to forecasters.

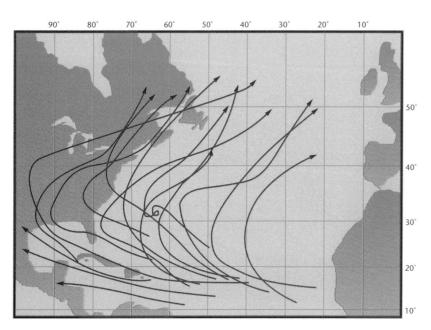

FIGURE 14.3 *Tracks of Some Typical Hurricanes in August. Notice that the storms follow the prevailing winds.*

Name	Date	Affected Area	No. of Deaths
*	Sept. 1900	coastal Texas	6000
*	Sept. 1926	Florida, Alabama	372
*	Oct. 1926	Cuba	600
*	Sept. 1928	West Indies, Florida	6000
*	Sept. 1930	Dominican Republic	2000
*	Sept. 1938	New England	600
*	Sept. 1944	Atlantic coast from North Carolina to Maine	389
Hazel	Oct. 1954	Eastern USA, Ontario	347
Diane	Aug. 1955	Eastern USA	400
Hilda	Sept. 1955	Mexican coast	200
Janet	Sept. 1955	Caribbean	500
Audrey	June 1957	Louisiana, Texas	430
Hattie	Oct. 1961	Belize	400
Flora	Oct. 1963	Cuba, Haiti	6000
Inez	Sept. 1966	Caribbean, Florida, Mexico	293
Camille	Aug. 1969	Mississippi, Louisiana	256
Fifi	Sept. 1974	Honduras	2000
David	Sept. 1979	Dominican Republic, Dominica, Florida	1200
Allen	Aug. 1980	Caribbean, Texas	272
Gilbert	Sept. 1988	Jamaica	318
Andrew	Aug. 1992	Bahamas, Florida, Louisiana	62

*Hurricanes were not named prior to the Second World War.

FIGURE 14.4 *Significant Hurricanes of the Twentieth Century*

Sometimes storms head for the mainland. Destruction can be terrible as waters rise in low-lying areas and winds destroy property. Frequently lives are lost. Hurricanes that move inland do not stay dangerous for long, however. They lose power because the land they are crossing does not build up energy in the storm the same way in which warm tropical oceans do. Once they move inland, these violent storms lose strength and become mid-latitude cyclones.

Called "the storm of the century," Hurricane Andrew hit the coast of the United States in late August and early September 1992. The storm had already devastated The Bahamas in the Caribbean. Meteorologists in the US warned people of the impending dangers. Residents and tourists heeded their warnings and evacuated low-lying areas all along the Florida and Gulf of Mexico coasts. Stores, homes, and other buildings were boarded up. When Hurricane Andrew screamed into Miami in the early morning of 31 August 1992, the people were ready. After battering southern Florida, the hurricane roared across the Gulf of Mexico, building up more power from the warm waters. By the time it hit the Mississippi Delta in Louisiana, wind gusts reached up to 270 km/h—as high as wind speeds recorded in some tornadoes! But this storm was much bigger and lasted much longer than a tornado. As Andrew moved inland up the Mississippi Valley, it quickly lost its energy. By the time it reached southern Ontario, it was just a bad storm.

In all, Hurricane Andrew claimed sixty-two lives—a low death toll for such a wild storm. But the destruction Andrew caused was unbelievable. The hurricane became the most costly natural disaster in American history. A quarter

In North America, hurricanes sometimes travel up the Atlantic coast of the United States. At other times they slash their way across the Caribbean Sea and the Gulf of Mexico. Storms that linger over the ocean are often long-lived because they continue to pick up energy from water that has been heated all summer by the tropical sun. Fortunately, however, they cause little damage unless they strike an island. Then, of course, the scale of the tragedy can be immense. Such was the case in September 1988 when Hurricane Gilbert hit Jamaica. Three hundred and eighteen people were killed. Thousands lost their homes, tourist hotels were devastated, and agricultural crops were destroyed.

of a million people were left homeless, thousands were injured, and property damage was estimated at over $35 billion (US). The clean-up took months and in some situations years. The storm was also an ecological disaster.

Stately palm trees were uprooted and citrus groves lost their crops. Parts of the Everglades swamp were inundated with sea water, which resulted in the destruction of many of the delicate ecosystems found there.

J	F	M	A	M	J	J	A	S	O	N	D	TOTAL
0	0	0	1	3	23	33	143	181	88	21	3	496

FIGURE 14.5 *Total Number of Hurricanes by Month, 1886-1986*

Things To Do

1. Describe the characteristics of hurricanes.

2. Use a data base or spreadsheet program with graphing capabilities to record the information listed in Figure 14.5. Determine the annual frequency of hurricanes for each month. Show your results on a bar graph. What generalization can you make from this? How could this affect tourism in the Caribbean?

3. Prepare a graded shading map using the data from Figure 14.4. Use one colour. Each time a place is mentioned in the organizer, colour that spot. If a place is mentioned more than once, it will have a darker shade of the colour because you have coloured it more than once. When you have finished, you will have a map that shows locations that are most prone to hurricanes.

4. a) What generalizations can you make about where hurricanes occur?
   b) On a world map, show other locations where hurricanes could occur.
   c) Explain why hurricanes do not form over land, in middle latitudes, or over the Equator.

5. Prepare a list of planning regulations that a town in a hurricane-prone region should follow.

Tornadoes

Tornadoes are similar to hurricanes in that both are violent storms with high winds and heavy rains. Both also tend to move with the prevailing winds; in Canada, this is usually from the southwest. Hurricanes, however, cause tremendous damage over large areas; entire states and even whole countries can be devastated by a hurricane. Tornadoes, on the other hand, cause more localized damage; their enormous power is confined to a relatively narrow band. Hurricanes are created over warm seas in late summer or early autumn. Tornadoes are most common in the spring or early summer and almost always form over land. The trade winds cause a tropical storm to intensify into a hurricane. **Wind shear** caused by a cold front moving through a maritime tropical air mass starts a tornado spinning. The different characteristics of each storm require people to take different precautions.

Tornadoes move along the ground about as fast as a car— between 50 and 100 km/h. You can usually see a "twister" coming; if you do, get out of the way! Tornadoes can pack wind

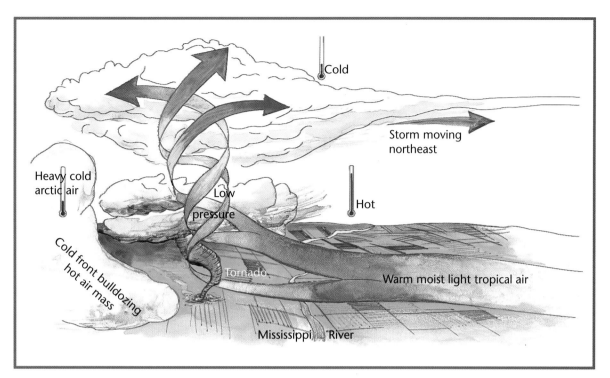

FIGURE 14.6 *How Tornadoes Are Formed*

speeds as high as 500 km/h. This is powerful enough to drive a wooden plank through a concrete wall or lift a car 20 m in the air.

Unlike hurricanes, tornadoes cannot be detected from satellite images. The cloud cover is so great that the characteristic funnel clouds are obscured. The only way to detect a tornado is from the ground. If a tornado is approaching, the sky turns greenish-grey, the temperature drops from a hot, humid condition, winds become very strong, and there is an exhilarating feel to the air.

When conditions are right for a tornado, a tornado watch alert is issued by the Atmospheric Environment Service. Trained volunteer weather observers keep a close watch and report any **funnel clouds** to the local weather office. When a dark, funnel-shaped cloud hanging from a stormy sky touches down, a tornado warning is issued. When this happens, people should take cover immediately.

A tornado will skip and hop across the countryside. Whenever it touches the ground, dust, debris, and anything not held down is sucked up the funnel just as if it were a giant vacuum cleaner! The event may last a few seconds or a few minutes at most, but the death and destruction left behind will be remembered for a lifetime.

Unlike hurricanes, tornadoes are almost impossible to study. While meteorologists can predict when they might occur, it is almost impossible to tell exactly where they will touch down. Wind speeds can be estimated after the fact by studying the damage. Engineers are able to determine the amount of force required to cause certain types of damage. From this data, it is possible to ascertain the amount of energy given off by the whirlwind. As far as forecasting tornadoes, however, the best meteorologists can do is study the atmospheric conditions that caused tornadoes in the past and assume that the same conditions will cause tornadoes in the future.

There are two conditions necessary for a tornado to occur. First, the weather must be so hot and humid that there are strong convection currents. Considerable cloud cover in the form of cumulonimbus clouds completely blocks the sky after a hot, humid morning and early afternoon. Second, a cold front must move into the area, usually from the northwest. The coriolis effect causes the convection currents to spin around the low pressure cell. The wind shear from the cold front creates extra energy in the atmosphere, which may form into funnel clouds. When these conditions exist, the weather office is on the lookout for these most violent of all storms.

Location	Wind Speed* (km/h)	Length† Travelled (km)	Width† (m)
Rush Cove	150	4	30
Hopeville	200	17	150
Corbetton	200	35	200
Borden	150	18	150
Essa	100	1	30
Barrie	250	10	250
Grand Valley	250	102	250
Wagner Lake	100	4	100
Reaboro	100	8	100
Alma	200	33	150
Ida	150	9	150
Rice Lake	200	14	200
Minto	100	1	13

* Wind speeds are estimates based on damage.
† Width and length estimates are based on aerial photographs.

FIGURE 14.7 *Ontario Tornadoes, 31 May 1985*
It was very unusual for so many tornadoes to touch down in a single day in a Canadian location.

FIGURE 14.8 *A Tornado Touches Down*

TORNADOES IN SOUTH-CENTRAL ONTARIO: A PERSONAL ACCOUNT

It had been a hot and humid summer day. As I drove north from Toronto, I could see the huge, fluffy cumulus clouds building like a fortress on the horizon over Newmarket. Occasional crackles on the car radio, and later flashes of lightning, warned me that we were in for a major thunderstorm. As I got closer to the storm, the clouds became so thick I could not see the sky at all. The wind picked up until it reached gale force. The rain started to fall in sheets. The weather office issued a tornado watch; a funnel cloud had been sighted in the area. It was 4:15. I should have pulled off the road, but I wanted to get home to make sure that my family was safe. Driving was treacherous. Cars were swerving all over the road because of the poor visibility and the strong winds. Trees started to come down. Finally, my driveway came into view. I hurried into the house and found that everyone was safely at home. They were all watching the storm from our big picture window on the second floor. Annie, our dog, was sleeping peacefully in the basement. We had been lucky. The tornado missed us this time.

Province/State	Tornadoes	Province/State	Tornadoes
Alberta	0.3	Nova Scotia	0.5
British Columbia	0	Ontario	0.2
Manitoba	0.1	Prince Edward Island	0
New Brunswick	0	Quebec	0
Newfoundland	0	Saskatchewan	0.5
Alabama	0.8	Montana	0.8
Alaska	0	Nebraska	0.9
Arizona	0.1	Nevada	0.7
Arkansas	1.0	New Hampshire	0
California	0.4	New Jersey	0
Colorado	2.8	New Mexico	0
Connecticut	0	New York	0
Delaware	0	North Carolina	0.8
Florida	4.0	North Dakota	0.9
Georgia	1.1	Ohio	0.7
Hawaii	0	Oklahoma	3.9
Idaho	0.5	Oregon	0.2
Illinois	2.1	Pennsylvania	0.6
Indiana	1.5	Rhode Island	0
Iowa	3.7	South Carolina	1.3
Kansas	5.5	South Dakota	1.7
Kentucky	0.8	Tennessee	0.8
Louisiana	3.1	Texas	2.8
Maine	0.1	Utah	0.1
Maryland	0.4	Vermont	0.4
Massachusetts	0.5	Virginia	0.2
Michigan	1.5	Washington	0.1
Minnesota	1.8	West Virginia	0.2
Mississippi	1.0	Wisconsin	0.7
Missouri	1.4	Wyoming	0.9

FIGURE 14.9 *Tornado Frequency per 10 000 km² by Province/State, 1991*

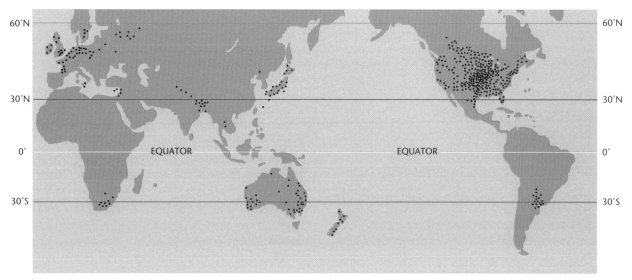

FIGURE 14.10 *Incidence of Tornadoes Worldwide*

Things To Do

1. Prepare an organizer comparing tornadoes with hurricanes, using headings and subheadings.

2. Try this simulation. Get two 2 L plastic soft drink containers. Fill one almost to the top with water. Attach the top of the second bottle to the top of the first using two large washers and strong duct tape. Invert the bottles so that the empty one is on the bottom. Rotate the bottles to the left so that the water in one flows down to the other one. What you have is a "tornado in a bottle."
   a) How is this simulation like a tornado?
   b) How is it different?
   c) Explain how tornadoes are formed using Figure 14.6 and this simulation.

3. Complete the following statistical study using Figure 14.7. Use a spreadsheet or data base with graphing capabilities if available.
   a) Calculate the average length, width, and wind speed of the tornadoes.
   b) Draw bar graphs of the length, width, and wind speed for each.
   c) Prepare scattergraphs for each of the following pairs of data: length and wind speed, length and width, width and wind speed.
   d) What correlations do you find among the data?
   e) How could this correlation help researchers studying tornadoes?

4. a) Prepare a graded shading map using the data in Figure 14.9 to show the distribution of tornadoes across North America.
   b) Explain why the organizer did not just list the number of tornadoes sighted by state or province but instead showed the number of tornadoes per 10 000 km^2.
   c) What generalizations can you make about where tornadoes are most commonly found?
   d) Explain why tornadoes are most common in the regions listed in part c).

5. Study Figure 14.10. List the regions of the world where tornadoes occur. What generalizations can you make about the locations of tornadoes?

6. a) Read the eyewitness account of a tornado on page 100. What did the eyewitness and the family do wrong? Did any member of the family do the right thing?
   b) Rank the following places from safest to most dangerous in a tornado and explain your ranking: a parked car, a soccer field, a basement, an attic, a ditch, a tree house, a shopping mall, a sailboat.
   c) Make a list of safety precautions people should take when there is a tornado watch.

Blizzards

Blizzards are severe winter storms. Environment Canada describes blizzards as having a temperature of -12°C or less, a wind speed of 40 km/h or more, visibility of less than 1 km, and a duration of no less than three hours. Canada is famous for them. The blizzard of 1959 is long remembered in Newfoundland. Six people died, 70 000 were left without power, and highways were blocked with giant snow drifts for days. Blizzards can cause havoc, especially when they hit regions unaccustomed to heavy snowfall. The temperate southwest coast of British Columbia was frozen by one of these wintry blasts on 4 December 1980. Victoria received over 20 cm of snow, accompanied by strong winds. Three people died.

Blizzards are caused when frigid arctic air comes in contact with warm, moist tropical air. Of course, the temperature has to be below freezing for the precipitation to fall as snow. Mid-latitude cyclones dip far south into temperate regions with the jet stream that flows over North America. If the path of the jet stream loops far to the south, there is a good chance that cold, dry air will come in contact with moist, tropical air from the southern United States. When this happens, look out! A blizzard is on the way.

Such a storm hit the southeastern United States in March 1993. It stranded thousands of vacationers as they travelled south for the traditional March school vacation.

FIGURE 14.11 *Blizzard Conditions in the BC Rockies*

25 killed; 3 million on coast without power

By Roger Petterson
Associated Press

A fierce storm bombarded the East Coast yesterday with record snow — including 4 m drifts in Virginia — wind exceeding 160 km/h and killer tornadoes. At least 25 people died, more than 3 million customers were without electricity and thousands of travelers were stranded....

Thirteen of the deaths were in tornado-ravaged Florida, still recovering from Hurricane Andrew last summer.

Snow depths by early evening included 1 m in western North Carolina; 60 cm in the mountains of West Virginia; 60 cm in Eastern Kentucky; 53 cm in eastern Tennessee; a record 38 cm at Birmingham, Ala.; and 40 cm in northern Georgia.

Farther north, 38 cm fell at Philadelphia and 51 cm elsewhere in Pennsylvania; 25 cm at New York City's Central Park; and 38 cm in parts of New Hampshire.

"It's turning into a record snowstorm for the East Coast," said National Weather Service meteorologist Mike Wyllie in New York City. But blizzard warnings were canceled along the southern New England coast as the precipitation turned to rain.

Shore residents were evacuated from Delaware to Maine....

Power outages affected nearly 2 million customers in Florida, and outages also were reported across the South and up the East Coast to Connecticut, utilities reported. "We have some people in trucks who can't go anywhere because the snow is too deep," said Alabama Power Co. spokesman Dave Rickey.

Up to 1 m of snow — before drifts — was forecast for hardest-hit areas from the mountains of West Virginia to northern New England.

"This is like a hurricane with snow," said Devin Dean, a forecaster at the Atmospheric Science Research Center of the State University of New York in Albany.

The atmospheric pressure at the storm's centre was lower than the extra-low pressures at the center of some hurricanes. By creating a contrast with surrounding air, the low pressures create high winds.

States of emergency were declared from Florida northward to Maine.

The storm closed key airports at New York City, Washington, Baltimore, Boston, Philadelphia, Pittsburgh and Atlanta, stopping hundreds of flights. Almost 3000 people were stranded at just New York's major airports, Port Authority officials said.

All interstate highways in Pennsylvania were ordered closed and motorists in West Virginia were told that interstates would not be plowed. Highways also were closed in North Carolina, Kentucky, Virginia, Tennessee and New York state....

Reprinted by permission of Associated Press.

FIGURE 14.12 *The Blizzard of '93*

The Blizzard of '93: One Family's Experience

We left London, Ontario, after school on Friday, hoping to get to Daytona Beach, Florida, by Saturday night. A severe storm was reported moving into the southeastern United States, but we had driven in bad weather before and we were eager to get to our vacation in the sun, so we decided to go for it.

Our plan was to cross the border at Niagara Falls and travel down Highway 90 to interstates 79, 77, and 95 in the eastern US. Instead we decided to drive west and take the I-75 from Detroit all the way to Orlando. Although this route is longer, we thought we would skirt the storm, which was forecast to move up the eastern seaboard.

We were making great time; after only ten hours we were in Kentucky. The weather was cold but clear. But things started to change once we entered the mountains of Kentucky. The weather took a turn for the worse and road conditions started to deteriorate. At first a light dusting of snow covered the highway. By the time we reached Lexington, however, the snow was really coming down. We decided to find a motel, but it was 3:00 in the morning and all the rooms were booked. We had no choice but to travel on.

Road conditions were really bad. There were no snow ploughs or salt trucks out, even though the snow was already 10 cm deep on the highway! Cars in the right lane were moving so slowly they could not make the mountain grades. Left-lane traffic was travelling too quickly and many cars were skidding out of control. When a tractor-trailer jack-knifed in front of us, we knew it was too dangerous to continue. We pulled off the highway at the first opportunity.

We ended up spending the next twenty-four hours sitting out the blizzard at a truck stop with 150 other weary travellers. We heard that over 1000 people were stranded on the interstate in Kentucky and Tennessee. We realized we were the lucky ones.

Sunday morning, after sleeping on the floor of the truck stop, we took a walk to the highway. The snow was 30 to 40 cm deep, with drifts up to 2 m, but the road was clear. The worst was over, we thought. We started driving, but after only 19 km the highway ahead was closed. This time we were lucky enough to get a room at a motel. From our window we could see the armoury across the road, where hundreds of vehicles were being towed from the interstate.

On Monday we were still determined to get to Florida. After what we had been through, we really deserved some rest and relaxation on the beach! In Tennessee the work crews had done an amazing job clearing the road. Although we heard that the highway was closed once again in Georgia, we figured it would be clear by the time we got there. But again things didn't go according to plan. In Georgia we got stuck in a seven-hour traffic jam.

We decided to leave the interstate and try our luck on a smaller highway running through the mountains. Axle-deep snow made the trip treacherous, but at least we were moving. Once we reached the Florida border we rejoined the interstate and finally made it to Daytona Beach—at 6:00 Tuesday morning! Our twenty-four hour drive had taken eighty-one hours. Next time we'll fly!

Things To Do

1. Summarize the problems the blizzard caused as described in the news story. What precautions do you think travellers should take in such storms?

2. a) Use a sketch map of Figure 14.13 to draw the two routes mentioned in the eyewitness account.
   b) On a road map of North America, determine the shortest distance from where you live to Daytona Beach, Florida.
   c) Estimate the travelling time on this route at an average speed of 100 km/h.

3. Evaluate the decisions made by the family driving to Florida. Which ones were good? Which ones were foolhardy?

4. Write a personal account of an incident when weather altered your plans.

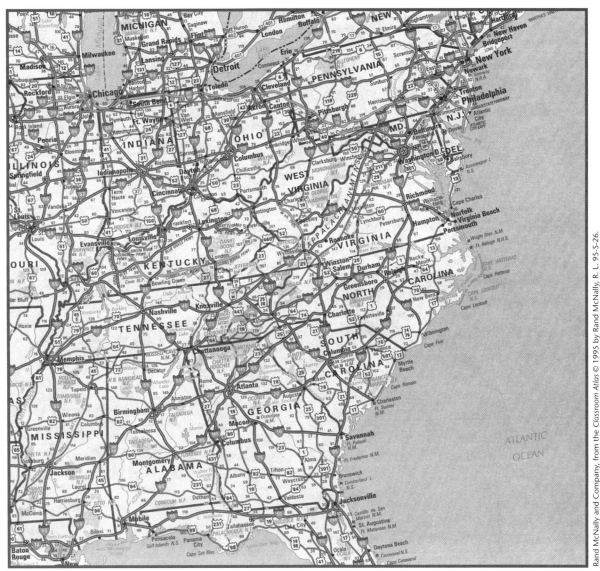

FIGURE 14.13 *Road Map of the Eastern United States*

Chapter 15
Climatology

INTRODUCTION

Long-term average weather, or **climate**, is essential to the study of physical geography. The concept of climate is really quite abstract since it is based on long-term averages. However, climate is central to much that happens on the planet. The biosphere develops the way it does because of the climate. Similarly, the biosphere determines, to some extent, what the climate is like. For example, when forest cover is stripped away, rainfall often decreases due to reduced transpiration.

Soils and other elements of the lithosphere are also affected by climate. If rainfall is heavy, soils are often **leached**, but if it is light, salt deposits may develop on the surface of the soil. The amount of precipitation and the temperature affect the way rocks **weather**, or break down.

The hydrosphere is intimately bound to climate. Precipitation patterns, ocean currents, and other aspects of the hydrosphere greatly affect climate and are in turn affected by it. Human activities are also influenced by climate. To study the climate of a place is to understand the basis for life in that region.

FIGURE 15.1 *British Columbia and Alberta in February*
What differences are there in the photo on the left, taken near Victoria, B.C., and the photo on the right, taken near Calgary, Alberta? Where would you rather be in February? Why?

Classifying Climate

The earth has many different climates. People living in Victoria have a slightly different climate than people living as close as Vancouver. The two cities are just 182 km apart by ferry across the Strait of Georgia. Local climate variations exist, however, because of differences in relief and exposure to prevailing winds. Yet there are similarities, too, at least when compared to the climate in another place, say Calgary. These similarities are the basis for climate classification.

City		J	F	M	A	M	J	J	A	S	O	N	D
Victoria	T [°C]	4	5	7	9	12	14	16	16	14	11	7	5
	P [mm]	114	76	58	31	25	23	10	15	38	71	109	119
Vancouver	T [°C]	2	4	6	9	13	16	18	17	14	10	6	4
	P [mm]	218	147	127	84	71	64	31	43	91	147	210	224
Calgary	T [°C]	-11	-8	-4	4	10	13	16	15	10	4	-3	-7
	P [mm]	13	13	20	25	58	79	64	58	38	18	18	15

FIGURE 15.2 *Monthly Average Temperatures and Precipitation Totals for Victoria, Vancouver, and Calgary*

Things To Do

In your atlas, locate Victoria, Vancouver, and Calgary, then examine the climate statistics in Figure 15.2.

1. Copy the organizer below into your notebok and complete the comparison of the three climate stations.

Criterion	Victoria	Vancouver	Calgary
Coldest temperature (with month)			
Warmest temperature (with month)			
Mean annual temperature			
Temperature range			
Months with temperature <0°C			
Total precipitation			
Driest month (with precip. total)			
Wettest month (with precip. total)			

2. a) Prepare a point-form summary outlining how the climates of Victoria and Vancouver are similar and different.
   b) What generalizations can you make about the climate of southwestern BC?

3. a) Prepare a point-form summary outlining how the climate in southwestern BC is different from the climate in Calgary.
   b) Outline how life in the two regions might be different because of their climatic differences.
   c) Based on climate, in which of the two regions would you prefer to live? Explain your answer.

4. Compare the climate of your region with that of southwestern BC. How are the regions similar? How are they different?

Temperature Zones and Climate Classification

The first people to come up with a climate classification were the ancient Greeks. They based climate on temperature. The region that was cold all year was called the frigid zone. This was the land of ice and snow where the mean monthly temperature never rose above 10°C and life was hard. In equatorial regions, the mean monthly temperature never dropped below 18°C. This torrid zone was also difficult to live in. The Greeks believed that the area where they lived was the best because there were four seasons. It was cooler than the torrid zone but not as cold as the frigid zone. They called this region the temperate zone. These three zones made up a simple classification system, but there were definite limitations.

Climate Classification with Two Variables

The system developed by the Greeks was a start, but it is obvious that it needed to be refined and expanded. The climates of Vancouver and Calgary are both temperate, but so is the climate of southern California and northern Siberia. To say that they are similar would be misleading. Obviously the temperate zone needed to be further divided.

Another problem with the system is that it does not take precipitation into consideration. Figures 15.3 and 15.4 both show tropical climates, but it is obvious that they are quite different. One is a tropical rainforest climate and the other is a desert. A thorough classification system would have to include both temperature and precipitation characteristics.

The Köppen Climate Classification System

In the previous activity, you probably discovered that the main problem with any classification system is definitions. How wet is wet? How cold is cold? Obviously there is a need for a more precise system of classification. Such a system was developed by Austrian cli-

FIGURE 15.3 *The Tropical Rainforest*

FIGURE 15.4 *The Hot Desert*
Both the tropical rainforest and the hot desert are tropical, but in what ways are they different from one another?

matologist Wladimir Köppen in 1918. Since that time this system has undergone many modifications, but it remains the standard used by climatologists the world over.

Köppen's classification system is based on temperature and precipitation. Instead of using descriptive terms like cold or wet, he added numerical values to define exactly what he meant by cold, wet, and so on. Each climate has a two- or three-letter code. By learning the definitions of each code a precise understanding of the climate can be determined. The classification system in this book is a simplification of the Köppen model.

Things To Do

1. Today the three zones created by the ancient Greeks are called polar, tropical, and temperate. Write mathematical expressions to define each. For example, polar would be $T<10°C$, where T is the temperature of the warmest month.

2. a) Study the climate statistics in Figure 15.6 on page 110. State the zone each climate station is in. Be prepared to justify your response.
   b) Use an atlas to find the location of each station. Label each station on a map of the world. Add the 10°C January isotherm and the 18°C July isotherm. Colour in the three temperature zones. (See Figure 15.5.)
   c) Study the differences within zones and determine the limitations of this system. How would you improve this classification system?

3. Write down all the possible climate combinations you could have using the following descriptive terms:

Temperature	Precipitation
Tropical	Wet all year
Temperate mild winter	Dry all year
Temperate cold winter	Very dry all year
Polar	Winter drought
	Summer drought

4. a) Work with a group to make climate graphs for each of the climate stations in Figure 15.7 on page 110.
   b) Match each graph to one of the climate types you established in activity 3.
   c) Share your group's results with other groups in the class. Did you all agree? What difficulties still exist with your classification system?

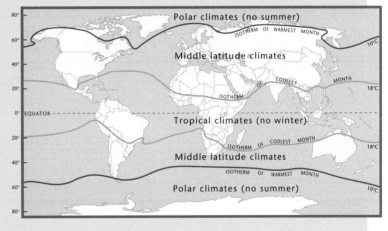

FIGURE 15.5 *The 10°C Summer Minimum Isotherm and the 18°C Winter Maximum Isotherm*

The Köppen Codes

The First Letter

Five major climate groups are identified by capital letters. Four relate to temperature, but one, the B climates, relates to precipitation.

A Tropical Climates: The mean monthly temperature of the coldest month is above 18°C.

B Dry Climates: Evaporation is greater than precipitation, so the ground is dry and vegetation is sparse. Precipitation is generally less than 500 mm per year in hot climates.

C Temperate Mild Winter Climates: The mean monthly temperature is over 10°C in the warmest month and under 18°C in the coldest month. The mean temperature of the coldest month never falls below -3°C.

D Temperate Cold Winter Climates: This is the same as the C climates except that the mean temperature of the coldest month does fall below -3°C.

E Polar: The warmest month has a mean temperature below 10°C.

Station	J	F	M	A	M	J	J	A	S	O	N	D
Toronto, Canada [43°N 79°W]	-5	-6	-1	6	12	18	21	19	16	9	3	3
Dakar, Senegal [15°N 17°W]	22	22	23	24	27	28	28	28	28	28	26	23
Arkhangel'sk, Russia [64°N 40°E]	-15	-14	-9	-1	4	10	14	12	7	0	-7	-12
Aklavik, Canada [68°N 135°W]	-29	-27	-22	-13	-1	4	10	14	12	-7	-19	-26
Tombouctou, Mali [17°N 3°E]	22	24	28	32	34	34	32	30	32	31	28	22
Thule, Greenland [77°N 69°W]	-22	-27	-23	-18	-6	2	5	3	-3	-9	-15	-23
Rio de Janeiro, Brazil [23°S 43°W]	26	26	25	22	21	21	21	21	21	22	23	24
Los Evangelistos, Chile [52°S 75°W]	8	8	8	7	6	4	4	4	4	5	6	7

FIGURE 15.6 *Selected Climate Statistics*

The Second Letter

In the case of A, C, and D climates, precipitation is greater than evaporation (over 500 mm annually). The following lower case letters are codes that refer to seasonal variations in precipitation. An effort has been made to help you remember the meaning of each code.

f: Precipitation is distributed evenly throughout the year. In the A climates no month receives less than 60 mm. In C and D climates no month receives less than 30 mm. (Think of "f" as standing for "full" rainfall.)

w: There is a summer rainy season and a winter dry season when at least one month has less than 60 mm of precipitation in A climates. In C and D climates at least one month receives less than 30 mm. ("w" is for winter drought.)

s: Precipitation is heavy in winter; drought occurs in summer. ("s" is for summer drought.)

m: A special type of A climate where rainfall is extremely heavy in summer and almost non-existent in winter. ("m" is for monsoon.)

In the case of B climates, two different codes, also relating to precipitation, are used. These are written in upper case letters.

S Steppe Climate: This is a semi-arid climate where precipitation is between 250 mm and 500 mm.

Station		J	F	M	A	M	J	J	A	S	O	N	D
Chongqing,	T [°C]	7	10	14	19	23	26	29	30	24	19	14	10
China	P [mm]	15	20	38	99	142	180	142	122	150	112	48	20
Georgetown,	T [°C]	26	26	26	27	27	27	27	27	28	28	27	26
Guyana	P [mm]	203	114	175	140	190	302	254	175	81	76	155	287
Cairo,	T [°C]	13	15	18	21	25	28	28	28	26	24	20	15
Egypt	P [mm]	5	5	5	3	3	0	0	0	0	0	3	3
Dongala,	T [°C]	19	21	24	29	32	34	34	34	33	29	26	20
Sudan	P [mm]	0	0	0	0	0	0	8	8	0	0	0	0
Baker Lake,	T [°C]	-35	-32	-25	-17	-6	2	9	9	2	-9	-19	-27
Canada	P [mm]	5	5	8	8	5	18	23	28	20	13	8	8
San Francisco,	T [°C]	10	12	13	13	14	15	15	15	17	16	14	11
United States	P [mm]	115	92	74	37	15	3	0	1	5	22	51	109
Medicine Hat,	T [°C]	-10	-9	-2	7	13	17	21	19	13	8	-2	-7
Canada	P [mm]	18	18	20	25	38	58	35	35	38	18	18	20
Lahore,	T [°C]	12	14	21	27	32	34	32	31	29	24	17	13
Pakistan	P [mm]	23	25	20	13	18	35	128	118	58	8	3	10
Montreal,	T [°C]	-10	-9	-3	6	13	19	21	20	16	9	2	-7
Canada	P [mm]	83	81	78	72	72	85	89	77	82	78	85	89
Bombay,	T [°C]	24	24	26	28	29	29	27	27	27	28	27	26
India	P [mm]	2	2	2	1	18	485	617	340	264	63	13	2
Deception Island,	T [°C]	1	1	0	-3	-5	-8	-9	-8	-6	-3	-2	1
South Shetlands, UK	P [mm]	58	53	69	51	5	8	15	25	23	109	97	51
Galveston,	T [°C]	12	13	17	21	24	28	29	28	27	23	17	14
United States	P [mm]	203	114	175	140	190	302	254	175	81	76	155	287

FIGURE 15.7 *Mean Monthly Temperatures and Precipitation for Selected Climate Stations*

W Desert Climate: This is a dry climate with total annual precipitation under 250 mm.

T Tundra Climate: The mean monthly temperature of the warmest month is between 0°C and 10°C.

F Icecap Climate: Mean monthly temperatures are all below 0°C and the area is normally ice-covered.

The Third Letter

The final code makes the climate zones even more distinct. This is especially necessary in temperate climates where there is so much variation. These lower case letters refer to temperature and are used only with C and D climates.

a: The warmest month is over 22°C.

b: The warmest month is below 22°C.
c: Fewer than four months have temperatures over 10°C.
d: Winters are so cold that the coldest month is below -38°C.

Two codes are used with B climates to differentiate temperature.
h: This is a hot climate with mean annual temperatures over 18°C.
k: This is a cold climate with mean annual temperatures under 18°C.

Köppen not only defined the climate regions, he theorized where each would be found on a hypothetical continent. (See Figure 15.8.) This hypothetical continent is a simplification but it serves as a theoretical model. It has three characteristics: it is perfectly flat, it is at sea level, and it has no outstanding coastal features. Once the locations of climate zones are found on the model, it is a simple task to determine where they occur on real continents with their varied relief, elevated landforms, and irregular coastlines. The hypothetical continent considers prevailing winds, ocean currents, and solar radiation at different latitudes. Like any model, it oversimplifies reality, but it serves to bring order to what would otherwise be a chaotic study.

MAJOR CLIMATES OF THE KÖPPEN CLASSIFICATION SYSTEM

The Köppen Classification System includes the following major climates:

Af Tropical Rainforest: Tropical and wet all year. (Example: Amazon Basin)

Aw Savanna: Tropical and wet in summer but dry in winter. (Example: Grasslands of Africa)

Am Monsoon: Tropical and wet in summer like the Aw but much more extreme. (Example: Southern India)

BWh Hot Desert: Under 250 mm of rain per year and mean annual temperature >18°C. (Example: Sahara Desert)

BWk Cold Desert: Under 250 mm of precipitation per year and mean annual temperature <18°C. (Example: Patagonia)

BSh Hot Steppe: Rainfall between 250 and 500 mm per year and mean annual temperature >18°C. (Example: Australian Outback)

BSk Cold Steppe: Precipitation between 250 and 500 mm per year and mean annual temperature <18°C. (Example: American Great Plains)

Cfa Subtropical: Temperature of coldest month <18°C and temperature of warmest month >10°C, with mild winters and precipitation all year. (Example: Florida)

Cfb/c Marine West Coast: Temperature of coldest month <18°C and temperature of warmest month >10°C, with mild winters and precipitation all year; cooler summers than Cfa. (Example: British Columbia)

Csa/b Mediterranean: Temperature of coldest month <18°C and temperature of warmest month >10°C, with mild winters and summer drought. (Example: California)

Dfa/b Continental: Temperature of coldest month <18°C and temperature of warmest month >10°C, with cold winters and regular precipitation all year. (Example: Southern Prairie Provinces)

Dfc/d; Dwc/d Subarctic: Temperature of coldest month <18°C and temperature of warmest month >10°C, with very cold winters and mild summers. (Example: Yukon)

ET Tundra: Mean monthly temperature <10°C and >0°C. (Example: Northern Siberia)

EF Icecap: Mean annual temperature <0°C. (Example: Antarctica)

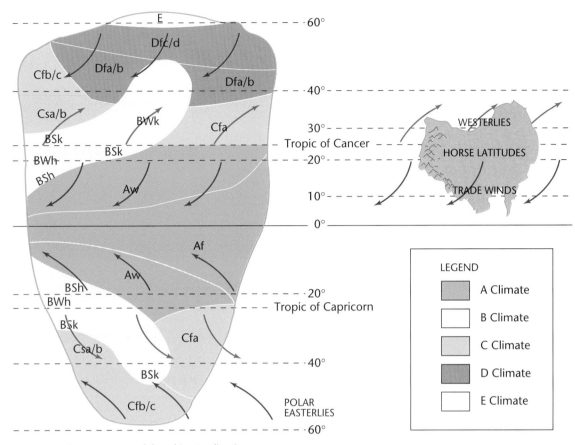

FIGURE 15.8 *The Köppen Model and its Application*

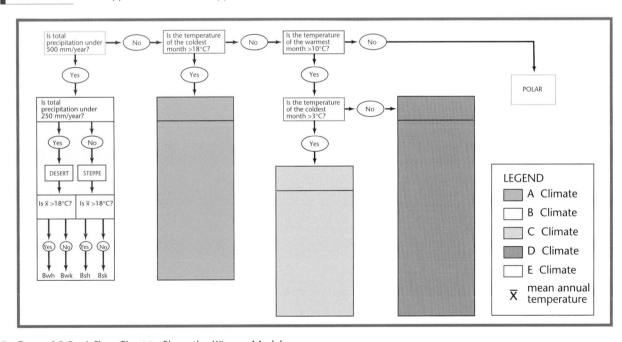

FIGURE 15.9 *A Flow Chart to Show the Köppen Model*
Given any set of climate statistics, it is possible to determine the climate region using a flow chart. The B climates are done as an example using the definitions on page 112. Copy this chart and complete it for A, C, D, and E climates.

Things To Do

1. Compare the Köppen codes to the classification system you developed in activity 3 on page 109.

2. Copy the organizer below into your notebook. Fill in the appropriate code and corresponding numerical values for each letter of the code. The first one has been done as an example.

3. Match the climate graphs you created in activity 4 on page 109 to a climate in the Köppen Classification System.

4. Prepare a flow chart summarizing the Köppen Classification System. Figure 15.9 will help get you started.

5. Develop a computer program that will determine the climate of a given set of climatic data using a series of input statements.

6. a) Copy Figure 15.8 showing the Köppen model and an imaginary continent on a piece of paper. Make sure you include the lines of latitude and the prevailing winds.
   b) Draw the climate regions on the imaginary continent using the Köppen model as a guide. Be sure to make allowances for relief and an irregular coastline.
   c) Work with a group to prepare similar climate maps for the following continents: North America, South America, Australia, Asia, Africa, and Europe. Where there are high mountain ranges like the Andes and the Himalayas, label these regions "H" for "highland climate."
   d) Which continent is most like the one in Figure 15.9? How are they different?

Climate	Code	First Letter	Second Letter	Third Letter
Tropical Rainforest	Af	coldest month >18°C	rainfall every month >60 mm	
Savanna				
Monsoon				
Hot Desert				
Cold Desert				
Hot Steppe				
Cold Steppe				
Subtropical				
Marine West Coast				
Mediterranean				
Continental				
Subarctic				
Polar				

Chapter 16
Climate Regions of the World

Introduction

A climate region is a collection of places with similar climate characteristics. By studying sample climate stations, we can obtain a general understanding of what each climate region is like. It is important to note that there may be considerable variation among climate stations within a single climate region. Why do these variations exist?

Figure 16.1 *The Tropical Rainforest Climate of South East Asia*
The tropical rainforest climate here is almost identical to that found in South America and Africa.

Things To Do

1. This chapter describes the different climate regions of the world. Read about each of these regions, then prepare a comparison organizer describing the various characteristics of each one.

2. Select one climate region and use the inquiry model below to explain the patterns found there.

Focusing
Make up a series of questions about the climate region you have chosen. For example: Why is there a winter drought in the savanna? Why is the desert often hotter than the rainforest? Why is the marine west coast climate mild all year? From your list, select one or two questions.

Listing Alternatives
List alternatives that might be a factor in answering each question. For example, wind, air pressure, latitude, the sun's position, and cloud cover could all relate to why there is winter drought in the savanna.

Gathering Data
Using this textbook and other sources, gather information about each alternative. For example, there is no cloud cover in the desert so the amount of sunlight reaching the ground there is greater than in the tropical rainforest.

Synthesizing
From your data determine the alternatives that answer each question and establish the relationship among alternatives. Sometimes the question is answered by only one alternative; other times several alternatives work together to create a pattern.

Concluding
Write a concise explanation to answer the question. Include maps and diagrams as necessary. Present your findings to your group or class.

Evaluating
Be prepared to support your conclusion with data. If necessary, change your conclusion to take into account errors in interpretation.

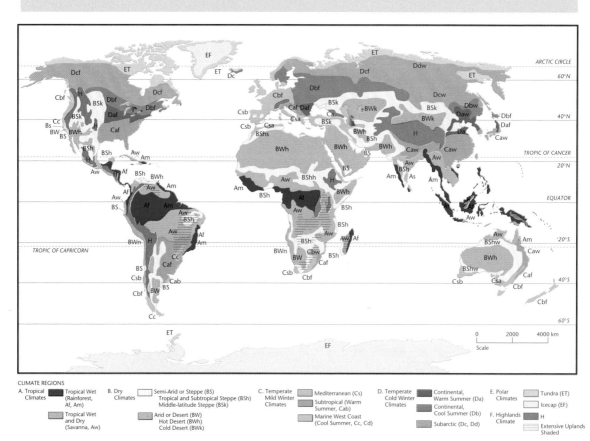

FIGURE 16.2 *Climate Regions of the World*

The Tropical Rainforest Climate (Af)

We already know two important characteristics of the tropical rainforest climate. The first is that temperatures are high all year round. The second is that heavy rainfall occurs throughout the year. In equatorial lands it is impossible to tell what month it is from the climate because every day is exactly the same as the one before.

The high temperatures are the result of the region's location on or near the Equator. On the east coast, the region extends towards the poles as far as the Tropics of Cancer and Capricorn. Warm ocean currents on the east coast and the constant trade winds blowing over the water bring warm temperatures to coastal regions. Therefore the Af climate extends closer to the poles on the east coast, but is centred around the Equator on the west coast where cold ocean currents limit the region's expanse. The tropical rainforest is found as far south as the Tropic of Capricorn on the large island nation of Madagascar, off the southeast coast of Africa. On the west coast of Africa, the tropical rainforest is restricted to a relatively narrow coastal region on either side of the Equator. In the Western Hemisphere, the tropical rainforest is found in southern Mexico around 20°N latitude, but once again it is confined to a narrow coastal belt along the west coast of South America. Local **anomalies** sometimes result in hot, wet climates occurring outside of the Af zone.

Although this climate is hot all year, daily maximums fall short of the incredibly high temperatures found in the hot desert climate (BWh). There are two reasons for this. The heavy cloud cover of the tropical rainforest acts as insulation against the equatorial sun. Thus daytime temperatures are lower than you might expect. The second reason is because energy is consumed by the processes of photosynthesis and evapotranspiration. Therefore there is less solar energy to heat the air than if the region were dry. Nevertheless, the temperature feels hotter than it is because of the extremely high humidity. Nights, though less humid, are also very uncomfortable. The **diurnal temperature range** is low because the humidity and cloud cover keep the heat from escaping into the upper atmosphere.

The high precipitation occurs because of the strong convection currents. A low pressure cell is found over equatorial regions for most, if not all, of the year. As the moist, hot air rises, water vapour condenses on dust particles in the air, producing rain. Daily thunderstorms are often violent. The storms usually come in the late afternoon. Evenings can be less oppressive as the humidity drops after the storms have ended. By the next morning, the humidity starts to rise again as another storm begins to form.

As you move away from the Equator, an area of slightly drier climate occurs. Some months receive much less rain than others. Consider Georgetown, Guyana, for example. Located 7° north of the Equator, it has two months when the rainfall drops significantly. (See Figure 16.3.) Singapore, which is much closer to the Equator, has little variation in monthly rainfall. (See Figure 16.4.) The reason for this involves seasonal changes in wind patterns. Moving away from the Equator, rainfall patterns become increasingly seasonal until monthly totals drop below 60 mm. At this point the region is defined as a savanna climate. Though the climate regions map shows the division between the two as a clean-cut line, in actual fact the transition is anything but precise. In reality, the transition between Af and Aw zones can be as much as several

	J	F	M	A	M	J	J	A	S	O	N	D
T [°C]	26	26	26	27	27	27	27	27	28	28	27	26
P [mm]	203	114	175	140	280	302	254	175	81	76	155	182

FIGURE 16.3 *Climate Statistics for Georgetown, Guyana*

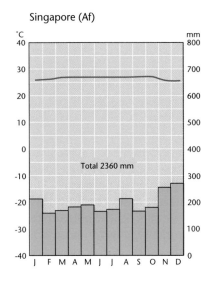

FIGURE 16.4 *Climate Graph for Singapore*

hundred kilometres and can shift as long-term averages change.

Tropical rainforests are exotic and beautiful, but the high temperatures, heavy rain, and constant humidity make living conditions uncomfortable. However, many coastal areas in this region are densely populated. The strong trade winds increase evaporation, making the region feel more comfortable than inland places more removed from the winds.

The Savanna Climate (Aw)

The savanna climate is different from the Af climate in two ways. It receives slightly less rainfall and has definite dry and rainy seasons. The best known example of the savanna

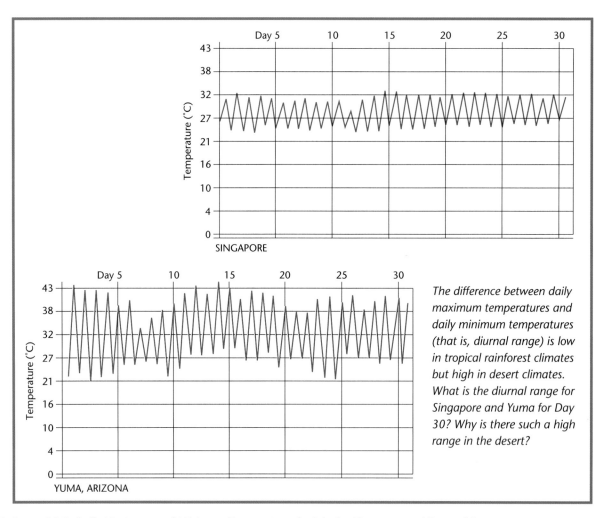

The difference between daily maximum temperatures and daily minimum temperatures (that is, diurnal range) is low in tropical rainforest climates but high in desert climates. What is the diurnal range for Singapore and Yuma for Day 30? Why is there such a high range in the desert?

FIGURE 16.5 *Daily Maximum and Minimum Temperatures for July, for Singapore and Yuma, Arizona*

climate is the vast grassland of Africa. The African savanna is the greatest expanse of tropical grassland in the world. The high altitude of the East African Highlands extends the savanna into equatorial regions where you would expect to find tropical rainforest. The savanna climate is also found in South America, Australia, and southern Asia. In each continent, the climate is almost identical, even though the regions may be on opposite sides of the earth.

The savanna climate lies between the tropical rainforest and the arid B climates. Because of its location, it has characteristics of both zones. In summer, the savanna experiences high temperatures and heavy rainfall like the tropical rainforest. The air pressure gradually rises as summer turns to autumn and autumn turns to winter. Temperatures remain high because the region is in the tropics. Rainfall patterns, however, change considerably.

Canadians determine changes in the seasons by differences in temperature. In the savanna, seasonal changes are measured in precipitation. The savanna takes on many of the characteristics of the hot desert during winter. Days are hot and sunny, streams dry up, plants lose their leaves, and animals migrate to more humid lands. The savannah is definitely a **transitional climate**.

The reason for seasonal differences in rainfall has to do with the apparent movement of

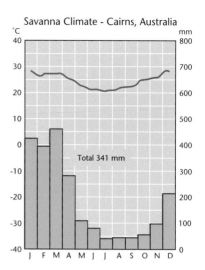

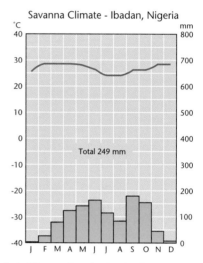

FIGURE 16.6 *Climate Graphs for Aw Savanna*

FIGURE 16.7 *The Wet and Dry Seasons of Aw Savanna*

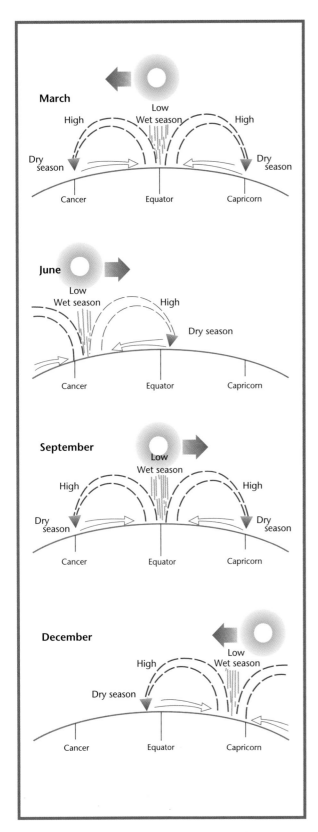

FIGURE 16.8 *Seasonality in Savanna Rainfall*

the sun. (Of course, the sun does not really move; it just seems to move because of the earth's revolution around the sun.) Convectional rain occurs wherever the sun is directly overhead. Upwellings of hot, moist air cool in the upper troposphere. The water vapour condenses and heavy rain results. A permanent low pressure cell follows the sun as it appears to move between the Tropic of Cancer and the Tropic of Capricorn. This equatorial low pressure cell, also known as the **thermal equator**, is responsible for the heavy rains in the tropical rainforest and the savanna.

Why is the savanna dry for part of the year while the tropical rainforest is always wet? Figure 16.8 shows this climate pattern. In March the thermal equator lies over the Equator; rains are heavy. As the earth revolves around the sun and the year progresses, the thermal equator moves north, bringing rains to the savanna in the Northern Hemisphere. For the months of June, July, and August, this region north of the Equator experiences heavy rains like those of the tropical rainforest. By September, however, the thermal equator is over the Equator once again. The northern savanna is now entering the dry season. As summer (December, January, and February) comes to the Southern Hemisphere so do the rains. The low pressure cell is now stationed over the southern savanna. By March the sun is again directly over the Equator. The rain cycle continues once again, just as it has for millions of years.

Of course, when it is the rainy season in the Southern Hemisphere it is the dry season in the Northern Hemisphere. The air over the equatorial low has cooled and lost its water vapour. As it descends, the temperature rises and the relative humidity drops even more. By the time this descending air reaches the earth, it is hot and bone dry. This tropical high pressure cell found over the Tropics of Cancer and Capricorn when the sun is directly over the Equator migrates as the sun appears to move. So when it is the rainy season in the Southern Hemisphere savanna, it is the dry season in the Northern Hemisphere savanna. Conversely, when the northern savannah is experiencing

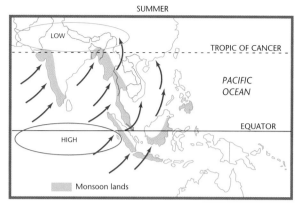

FIGURE 16.9 *Wind Directions in Winter and Summer for Monsoon Regions*

heavy rain, the southern region is dry. This is why the savanna has such definite rainy and dry seasons.

The Monsoon Climate (Am)

The monsoon climate is an extreme version of the savanna climate. It is most commonly found in South East Asia. The summer rainy season here is so severe that extreme flooding can occur. In Bombay, for example, water on the streets can be knee-high at the height of the monsoon. But during the winter, weeks may pass with absolutely no precipitation. Despite this tremendous seasonal variation, monsoon lands often have rainfall totals as high as those in Af climates.

The seasonal rainfall pattern of the monsoon climate is exaggerated by landform patterns that influence tropical air flows. The Himalayas, the highest mountains in the world, separate South East Asia from the rest of the continent. The high altitude causes mountain glaciers and extremely low temperatures. South of the land mass lies the super heated waters of the Indian Ocean. In the winter, the equatorial low is south of the Equator. A high pressure cell develops over the mountains. Prevailing winds flow from the Himalayas to the Indian Ocean. This dry, cold air gradually expands as it descends the mountain slopes. Coupled with the strong tropical sun, the air temperature rises and the relative humidity drops. This is the season of blistering dry winds.

In the summer the flow reverses. The land between the mountains and the ocean warms up rapidly from the nearly vertical rays of the sun directly overhead. The sea, while warm, is relatively cool compared to the land. Now the equatorial low is over the land, and strong convectional currents as well as onshore winds bring a deluge unlike anything found elsewhere on the planet! The monsoons have come once again.

The Desert (BW) and Steppe (BS) Climates

The main characteristic of the B climates is the small amount of precipitation. Köppen defined B climates as regions where evaporation exceeded precipitation. Even though the rate of evaporation varies with the temperature, here we will assume that desert climates have under 250 mm of precipitation and steppe climates have between 250 and 500 mm no matter what the temperature. Deserts are bordered by the steppe climate, which serves as a transition zone between the desert and savanna climate regions.

There are three general locations where deserts occur: in a band 10° to 30° in latitude north and south of the Equator; on westerly coasts of continents between 10° and 30° north and south of the Equator; and on the leeward side of mountain barriers. Each type

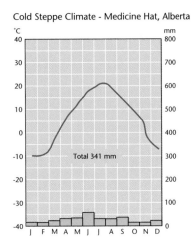
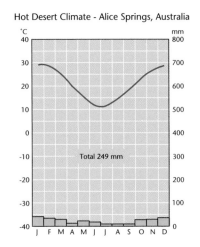
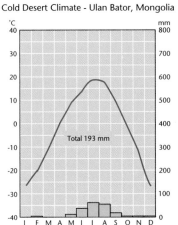
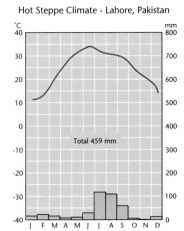

FIGURE 16.10 *Climate Graphs for Cold Steppe, Cold Desert, Hot Desert, and Hot Steppe Climate Regions*

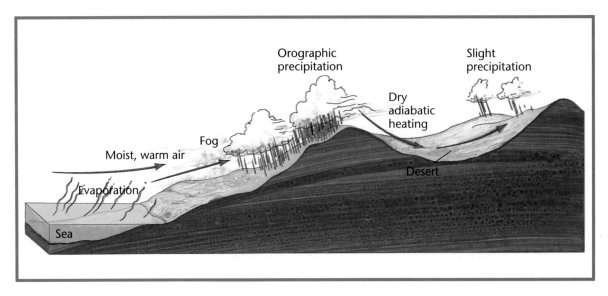

FIGURE 16.11 *How the Rainshadow Causes a Desert*

of desert has slightly different characteristics, but all have one thing in common: they are very dry all year round.

Deserts are located on the west coasts of continents, extending towards the poles through the interior of the continents. Mountain barriers often increase the size of deserts by blocking moisture-laden winds. These mountain barriers cause rainshadow—areas of relatively low rainfall on the lee side of the mountains. (See Figure 16.11.) Rainshadow deserts are found in Alberta and the American Great Plains, in central Asia, and in southern Argentina. Deserts that form behind great mountain barriers have much lower temperatures than deserts in the tropics. During the winter, they are frequently very cold and snow may occur. When it melts there is a brief period when the desert does not seem to be so arid. The moisture that is trapped as snow and the low evaporation rate make these regions better suited for pasturing livestock than the hot, dry tropical deserts closer to the Equator.

The general circulation of air in the tropics results in such great hot deserts as the Sahara and the Kalahari in Africa, the Arabian and the Thar in Asia, the Great Australian Desert, and the Mexican Desert. As you know, the air over the Equator loses enormous amounts of moisture as it rises. In a band between 10° and 30° north and south of the Equator, the resulting cool, dry air descends to earth. But the air does not stay cool for long. As it falls, the molecules are compacted together, thereby increasing the temperature. The adiabatic heating of the air, coupled with the constant bright sunshine, results in very high temperatures during the day. The relative humidity is so low that surface water is quickly evaporated, leaving behind minerals that had been dissolved in the water. The hot, dry air blows out of the deserts either back into the Equator as the trade winds or in a westerly direction in the middle latitudes. The cycle starts all over again, bringing rains to the low latitudes and drought to the tropics.

These tropical deserts have the highest temperatures of any climate. But they also have very low nighttime temperatures. The diurnal range of hot deserts can be surprisingly high. Some observers have called the desert night "the winter of the desert." Temperatures have been known to drop 10° to 20°C in just three hours after the sun goes down! What sound like gunshots in the Arizona desert are not from the ghosts of cowboys past. They are in fact the sound of the desert rocks contracting

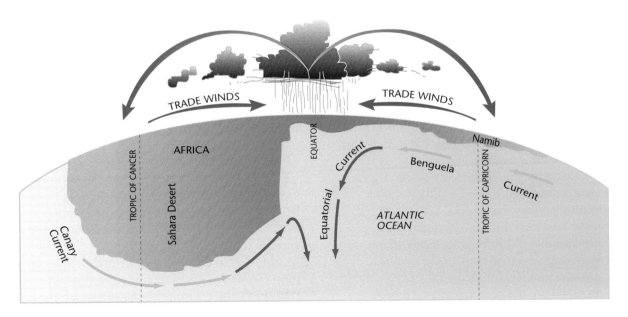

FIGURE 16.12 *Convection Currents Over Deserts*

rapidly as the temperature drops at night. These low temperatures are caused by the lack of cloud cover and low humidity. There is nothing to prevent the daytime heat from radiating into space once the sun sets.

It seems curious that some of the world's great deserts lie right along the coast. The Atacama Desert of Peru, the Namib Desert of South West Africa, Baja California, and portions of the Sahara and Great Australian deserts are all located on the west coasts of continents. You would think that these coastal areas would receive orographic precipitation off the ocean. Yet they are the driest of all the world's deserts.

What explains this phenomenon? The air over the oceans has the same characteristics as the air over the land. Rainfall is lower than the rate of evaporation, so there are actually desert conditions over the ocean. The same pattern of adiabatically heated dry air occurs over the ocean as well as over the land. Simply put, the air over these tropical waters is heated because it is descending. Cold currents off the coast of these deserts further increase the stability of the air. Interestingly, many of these coastal deserts experience heavy fog and are often cloudy as the warm tropical air drifts over the cool ocean currents that flow to the coast. It does not take long for the dry, hot air over the land to evaporate whatever moisture drifts over the land, however. Characteristically, these coastal deserts have lower mean monthly temperatures and lower diurnal temperature ranges due to the cold offshore currents.

The steppe climate is found bordering the desert. It has many of the same characteristics as the desert: it is hot all year, has a high diurnal range, and receives little rainfall. However, it does get more rain than the desert. It often has a definite rainy season in the summer, especially when the steppe lies between the desert and the Equator. This marginal region is influenced by the same migrating low pressure that brings summer rains to the savanna.

In the middle latitudes, cold steppe climates surround the temperate deserts. These zones also receive more rain than the desert and often support vast prairies of natural grass. The Russian Steppes, the Prairies of North America, and the Great Australian Outback are all examples of this transitional zone between deserts and more humid lands.

The Mediterranean Climate (Csa, Csb)

The Mediterranean climate is one of the nicest climates in which to live. It has warm to hot summers, unusually mild winters, and bright sunny days all summer. While the summers are dry, a moderate amount of rain falls in the winter. This is the only climate zone that experiences summer drought and winter rain. Snow and frost are almost unheard of. It is little wonder that the Mediterranean climate supports some of the densest populations in the world.

The climate is named after the Mediterranean Sea located between Africa and Europe. It is here that the largest example of this type of climate region can be found. Generally this region lies on the west coasts of continents between 30° and 40° north and south of the Equator. A large Mediterranean zone is found on the west coast of the United States. Smaller regions exist in Chile, South Africa, and Australia. No matter what continent the region is found in, however, it is always on the west coast just north of the Tropic of Cancer and south of the Tropic of Capricorn extending into the middle latitudes.

The unusual characteristics of this climate are the result of its location. On the border closest to the Equator are dry tropical steppes, while on the side closest to the poles are wet marine west coast climates. Generally rainfall increases moving towards the poles. The position of the climate region is why it has a summer drought and mild temperatures all year.

The prevailing winds move north and south as the equatorial low migrates each year. Because of its location, this climate is influenced by the westerlies of the middle latitudes and the trade winds of the tropics. In the winter, the westerlies blow over the ocean and onto the land. These warm maritime winds pick up moisture from the cool water off the west coast. When they reach the relatively cool air over the land, water vapour condenses,

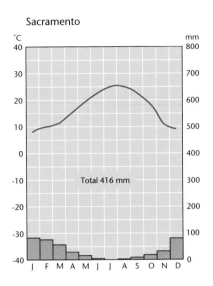
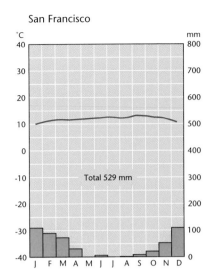

FIGURE 16.13 *Climate Graphs for Sacramento and San Francisco, California*

forming fog, drizzle, and light rain. These winter rains are all the moisture the region will receive all year. The pattern of prevailing winds reverses in the summer. In the Northern Hemisphere, the equatorial low moves north towards the Tropic of Cancer. The easterly trade winds move north with the equatorial low. When the low is near the Tropic of Cancer, the winds blow from the east. Because the winds are blowing out from dry desert and steppe regions to the southeast, they contain little water vapour. Consequently there is virtually no precipitation in summer. Of course, the same pattern of precipitation occurs in the Southern Hemisphere.

Temperatures in the Mediterranean region are moderate all year. The high summer temperatures are due to the subtropical location fairly close to the Equator. Mild temperatures in winter are caused by the moderating influence of the ocean currents. This climate is affected by maritime air masses in winter and continental air masses in summer.

The maritime influence is obviously much greater on the coast than it is inland. Coastal cities like San Francisco never experience the summer heat of cities like Sacramento, which is located further inland. Although these two cities are less than 200 km apart, the July temperature is significantly hotter inland. While San Francisco is a damp, cool 15°C Sacramento is a hot, dry 23°C! (See Figure 16.13.) It is little wonder that many Californians flock to the beach in the summer to get away from the inland heat.

The Marine West Coast Climate (Cfb, Cfc)

The marine west coast climate is also found on the west coasts of continents. It extends from the Mediterranean region well into the upper latitudes. In Norway and Alaska it extends as far north as the Arctic Circle. This climate zone is found throughout much of western Europe, New Zealand, and coastal British Columbia, as well as small sections of the coastlines of southern Chile and Iceland. Location is the key factor in this climate's characteristics.

As the name implies, the marine west coast climate is influenced by the oceans. Warm ocean currents heat the region so much that it is surprisingly warm in the winter considering its latitude. The Gulf Stream and North Atlantic Drift carry warm water all the way from the Caribbean to northern Europe. As a result, winters here are mild. Although snow does fall, the most common form of winter

FIGURE 16.14 *The Marine West Coast Climate in New Zealand*

precipitation is rain. Westerly winds pick up moisture from the relatively warm ocean and dump it along the coast. Fog and rain are the common characteristics of winter. But the climate provides plenty of moisture for crops and pastures. In the summer, the westerlies bring abundant rain. Summers are characteristically cool, moderated by offshore ocean currents. Overall, the marine west coast has a low temperature range, with mild winters and cool summers. As you would expect, the temperature drops moving away from the Equator. In the higher latitudes the Cfc climate is found, while closer to the Equator the Köppen code is Cfb. A Cfb climate has more than four months *over* 10°C, while the Cfc is cooler with more than four months *below* 10°C. This climate region is a comfortable one in which to live and is one of the most densely populated regions in the world.

The Humid Subtropical Climate (Cfa)

The humid subtropical region is another heavily populated climate zone. Unlike the other two C climates, this region is found on east coasts. It receives much more rainfall than the other C climates, and there is no summer dry season like that in the Mediterranean region that lies at the same latitude on the west coast. The region extends from about 25° to 40° north and south of the Equator. Because of its low latitude, summers are hot and winters are cool to mild. The influence of the ocean on temperature is much less than on the west coast, but it still has an impact on precipitation.

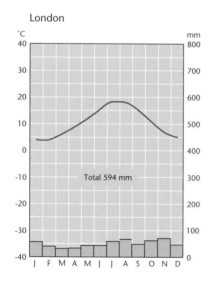
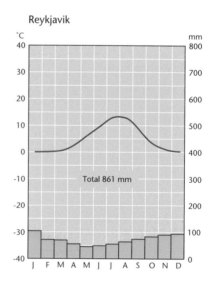

FIGURE 16.15 *Climate Graphs for London, England and Reykjavik, Iceland*

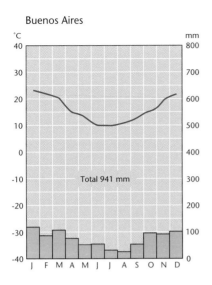
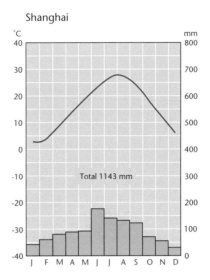

FIGURE 16.16 *Climate Graphs for Buenos Aires, Argentina, and Shanghai, China*

There are vast regions of humid subtropical climate on most continents. One example is the southeastern United States, stretching from the mid-Atlantic states to the Gulf of Mexico and inland to the vast cold steppes of the Great Plains. In Asia, the most populous part of China is a humid subtropical region. There is also a large humid subtropical region in South America. Most of the major cities of Argentina, Uruguay, and southern Brazil are located in this temperate warm, moist climate.

The regular precipitation is caused by two mechanisms. In winter, frontal precipitation occurs as tropical maritime and polar continental air masses collide over the area. Depending on the air temperature, the precipitation can take many forms, from snow to sleet to freezing rain to drizzle and even to heavy downpours. Latitude obviously has a great effect on the type of precipitation. But tropical air can extend well into the upper latitudes and polar air can dominate regions close to the tropics.

In summer, the jet stream moves north and polar incursions of cold air are less common. However, precipitation in the form of thunderstorms is still common. The wind blows from the sea across the land in a similar fashion to what happens during the tropical monsoon. A low pressure cell often forms over the land because of the high daytime temperatures. Anticyclones offshore pump moist, cool air over the land. A convectional flow occurs, making heavy rains common on hot afternoons.

The Continental Climate (Dfa, Dfb, Dwa, Dwb)

Continental climates are found north of the humid subtropical climate in the Northern Hemisphere. These two climates are similar in many ways except that continental climates are colder in winter. In North America, the region is located on the east coast, extending into the continent all the way to the marine west coast. It is found throughout most of southern Canada, all of New England, and most of the American Midwest. A belt of continental climate is also found in Poland and western Russia. It does not extend as far east into continental Asia because of the rainshadow caused by the Himalayas to the south. A small region of continental climate also exists on the northeast coast of Asia but it experiences a winter drought. The nature of this climate zone is determined by its location on large continental land masses. There are no examples of continental climates in the Southern Hemisphere because there are no major mid-latitude land masses south of 40°.

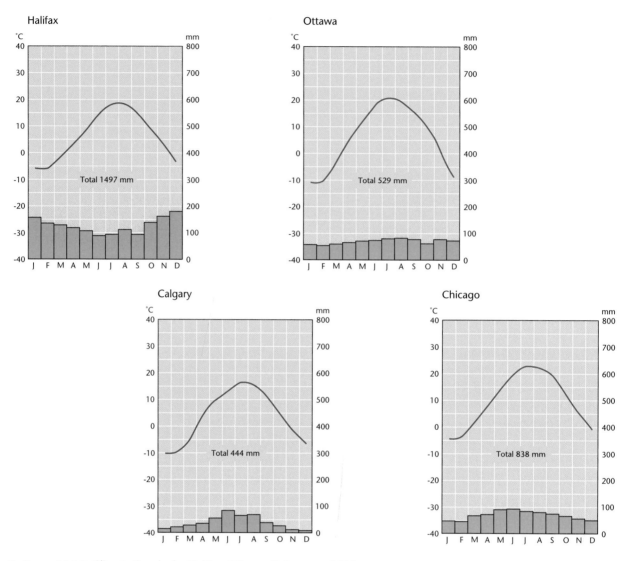

FIGURE 16.17 *Climate Graphs for Halifax, Ottawa, Calgary, and Chicago*

The temperature characteristics vary with distance inland and from the Equator. Climate stations on the coast, such as Halifax, have more moderate temperatures than climate stations inland, such as Ottawa. As you move away from the moderating influence of the oceans, the temperatures become colder in winter and warmer in summer. The cold Labrador Current that flows down the east coast of Canada keeps summer temperatures cool in the Maritimes. Temperatures also vary with latitude. Chicago has a mean annual temperature of 10°C while Calgary's average is only 5°C.

Summers can be remarkably hot in the interior of the continent, even though this region is far away from the Equator. The sun may be at a low angle on the horizon, but summer days are very long. The sun shines for as many as fifteen hours a day, making up for its low intensity. Hot temperatures often trigger thunderstorms. Mid-latitude cyclones occur during any season as polar continental air meets tropical maritime air from the south. In the winter, mid-latitude cyclones carry bitterly cold weather from the northern polar regions. Weeks of sub-zero temperatures can make the winter seem unbearably long.

As with the humid subtropical climate, this zone receives plenty of precipitation all year round. This temperate climate offers great variety. It is well suited to agriculture and is therefore quite heavily populated.

The Subarctic Climate (Dfc, Dfd, Dwc, Dwd)

Unlike the other temperate climates, the subarctic climate does not have many people living in it. It is found in enormous stretches of Canada, Alaska, Scandinavia, and Russia, 50° to 70° north of the Equator. This climate doesn't exist in the Southern Hemisphere because there are no continental land masses at the same latitudes.

Winters are long and bitterly cold. The cool summers are often too short for agriculture, although occasionally summer temperatures can get quite high. This is because the long summer days, lasting sixteen hours or more, allow the air to heat up even though the sun's rays are low on the horizon. The annual temperature range is therefore quite high compared with other climates.

Rainfall is often less than it is further south, but it is more effective for plant life because the rate of evaporation is so low. Despite what many people think, there is little snowfall in the subarctic region. People often find all-terrain vehicles more practical for winter travel than snowmobiles!

The inland regions of this climate are the source regions for polar continental air masses. Stable air is formed in massive high pressure cells over the frozen continents. Because the air is so stable, convectional rainfall is almost unheard of, especially in winter. Cyclonic storms in the form of blizzards occur when incursions of milder maritime air move into the region. Tropical air masses seldom reach these upper latitudes, so the snow that does fall seldom has the opportunity to melt until spring. Thus after six months of below-freezing temperatures, there may be quite an accumulation of snow. Once the thaw takes place in late April or May, the region seems wetter than it really is. There is plenty of surface water. Low rates of evaporation and the

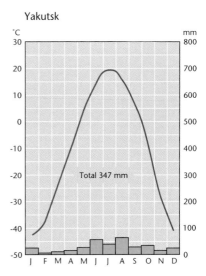

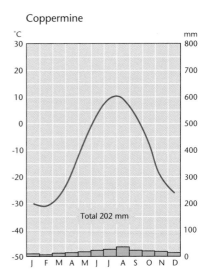

FIGURE 16.18 *Climate Graphs for Yakutsk, Russia and Coppermine, Canada*

permanently frozen ground (**permafrost**) make the earth muddy and water-logged.

In summer, cyclonic storms add light rain to the already soaked terrain. It is only by late summer that the landscape starts to dry out. The moist ground, long hours of sunlight, and relatively warm temperatures are ideal for trees and poorly drained bogs. Dense **boreal forests** cover vast regions of these northern lands. Animals such as deer and wolves and mosquitoes by the billions dominate the region.

Polar Climates (ET, EF)

The polar climates can be subdivided into two groups, tundra and icecaps. Lying between the poles and the polar circles, these climates are extremely severe; they have virtually no summer. The tundra regions are slightly moderated by the oceans, especially in the Southern Hemisphere where there is more ocean than land.

Both polar climates are extremely cold. The sun does not shine for weeks and even months at a time during the winter. In summer, days are long but the region is so far from the Equator that the mean monthly temperature never rises above 10°C. As with the subarctic climates, there is little precipitation. On the icecaps, snow is the predominant precipitation. Some rain falls in the summer in the tundra, mainly due to cyclonic storms formed between polar and arctic air masses. As with the subarctic climates, few people live in these vast frozen regions.

FIGURE 16.19 *Antarctica*
This polar climate is cold year round.

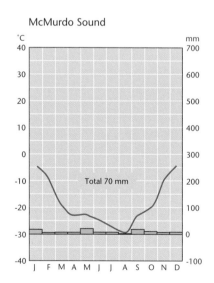

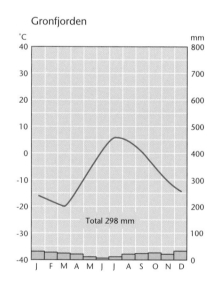

FIGURE 16.20 *Climate Graphs for McMurdo Sound, Antarctica and Gronfjorden, Spitsbergen*

Things To Do

1. a) In a group, brainstorm the characteristics of the ideal climate. Your criteria should be extensive and should include such factors as sunny summers, reliable rainfall for agriculture, and so on. The more criteria you establish, the more accurate your analysis will be.
   b) As a group, decide which five climates you think have the greatest appeal.
   c) Prepare an organizer to compare the criteria for each climate zone selected in b). The organizer below provides an example.
   d) Rank each of the climate zones for each criterion. Where necessary, weight criteria that you think are more important. For example, you may think that a broad temperature range (that is, four seasons) is good because it adds variety to the climate, so you weight it at two times what the other criteria are worth.
   e) Add up your ranking totals; the lowest total indicates the highest ranking. Rank the five climates from best to worst.
   f) Compare the climate map in Figure 16.2 to a population distribution map in your atlas. Is the population density the greatest in your "best" climate?
   g) Compare your group's results with the results obtained by other groups. Did you all agree?
   h) Prepare a presentation defending your choice of the world's best climate. This may take the form of a TV commercial, an advertisement, an information brochure, or a formal report.

2. a) What are the shortcomings of the Köppen Climate Classification System? What things does it leave out?
   b) How would you improve the Köppen system?
   c) Explain how modern technology could help you develop a better system.

3. a) List the characteristics of the climate where you live.
   b) Identify those aspects of the climate that you like and those aspects that you do not like.
   c) Write an essay describing your climate and why you like or dislike it.

4. a) On flash cards, write questions and answers to review the different climates. For example, one card might read "What climate experiences summer drought?" *Answer:* Cs Mediterranean.
   b) Make up a game using the flash cards to review the climate types.

5. Explain how each of the following factors determines climate and give a specific climate zone as an example.
   - latitude and solar radiation
   - global wind systems
   - ocean currents
   - water bodies
   - landforms (mountains)
   - altitude

Criteria	Weight	Af		Aw		BWh		Csb		Dfa	
		Data	Rank	Data	Rank	Data	Rank	Data	Rank	Data	Rank
Broad temperature range	2x	1 season	5	2 seasons	3	1 season	4	2 seasons	2	4 seasons	1
Sunny summers											
Reliable rainfall											

Chapter 17

Microclimates

INTRODUCTION

One of the main problems with climate classification is that it makes tremendous generalizations. This is because it is the study of climate regions on a global scale. Most Canadians live in one of the temperate C or D climates. Yet we all know that there are considerable variations in climate across the country. The study of climate variations on a small scale is called **microclimatology**.

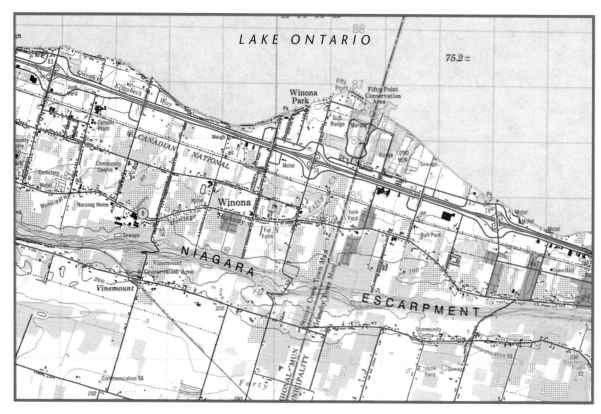

FIGURE 17.1 *The Niagara Escarpment*
Study this topographic map of the Niagara Escarpment. Why are most orchards located below the escarpment close to the lake?

Factors Affecting Microclimates

Many of the same global climate controls determine microclimates. Local winds blowing inland from bodies of water on warm summer days create low pressure cells. **Aspect** is also a control. The side of a hill or building facing the sun is warmer than the northern side. Ski resorts in North America are usually located on northern slopes so that the sun will not melt the snow as fast as it does on the south-facing slopes.

Altitude too can make a big difference to temperature. In the Okanagan Valley of British Columbia, orchards and vineyards thrive. It would be impossible to grow these fragile crops in the mountains.

Many microclimates are created by human developments. In cities, winds are often forced to flow between tall buildings. Thus they become concentrated in a small area, often creating violent winds and causing discomfort for people. Cities also modify the climate by increasing the temperature in the immediate area. The heat from vehicles and buildings combined with pollution make cities considerably warmer than the surrounding countryside. Poor air quality, while not always considered a part of climatology, is increasingly becoming important as cities grow. Urban areas also release less moisture into the atmosphere. The lack of vegetation results in lower transpiration. In addition, urban surfaces quickly drain into storm sewers after a rainfall so there is little surface water to evaporate.

CAREER PROFILE: CLIMATOLOGIST

Are you interested in learning how climate affects people and how people are changing climate? Are you an environmentalist? Do you like working with computers and statistics? If you answered "yes" to these questions, you might consider a career in climatology.

The Nature of the Work
Pamela Kertland is a climatologist with Environment Canada. She monitors how climate is changing in Canada. Kertland reads scientific papers and analyses statistical data on climate change. She writes summaries of what she reads in language that the general public can understand. Kertland also advises scientists in other fields as well as government officials and industry representatives about how climate is creating changes in other aspects of the Canadian environment and economy.

Unlike meteorologists, who are specialists, climatologists are generalists. They study the interaction and relationships between climate and people. They use geography to bring together information from many fields.

Qualifications
Students who are interested in climatology should like science and geography. They should also enjoy statistical studies and have an aptitude for computer applications. To become a climatologist, an undergraduate or graduate degree in science is required. Your guidance counsellor can advise you on the appropriate courses.

FIELD STUDY

1. Take careful measurements of such elements as temperature, precipitation, sunlight, wind speed, and wind direction at selected spots at your field study site. Choose at least ten spots.

2. Draw sketch maps showing the spots and readings of the measurements you took.

3. Write an explanation of the variations in microclimates for your site. Include sketches and photographs.

4. Outline how variations in the microclimate could affect people, plants, and animals.

Chapter 18

Global Warming: An Integrative Study

INTRODUCTION

One environmental issue that concerns many people around the world is **global warming**. Some scientists predict that the increasing amount of **greenhouse gases** in the atmosphere will cause world temperatures to rise over the next two to three centuries.

A change in global temperatures could radically alter natural and human systems on a global scale. Sea level would rise as continental icecaps melt. People living along low-lying coastal plains and river deltas and on islands would be forced to migrate to higher land. Island nations like Vanuatu in the South Pacific could disappear altogether. Countries like Bangladesh and American states like Mississippi and Louisiana would lose so much of their coastlines to flooding that enormous resettlement programs would be necessary.

FIGURE 18.1 *Hurricane Andrew*
Scientists predict an increase in violent storms as a result of global warming.

The damage would not be confined to coastal regions, however. It is predicted that continental interiors would become drier because of an increase in the rate of evaporation. (As you know, water evaporates faster as the temperature rises.) The midwestern United States, Ukraine, central Europe, and the Pampas of South America are some of the richest farmland in the world. These areas could become deserts as rainfall drops and temperatures rise. This alone could cause famine and hunger on a scale unknown in history.

More violent storms are also predicted as a result of global warming. Typhoons in Asia, hurricanes and tornadoes in the United States, and blizzards in Europe and North America are more common now than they used to be. (See Chapter 14 on pages 94-105 for details about violent weather.) Will these storms become commonplace in the years ahead?

Not all countries stand to lose from global warming, however. Northern regions, such as most of Canada, Alaska, and Siberia, may benefit from warmer temperatures. Tundra could melt and **soil profiles** could develop that would allow commercial farming to flourish. Countries like Great Britain and Norway could experience a longer growing season. This would enable them to expand the amount and variety of crops they can grow. In the higher latitudes, cattle may be able to winter outside without the need for expensive barns and hay storage facilities. In Canada, it is likely that the treeline would expand further north. Over the long term, the forestry industry could spread into new regions.

Global warming could also lead to the greatest human migration ever known. In the last ice age people migrated from cold lands to more temperate regions further south. The next migration may be in the opposite direction. People moving north to regions that are cooler may settle in vast northern territories like those found in Canada and Russia.

The Greenhouse Effect

We have probably all read some frightening things about the **greenhouse effect** and global warming. It is important to understand, however, that the greenhouse effect in itself is not a bad thing. In fact, without it, life on this planet as we know it would not exist.

Have you ever wondered why the temperature does not drop rapidly at night? On the moon, the temperature drops to absolute zero on the dark side and rises to extremely high temperatures on the bright side. Terrestrial life could not survive on the moon. On earth, however, the atmosphere moderates the temperature. At night, greenhouse gases retain the heat that the sun provided during the daylight hours. In nature's delicate balance, the amount of heat entering the atmosphere is the same as the amount of heat leaving it.

It is not the greenhouse effect itself that is responsible for global warming. It is *change* in the amount of heat that is being retained by greenhouse gases that is causing concern. Scientists predict that the delicate equilibrium between energy entering the atmosphere and energy leaving it is upset if there is an increase in greenhouse gases. Less heat is radiated into space, more heat is retained because there are more greenhouse gases, and global temperatures rise. Climate models predict warming of 1.5°C to 4.5°C over the next century. Although this may not seem like much, it is the largest temperature change in human history.

Greenhouse Gases

There are several different greenhouse gases, and all are increasing rapidly. The best known greenhouse gas is carbon dioxide. This gas is constantly being removed from the air by plants during photosynthesis. It is returned to the air through five natural processes: animal respiration (breathing); biomass decomposition (rotting plant and animal waste); combustion of organic substances (burning things that once lived); outgassing from water surfaces (free exchange of gas from water); and volcanic eruptions. In the past there was a

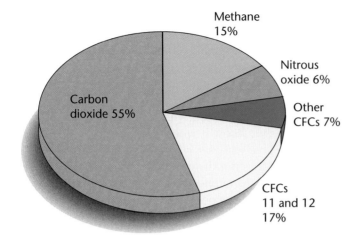

FIGURE 18.2 *Greenhouse Gases Compared*

balance in these exchanges. The amount of carbon dioxide entering the atmosphere was the same as the amount leaving it. Human activity has upset this balance, however. The burning of fossil fuels has increased carbon dioxide emissions. Thus whenever we drive cars, turn up the thermostat in our homes, or cook steaks on the barbecue, we are contributing to the threat of global warming.

Millions of years ago, prehistoric plants and animals trapped carbon dioxide, which eventually formed fossil fuels. When we burn these fuels we release all of this stored carbon dioxide. People living in today's industrialized societies have an enormous appetite for the energy created by fossil fuels. We use them to run our cars, generate our electricity, and manufacture the many consumer goods we use.

People also increase carbon dioxide levels when they burn forests. Many developing countries, such as Brazil and Indonesia, are clearing rainforest for human settlement and industrial development. As the forests are burned, huge amounts of carbon dioxide are sent into the air. The harmful effects are compounded by the fact that the number of trees has been drastically reduced. This means that less carbon dioxide is taken out of the air because the amount of photosynthesis is reduced. Many developing countries also rely on wood and animal manure for fuels.

Burning this biomass also contributes to carbon dioxide emissions. As the global population continues to grow, the amount of carbon dioxide being pumped into the atmosphere will increase unless action is taken.

Carbon dioxide is increasing at the rate of 0.5 per cent a year. Although this does not seem like much, it is cumulative and the rate is increasing. From 1960 to 1990, carbon dioxide levels increased by 11 per cent. Is there a limit to how much carbon dioxide the atmosphere can handle? It is the responsibility of each of us to make sure that our carbon dioxide emissions are kept to a minimum by burning fuels sparingly.

Another greenhouse gas is methane (CH_4). It is produced in wetlands when the biomass decomposes without the presence of oxygen. It is also produced through the digestive processes of termites and ruminating animals like cows. Human activity has increased the amount of methane in the atmosphere. Food waste and lawn clippings, among other organic wastes, are buried in landfill sites. As these materials rot, they produce methane gas. In parts of South East Asia and the United States, rice farming has increased steadily over the past thirty years. Methane gas is produced in rice paddies much like it is in natural wetlands. The amount of methane produced by domesticated cattle has also increased. This is because of the growing demand for meat and dairy products created by a rapidly expanding world population. Today there is twice as much methane in the atmosphere as there was before the Industrial Revolution. It continues to increase at a rate of 1 per cent a year. This rate of increase is faster than the rate for carbon dioxide, but there is much less methane—only 0.5 per cent as much as there is carbon dioxide. But one molecule of methane absorbs twenty-nine times as much heat as one molecule of carbon dioxide.

We are all aware of the connection between chlorofluorocarbons and ozone depletion. But CFCs are also potent greenhouse gases. They represent a relatively small part of the atmosphere but they are increasing at a rate of 4 per cent a year and they absorb up to 300 times more heat than carbon dioxide. Unlike natural greenhouse gases, chlorofluorocarbons do not readily leave the atmosphere. Instead they remain for many years, steadily increasing all the while.

Things To Do

1. Explain how global warming could make the lifestyle of your grandchildren significantly different from the one you enjoy today. Consider each of the following: clothing, leisure activities, housing, transportation, food, vacations.

2. For each of the following economic activities, determine if global warming is good or bad. Explain your answer for each.
   a) a ski resort in Quebec
   b) shrimp fishing in the Mississippi delta
   c) a theme park in central Florida
   d) farming in southern Saskatchewan
   e) farming in northern Saskatchewan
   f) living in London, England
   g) owning a home in Calgary
   h) snow removal in Winnipeg

3. Prepare an organizer comparing the various greenhouse gases. Include the following criteria: rate of increase per year; heat absorption compared with carbon dioxide; proportion of the atmosphere compared with carbon dioxide; natural and human sources; controls.

4. Write an argument either supporting or refuting the resolution that global warming is a good thing.

5. Prepare a world map showing areas that would benefit and areas that would suffer from global warming.

Scientific Uncertainty

There is some evidence to support those who argue that global warming is not as serious a threat as some scientists predict. There are so many natural systems that moderate global climates, it is possible that an increase in greenhouse gases would be offset by one of these factors. Volcanic eruptions and meteors hitting the planet create stratospheric dust clouds that reduce the amount of solar radiation entering the troposphere. These cataclysmic explosions could counter the effects of global warming, at least for a few years. Recent evidence suggests that the dust that resulted when a meteor struck southern Mexico 65 000 000 years ago plunged the planet into darkness. The explosion is credited with the extinction of vast numbers of animals, including dinosaurs. If a meteor of this magnitude were to hit the earth, global warming would seem like a minor problem!

There is also the moderating influence of the oceans. Not only do oceans store enormous amounts of solar energy and release it slowly, but sea water absorbs carbon dioxide. If sea level rises as the result of melting icecaps, there is more salt water to absorb carbon dioxide. So higher sea levels could help reduce the amount of carbon dioxide and thereby moderate global warming.

In New Zealand, some climatologists have predicted that a higher sea level and warmer temperatures may cause the Ross Ice Shelf—25 per cent of the icebound continent of Antarctica—to float north into the South Pacific. The incredible cooling of the southern mid-latitudes could be so great that global temperatures may drop as a consequence. Another ice age could result.

If global warming increased the rate of evaporation, there would be more cloud cover. Some clouds absorb heat and reradiate it to

(continued on page 140)

THE DEATH OF THE DINOSAURS
A Canadian scientist finds the smoking gun

By Wayne Campbell
Special to The
Globe and Mail
Ottawa

GEOLOGISTS HAVE LONG THEORIZED A COMET WAS THE VILLAIN. NOW AN OTTAWA RESEARCHER HAS CONVINCED SKEPTICS HE HAS FOUND THE IMPACT SITE

IT was as Alan Hildebrand puts it, a very bad day for the biosphere: 65 million years ago, an event occurred that brought an abrupt close to an entire chapter of life on Earth.

Now Dr. Hildebrand, 37, a planetary scientist with the Canadian Geological Survey, has documented to the satisfaction of just about everyone where this catastrophic event occurred.

The extinction of the dinosaurs — indeed, the end of 70 per cent of animal life on Earth — has been one of the most fascinating and hotly disputed issues among scientists for decades. Were they annihilated by the Earth's collision with an asteroid or comet? Or was the cause more down to Earth — a volcanic upheaval that obscured the sun and sent the planet into a deep freeze?

The first effect on dinosaurs from the collision was probably a tidal wave.

The case for a comet catastrophe has been compelling, but lacked an essential piece of evidence: a culprit crater — until last year, when Dr. Hildebrand and two colleagues published research claiming they had found evidence of the collision at a huge buried scar in Chicxulub on Mexico's northeast Yucatan Peninsula. "It's the impact crater of a comet," says the Ottawa-based scientist. The comet "was huge, about 20 kilometres across, and fast, about 180 000 kilometres per hour when it hit."

Though initially greeted with skepticism, Dr. Hildebrand's claims are now widely accepted by his peers — with the notable exception of vulcanologists (volcano experts.) The last roadblock to be removed was the objection of University of California geologist Dr. Walter Alvarez, a highly respected figure in his field, who recently endorsed the Canadian scientist's position.

THE search for the site of a comet collision began in the late 1970s. Dr. Alvarez, son of Luis Alvarez, the 1968 Nobel laureate in physics, discovered that a pencil-thin layer of iridium-rich clay (called the K-T boundary) always turned up sandwiched between the rock strata representing the Cretaceous period (the dinosaur era) and the more recent Tertiary (the mammalian era).

Iridium is extremely rare in the Earth's outer crust, but abundant in meteorites and comets. "It is at this K-T boundary, 65 million years ago, that all fossil evidence of the dinosaurs suddenly ceases," says Dr. Hildebrand.

The senior Alvarez, blessed with a protean imagination and unencumbered by geological and biological paradigms that viewed evolution as grindingly slow, soon proposed a startling idea. He suggested the iridium layer was the settled dust from the fireball of a meteorite impact. This suggested the mass extinctions could be explained by the collision.

Like the butterfly of chaos, whose wing-beat amplifies into a typhoon, the Alvarez hypothesis grew into a full-blown scientific controversy. Vulcanologists contended there was a more prosaic explanation for the iridium deep within the Earth. They said the dust of eruptions blocked the sun, lowered temperatures and, over thousands of years, led to the extinctions. Most scientists who study the evolution of life agreed.

Dr. Hildebrand was an impact proponent. There were features about the iridium clay that volcanoes couldn't explain: "The presence of BB-shot-sized stones called textites and quartz grains scored by shock striations," he says. "These are only found at meteor impacts."

FIGURE 18.3 *New Evidence of a Comet Catastrophe*

The day of the comet

The Chicxulub comet virtually collapsed the network of life on Earth — not unlike the effect that would be produced by a nearby supernova or a drastic alteration in the processes that drive the sun.

The comet was huge, some 20 kilometres in diameter, and slammed into the Yucatan at a time when it lay under a shallow sea. The size and depth of the crater, plus the distribution of ejecta, suggest the comet was moving at 180 000 kilometres an hour, and that the Earth ran right into it. In seconds, the comet punched down 40 kilometres, passing through a five-kilometre, limestone sea floor and the continental crust to the viscous mantle, the semi-molten interior.

The fireball blew about 3000 cubic kilometres of dust high into the atmosphere, which plunged the planet into darkness for more than three months, hampering life-sustaining photosynthesis. More than 25 000 cubic kilometres of melted and crushed rock exploded out of the crate, raining down for 5000 kilometres in all directions. The skies in the Americas were alight with molten debris, and a killing heat pulse burned the forests of both South and North America to the ground.

The effects were particularly catastrophic because the comet struck a continental shelf. Carbon dioxide levels jumped more than 10 times as seabed limestone vaporized. The resulting greenhouse effect lasted thousands of years. Sulphuric acid created from the gypsum in the limestone caused acid rain that scoured life from the upper reaches of the oceans.

"It was an Old Testament vision of hell," says Alan Hildebrand, "with darkness, terrible heat and sulphurous burning."

WAYNE CAMPBELL

Then in March, 1990, he heard of two geophysicists who worked for Pemex, the Mexican national oil company, who a decade before had discovered a circular magnetic and gravity anomaly in the Yucatan. Unfortunately, Pemex considered that it owned the geophysical data, and drill core samples of the area had been lost.

It was at this point that serendipity played a hand. Dr. Hildebrand applied for a job at the Geological Survey in Ottawa, for whom he had worked before. When he walked into the office for his interview he was stunned: On the wall was a map of minor fluctuations in the Earth's gravitational field, including "the circular gravimetric fingerprint of the impact site in the northern Yucatan." The map also supported his contention of an impact on land and sea. "The Yucatan was submerged under a shallow sea at the time."

The data from the map gave him the evidence he needed for publishing a paper on Chicxulub. He set to work studying the rocks of the Yucatan with Glen Penfield, one of the Pemex geophysicists.

Conventional wisdom had held that the area was volcanic, but the circular shape of the crater and the sequence of melted and broken rock supported the impact thesis. More important, Antonio Carmago, the other Pemex geophysicist, helped track down some of the drillhole samples, which showed clear evidence of the giant shock wave produced by the impact.

Though still dogged by skepticism, the trio published in the September, 1991, issue of *Geology*. Since then, supporting evidence from others has come in steadily, even from some who originally scoffed at the idea of the Chicxulub impact crater.

The final piece of evidence came in August this year. A paper in *Science* by Dr. Alvarez, following a carbon-dating analysis of the crater, pinpointed the age of the Yucatan site at 65 million years, the precise time of the extinctions. It confirmed the crater was the smoking gun of dinosaur extinction.

"The probability that the destruction of Cretaceous life came from the heavens has now risen above 99 per cent," Dr. Hildebrand said. "All the evidence was described as early as 1981. The impact wave deposits in Texas, the thick Haitian ejecta layer, the large, buried Yucatan crater. But no one put it together. The problem was that scientific prejudices favoured other explanations."

But where was the impact crater? Over the years, scientists had looked at several possible impact sites, but none fitted the evidence from the boundary layer.

Dr. Hildebrand joined the search in the mid-'80s, around the time a second, thicker clay layer was discovered under the world-wide fireball layer at K-T boundary sites in North and South America. "This is just what you would expect in the vicinity of the impact," he says. "The fireball would send its iridium-rich dust high into the atmosphere where it stayed for months, before settling and coating the entire globe. But near the impact site itself...coarser material would be ejected outward from the crater, falling back to ground within hours."

His analysis of fireball clay using a sensitive technique called neutronactivation analysis convinced him "that though the material seemed to be a mix of continent and ocean floor, the ocean floor predominated." Few gave this idea credence.

In 1990, he went to Haiti to examine a half-metre thick clay layer from the K-T boundary that underlay the iridium clay. "I knew immediately that it was ejecta," says Dr. Hildebrand. "It was almost completely composed of altered textites, which only occur at impact sites."

When news of this find hit the K-T community, the rush was on to Haiti and other Caribbean sites. Dr. Hildebrand, for his part, still thought the crater would be found on the ocean floor, the most likely candidate being a 300-kilometre wide depression in the Colombian basin, the body of water on the sea floor between Haiti and Colombia.

Wayne Campbell is a Montreal-based science writer.

the earth's surface. Other cloud types block the sunlight from reaching the ground, thereby reducing temperatures. So if there is more cloud cover, the planet could be warmer—or it could be colder!

As you have learned, weather forecasting is an imprecise science. The further into the future you make a forecast, the less accurate it is. Although methods have improved greatly, it is still impossible to make accurate forecasts for periods longer than a week. What makes climatologists think they can make long-term climate predictions for the next century?

Studying past climates helps scientists understand how climate could change in the future. We have statistical records dating back over 100 years, but that is not far enough to get an accurate understanding of long-term cycles.

It is possible to get a good idea what the climate was like in past eras by studying the biosphere and lithosphere. Plant pollen, lake sediments, tree rings, and air bubbles in icecaps, among other things, reveal what the planet was like hundreds, thousands, and even millions of years ago. We know, for example, that there is usually a period of global cooling every 100 000 years. This pattern has repeated itself at least eight times over the past 800 000 years. The last ice age ended in Canada about 10 000 years ago. Between each glacial period there is a time when temperatures rise 4° to 6°C. We are currently in one of these interglacial periods, but we are due for a new ice age to begin. But don't get out your overcoat too soon! It could be 10 000 years before the effects are widely felt.

Temperatures have been cooling over the past 5000 to 6000 years. From 1500 to 1900, temperatures dropped so low that the period has been dubbed "the little ice age." Norse settlements in Greenland and Labrador were abandoned, probably because it was too cold to continue growing crops. Twenty years ago, there was a lot of talk about the return of the ice age. Today we are more worried about the opposite extreme—global warming. Since the beginning of this century, temperatures have risen so much that they are now about where they were 1000 years ago, before the little ice age.

There are many theories about why these glacial periods occur. One suggests it has to do

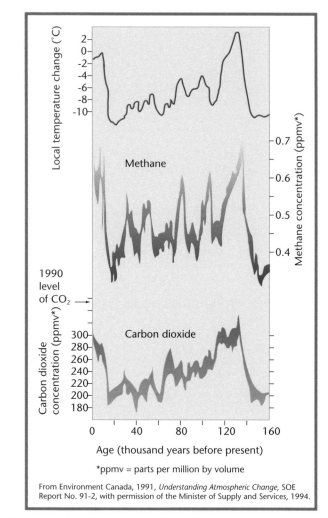

FIGURE 18.4 *Comparison of Local Temperatures and Atmospheric Concentrations of Methane and Carbon Dioxide in Antarctica Over the Past 160 000 Years*
These graphs show the positive correlation that exists between CO_2, methane, and temperature. Notice that all three were much higher in the past (about 125×10^3 ybp) and much lower (18×10^3 ybp) during the last ice age.

with fluctuations in the earth's orbit that occur every 100 000 years. Just as a spinning top wobbles from time to time, the earth could wobble as it spins on its axis. This would be enough to disrupt solar energy entering the planet so much that an ice age could result. Another theory suggests that there is a connection between the amount of greenhouse gases in the atmosphere and global temperatures. When there are a lot of greenhouse gases, there is an interglacial period. When there is a drop in greenhouse gases, then global temperatures drop. We don't yet know what causes these natural fluctuations. They could result from increased volcanic activity that releases more carbon dioxide into the atmosphere. Or they could be caused by increased methane gas that is released as ocean levels drop and continental shelves are exposed. Whatever the cause, it is likely that the planet will experience more ice ages in the future.

Predictions about future climate changes are hard to make. Yet it is important to try to determine the impact people have on the atmosphere and climate. Past climate changes occurred over such long periods of time that it was possible for living things to adapt. Other data suggest rapid climate change occurred several times in the past and led to large-scale extinctions. In the future climate change may happen quickly. Large mammals and other life forms may not have enough time to adapt.

Despite the uncertainty about global warming, we have to find out what changes our activities have brought about and how quickly they are happening. In the future, it may be too late to modify our behaviour. We must take action now!

TECHNOLOGY UPDATE: MATHEMATICAL CLIMATE MODELS

Sophisticated computers have revolutionized the study of climate change. Most of the processes that occur in the atmosphere can be measured. Temperature, humidity, wind speed, soil moisture, and many other measurements can be input for innumerable points on the earth's surface. Through the study of fluid physics, we know that there are relationships among these statistics. Mathematical expressions have been developed to show these relationships. Once the statistics are fed into the computer, it can predict what the climate will be like in the future. The mathematical expressions simulate how climate conditions will change with time. Scientists can alter one or more sets of statistics to see how a change in one variable will affect the climate in the future. In this way climate models are able to show how increased greenhouse gases will affect global climates through time.

The predictions of global warming are based almost entirely on climate models, but these models have their limitations. They cannot possibly replicate the earth's atmosphere; it is far too complex for even the most advanced computers. Called general circulation models (GCMs), these computer programs create a detailed three-dimensional model of the oceans and atmosphere. They can simulate how atmospheric and ground conditions change over time as various parameters are changed. But even though there are tens of thousands of measurement points and the model uses over 200 000 equations for each simulation, the computers cannot accurately forecast climate change. There are just too many variables and dynamics in the world that we do not fully understand.

The **resolution** of climate models is also a shortcoming. GCMs break the earth's surface into a three-dimensional grid, or series of boxes. The smaller the grid, the more accurate the model. The most advanced computer models are only able to break down the earth's surface into blocks of about 90 000 km^2 (about the size of Portugal or Israel). Many atmospheric processes take place on a much smaller scale. It is likely that computers may miss the minute changes that can cause tremendous differences in climate patterns.

One of the most advanced GCMs in the world was developed at the Atmospheric Environment Service in Toronto. It uses a higher resolution, or finer grid, than other models. It provides a detailed analysis of cloud cover and determines if the clouds are reflecting or absorbing solar radiation. However, like most climate models, it lacks inputs to record the influence of oceans on climate. This limitation is significant because the oceans store enormous amounts of solar energy and release it very slowly. The increased greenhouse gases may not cause the planet to warm up as fast as the model predicts.

In 1989, a simulation was conducted on the Canadian GCM. The amount of carbon dioxide was doubled. (The results are shown in Figure 18.5.) The simulation predicted that if carbon dioxide were to double, the average temperature of the earth would increase 3.5°C and the rate of evaporation would increase 3.8 per cent. The temperature in the high Arctic would increase 8° to 12°C during the winter, while the southern half of the country would become 20 per cent drier. Such a drastic change would create serious problems for Canadians from coast to coast. Why do you think this would be so?

Things To Do

1. Outline how climates have changed over the past 160 000 years. (See Figure 18.4.)

2. a) List the factors, past and present, that cause climates to change.
   b) Determine whether each factor causes temperatures to increase or decrease.
   c) Explain how each factor affects global temperature.

3. a) Explain climate mathematical models.
   b) Outline the reasons why these models are limited in their ability to predict future climatic conditions.

4. a) What predictions have been made about climate change in Canada if carbon dioxide emissions were to double?
   b) What impact would these changes have on the Canadian environment and people? Consider such things as agriculture, settlement patterns, permafrost, trees, wildlife, water resources, tourism, etc.
   c) Which regions of Canada could benefit from global warming? Which regions would be adversely affected?
   d) Explain the effects of global warming on the regions you listed in c).
   e) Debate whether *overall* global warming will be an advantage or a disadvantage to Canadians.

5. Read the article "The Death of the Dinosaurs" on page 138. How did the meteor affect the lithosphere, hydrosphere, atmosphere, and biosphere? Could such an event happen again?

Solving the Problem of Global Warming

Global warming has been widely publicized in television documentaries, the popular press, and scientific journals. Even though we do not fully understand why climates change, scientists, government officials, and the public are beginning to understand how we affect the environment. Attitudes are changing. There is much that individuals and governments can do to reduce emissions of greenhouse gases.

Unfortunately, Canadians are among the world's highest contributors to global warming. We use more fossil fuels than almost any other country. Cold winters require us to burn tremendous amounts of fossil fuels. It is also dark for long periods of time during the winter, so we use more electricity for lighting. In summer, of course, it is often uncomfortably hot in many parts of the country. So we use air conditioners that consume vast amounts of electricity. All of these factors contribute to the amount of carbon dioxide Canadians send into the atmosphere. Of course, more than half of the electricity generated in Canada comes from the burning of coal, oil, and natural gas. The size of the country also contributes to our carbon dioxide emissions. It requires a lot of energy to travel and move goods around this vast nation. Whether we are flying across the country or driving to school, lighting our houses or air conditioning our cars, we are all contributing to increased carbon dioxide levels and ultimately to global warming.

Everything we do can be analysed to determine the best environmentally sound alternative. Recycle. Turn off unnecessary lights. Keep room temperatures low. What other aspects of our daily routines could we change to help reduce the threat of global warming?

Governments are doing their part. Recycling campaigns have become common throughout much of North America. In fact, they have become one of the most successful environmental programs ever undertaken. Newspapers, bottles, and cans are recyclable.

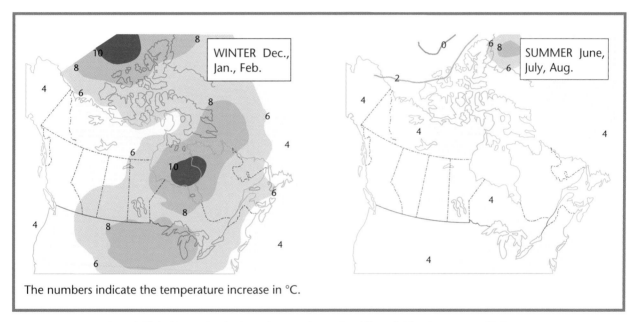

The numbers indicate the temperature increase in °C.

FIGURE 18.5 *Projected Global Warming in Canada, Based on a Doubling of CO_2, Summer and Winter Projections show that a doubling of CO_2 would increase winter temperatures substantially in Canada, especially in the high Arctic and eastern Hudson Bay where winter temperatures could increase by 10°C. What increase would occur where you live?*

So what can we do? There are many ways we can reduce our carbon dioxide emissions. Take commuting, for example. There are alternatives to driving long distances to work or school each day. It may be possible to find a job closer to home or to move closer to your work. Another solution is to car pool; this saves money and reduces the amount of fossil fuels being consumed by several cars. Public transit is also a good idea. Buses and trains emit less carbon dioxide per passenger than most cars. Cycling is a healthy and economical alternative to driving a car.

The energy used to produce these items is saved when the material is recycled. For example, it takes the same amount of energy to make an aluminum soft drink can as it does to run a television set for three hours. When the can is recycled, less energy is needed to make a new can than to make a new one from raw materials. An additional benefit of recycling is that it reduces the amount of garbage that ends up in landfill sites.

In 1987, many industrial countries signed the Montreal Protocol, which agreed to eliminate chlorofluorocarbon emissions by the year 2000.

By 1992, Canadians had reduced their dependence on this dangerous greenhouse gas by over 50 per cent. Public support will help government officials to make environmentally correct decisions.

The business community has also become environmentally aware. Grocery stores sell "green" products. These range from paper products made from recycled newspaper to "enviropac" containers made from less plastic. Industrial supply companies provide "zero waste" products where all components—bottles, cartons, wrapping, and so on—are returned for recycling. In this way no waste is produced, packaging costs are kept low, and less energy is consumed. As trustees of our beautiful planet, it is everyone's job to see that we do not destroy it. Encourage your family, friends, and elected representatives to act responsibly—act environmentally!

Things To Do

1. How have people and governments become more environmentally aware?

2. Study each of the following choices. Prepare a decision-making organizer to determine which choice is best from an environmental point of view. Include at least six criteria, including such things as carbon dioxide produced, waste produced, convenience, cost, peer image, etc. Weight each criterion so that the more important considerations are worth more.
   a) driving, walking, biking, or bussing to school
   b) using a lunch bag, a lunch box, or buying lunch
   c) shaving with an electric razor, a straight razor, or a disposable razor
   d) using disposable diapers, cloth diapers, or a diaper service
   e) using natural gas, oil, wood, coal, or solar energy to heat your home.

3. Develop your own Environmental Action Plan.
   a) List all the times you used energy either directly or indirectly over the past twenty-four hours.
   b) Circle all the times you could have saved energy by modifying your behaviour.
   c) Prepare an organizer to outline how your life could be modified to reduce energy consumption and thereby reduce greenhouse gas emissions.
   d) Take action on your environmental plan! Keep a log to show how you have conserved energy.
   e) Encourage others to act responsibly. Get your school to reduce the amount of lighting in classrooms. Get your family to turn off lights and electrical appliances when they are not in use. Encourage your parents to walk or cycle to the store instead of driving. What else can you do to be environmentally responsible?

PART 4

The Lithosphere:
Building Up the Land

Chapter 19
Geological Time

INTRODUCTION

When you visualize the earth in your mind's eye, what do you see? Great mountain chains marching down coastlines? River valleys snaking their way to the oceans? Vast plains sprawling across huge land masses? What you are actually visualizing is the lithosphere—the upper layer of the earth. But the lithosphere is more than just mountain chains, river valleys, and vast plains. It forms the foundations of the continents and the floors of the oceans. It includes minerals and precious gems, fuels like oil and natural gas, and the fossilized remains of creatures that roamed the planet millions of years ago. The lithosphere is perhaps the most fascinating of all aspects of physical geography. In this unit, we explore the forces of nature that have formed the lithosphere and that continue to change it today.

FIGURE 19.1 *The Grand Canyon*
The layers of rock in the Grand Canyon were laid down over millions of years, uplifted, and dissected by the Colorado River.

Geological Change

Change is one of the most important concepts in the study of geology. Change occurs when an object becomes different over time. Some of these changes happen over a short period; others take much longer. Think of human hair as an example. Our hair grows a little each day, but it takes many years for our hair to turn grey. In either case, however, it is easy to see how people change because even long-term change occurs within the **human time scale**.

The **geological time scale**, however, is something quite different. If you told a friend that something happened recently, your friend would probably think it happened within the last week or two. But in geological terms, a recent event may be something that happened 100 000 years ago! For most of us, a long time is 200 or 300 years. But to a geologist a long time may be 2 or 3 billion years!

The earth is probably about 4.6 billion years old. A lot has happened during this time. Mountains have risen and been washed into the sea. Continents have shifted over thousands of kilometres. Oceans have been filled in with sediment. Coastlines have risen and swamps have dried out. The changes that have taken place are almost unimaginable. Yet the physical features seen on modern maps have not changed significantly for hundreds of years. Because change often takes millions of years, it goes unnoticed.

Not all change happens slowly, however. Some changes occur very quickly. Volcanic eruptions, earthquakes, and mud slides happen in seconds or minutes and can radically alter the lithosphere. When Mount St. Helens erupted in 1980, 2.75 km^3 of rock, ice, and trapped air cascaded down the mountain at a speed of 250 km/h! This was followed by incredibly hot, steam-filled ash that shot out of the volcano at 400 km/h. These two events wiped out every living thing on the north side of the mountain for a distance of 30 km. The mountain lost 418 m of its original height in a matter of minutes. The eruption of Mount St. Helens was not unusual. Natural disasters of this magnitude happen frequently.

The Geological Record

Written records tell us when geologic events occurred in the past. For instance, we know that Mt. Vesuvius erupted in the year 79. Roman histories tell of the devastation as hundreds of people were buried alive under a searing blanket of volcanic ash. Written records are only available for the past 3000 to 4000 years, however. To discover what happened before people started keeping records, geologists have to look for **prehistoric** records. The layers of rock that make up the lithosphere provide just such information. They are as clear and concise as any records written by people. From the lithosphere, geologists can tell what happened and when it happened relative to other geological events.

There are two types of time—**relative time** and **absolute time**. Relative time is when an event occurs relative to other events. For example, you wake up *before* you go to school. The absolute time of each of these events is not determined but the sequence of events is. If you had a clock, however, you could measure the time when each event occurred. You may wake up at 7:00 a.m. and go to school at 8:30 a.m. This is absolute time.

Three principles help geologists use relative time to find out when geological events occurred. These are the principles of **superposition, uniformitarianism,** and **fossil correlation**. Scottish nat-

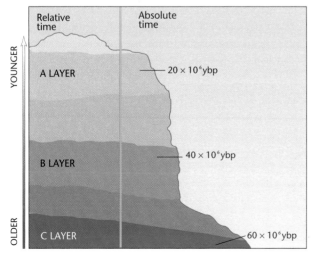

FIGURE 19.2 *Relative vs. Absolute Time*

ural philosopher James Hutton studied sediments as they were deposited along the seashore. He observed that each day fresh sediments laid down on top of older ones. This led him to form the principle of superposition—that is, the lower the layer of rock, the older it is. Figure 19.2 illustrates this principle.

Hutton also discovered the principle of uniformitarianism. This states that things change today exactly the same way they did a billion or more years ago. Under similar conditions, the lithosphere always reacts in the same way. So mountains are built in the same way, rivers meander in the same way, and continents move across the face of the earth in the same way. The principle of uniformitarianism can even be applied to other planets. Landforms on Mars, for example, indicate that there must have been surface water present at some point in time. Only running water could have created the channels that we can still see today on the planet's surface. The principle of uniformitarianism is an important one. It enables geologists to explain what happened in the past and to predict what could happen in the future.

The principle of fossil correlation is also useful in determining the relative age of a layer of rock. Fossils are formed when minerals in solution replace the organic material of decomposed plants and animals. Deposits that contain the same fossils are the same age. If the deposit was laid down before the plant or animal evolved, then its fossilized remains would not be present. Similarly, if the life form was extinct, the remains also would not be there. In this way sediments with the same fossils can be dated relative to other layers.

> **GEO-Fact**
>
> Fossils are formed when plants or animals are buried and preserved, usually in fine-grain soft sediments. Fossils include the shells of clams or snails, carbon from individual plant stems or leaves, and impressions made by soft-bodied creatures that have decayed. Dinosaur and bird tracks are also fossils. Sometimes organic material, such as wood, may be replaced by silica and appear to have "turned to stone." But most fossils are not made of rock.

The Geological Time Scale

Geologists read the geological record much like you read a book. It tells them what animals and plants lived, what the climate was like, what the conditions of the atmosphere were, and what geological formations were created. Over the past 200 years, the information contained in the geological record has been organized into the geological time scale. (See Figure 19.4.) This scale is the result of years of work by many scientists. It is divided into five eras based on the fossils found in each one.

The Archaeozoic Era

The Archaeozoic ("beginning life") Era is the oldest period in the geological time scale. Life was beginning as the lithosphere cooled after its creation. No fossils are contained in these ancient layers. It is likely that one-celled plants and animals may have existed, but they were too small and frail to be fossilized.

The Proterozoic Era

Next came the Proterozoic, or "early life," Era. The fossil record shows simple

FIGURE 19.3 *Proterozoic Era*

ERA	PERIOD	EPOCH	10^6 YBP*	FOSSILS	GEOLOGICAL EVENTS
Cenozoic	Quaternary	Recent	—	• human civilizations • domesticated animals	• minor postglacial gradation • western North America rising
		Pleistocene	0.01 2	• primitive humans • large mammals die out	• worldwide glaciation • ice ages and interglacial periods • Western Cordillera rises
	Tertiary	Pliocene	5	• earliest humans • large carnivores	• continents in present shape • Appalachians uplifted
		Miocene	23	• grazing animals thrive • elephants migrate to America • horses migrate to Asia	• volcanic activity in Rockies • North America joined to Asia
		Oligocene	38	• development of whales, bats, and monkeys	• India collides with Asia
		Eocene	53	• pygmy horses • flowering plants thrive	• mountain building in Rockies, Andes, Himalayas
		Paleocene	65	• many new mammals	• Rockies, Andes, and Himalayas rising
Mesozoic	Cretaceous		135	• first flowering plants • dinosaurs die out at the end of the period	• Laramide orogeny (Rockies, Andes, Himalayas start forming) • marine deposits on east coast of North America
	Jurassic		192	• first birds appear • dinosaurs thrive	• Nevadan orogeny (volcanoes in Rockies, Juras Mountains formed) • shallow seas cover interior of North America and Europe
	Triassic		230	• first mammals • first dinosaurs	• widespread volcanism • red sedimentary deposits
Palaeozoic	Permian		290	• conifers thrive • insects thrive • many reptiles and amphibians • trilobites extinct	• Appalachian orogeny • ice age in South America
	Carboniferous		350	• first reptiles • giant forests of sporebearing plants	• coal deposits form in eastern and central North America
	Devonian		410	• fish thrive • emergence of amphibians • earliest forests	• Acadian orogeny (eastern North American mountains)
	Silurian		438	• first appearance of plants and animals on land	• widespread marine deposits
	Ordovician		485	• first vertebrates (fish) • marine invertebrates thrive	• Taconic orogeny (mountains in New England)
	Cambrian		560	• marine invertebrates (trilobites)	• invasion of shallow seas and marine deposits
Proterozoic			2500	• first marine plants, worms, jellyfish evolve • stromatolites dominant • little fossil evidence	• great volcanism, lava flows, metamorphism of rock • ferrous metals formed
Archaeozoic			4600	• no fossil evidence • first life (bacteria)	• earth cooling • oldest known rocks

*Date when the period/epoch started in years before present in millions of years

FIGURE 19.4 *The Geological Time Scale*

plants and worms in the oceans. The oldest organic structures, **stromatolites**, date back to this era. Colonies of algae built up sedimentary deposits made of calcium carbonate and magnesium carbonate. These strange features are still found in the Canadian arctic and coastal Australia. No life existed on land. The atmosphere was too inhospitable and life had not evolved sufficiently to live out of water. There was a lot of volcanic activity and lava flowed over the continents. It was as if the land was preparing itself for life. Together, the Archaeozoic and Proterozoic eras are called the Precambrian Era.

The Palaeozoic Era

The Palaeozoic, or "ancient life," Era is broken down into seven periods. In the Cambrian Period, there was an explosion of many different marine species. Primitive marine plants and marine invertebrates, such as corals and molluscs, dominated the layers. Higher up, in the Ordovician Period, fish appeared for the first time. Spiders and scorpions were the first creatures to venture on land during the Silurian Period, while at this time life was thriving in the oceans. Land plants and vertebrates first appeared in the Devonian Period. Amphibians still spent most of their lives in the water but came on land from time to time. By the Carboniferous and Permian Periods, life abounded, not only in the seas, but also on land.

FIGURE 19.5 *Palaeozoic Era*

The Mesozoic Era

The Mesozoic, or "middle life," Era was the age of the reptiles. It occurred between the age of the invertebrates (the Palaeozoic Era) and the age of the mammals (the Cenozoic Era) and is made up of three periods. During the Triassic Period, reptiles flourished and the first mammals appeared. The Jurassic Period saw dinosaurs dominating land and sea, as well as the evolution of the first birds. By the Cretaceous Period, dinosaurs had declined as birds and mammals flourished.

The Cenozoic Era

Meaning "new life," the most recent age, the Cenozoic Era, is divided into two periods and

FIGURE 19.6 *Mesozoic Era*

FIGURE 19.7 *Cenozoic Era*

not found in the fossil record until the Quaternary Period. These two periods are divided into seven epochs that trace the evolution of different species of mammals until the present day.

Absolute Time

The geological time scale began as a relative scale. Each layer is older than the one above it, but no dates could be given for the individual time periods. It was not until **radiometric dating** was discovered in 1907 that an absolute time scale could be established. (See Technology Update: Radiometric Dating below.) Various methods of absolute dating provide geologists with the scientific means of measuring how long ago geological events occurred.

Because geological time periods are so long, the unit of measure is **years before the present** (ybp). Actual dates would not be practical. In

seven epochs. We know more about these sediments because they have been less affected by **metamorphism**, erosion, and other geological forces than older layers. The Tertiary Period was the age of mammals, although humans are

TECHNOLOGY UPDATE: RADIOMETRIC DATING

How are geologists able to date the fossil record? Radiometric dating is the process used to find the **absolute age** of rocks formed before the Cenozoic Era. Radioactive elements found in minerals provide the key. The element emits **alpha particles**. As each particle is given off, the element gradually becomes less radioactive. It becomes a new element, called an **isotope**, when half the elements have changed. This is called the isotope's **half-life**.

This **radioactive decay** occurs at a constant rate. It is not affected by heat, pressure, or any other external force. Based on the principle of uniformitarianism, it is assumed that the rate of decomposition has not changed since the earth was formed.

Uranium-238 is a common radioactive element used in radiometric dating. It is so named because it has an atomic mass of 238. When alpha particles are emitted, the isotope eventually decays until all that is left is lead, a non-radioactive element with an atomic mass of only 206. The half-life of uranium-238 is 4.5 billion years.

To establish the absolute date a layer of rock was formed, geologists compare the amount of uranium-238 in the rock to the amount of lead-206. The more lead there is the older the layer must be. For example, if a sample contained 4 g of uranium-238 and 0 g of lead-206, then the layer has just been formed. But if a sample contains 2 g of uranium-238 and 2 g of lead-206, then the isotope's half-life has been reached. The sample is therefore 4.5 billion years old.

Obviously uranium-238 is only good for measuring extremely old materials. But other isotopes have shorter half-lives. Potassium-40 has a half-life of 1.3 billion years. It changes into argon and is commonly found in many rocks. Still, many geological processes have occurred during the more recent Cenozoic Era. Another method is needed for dating these layers of rock.

Radiocarbon dating enables scientists to date organic substances—that is, those containing carbon. It is accurate from 1000 to 60 000 ybp. Carbon-14 is an isotope found in all things containing carbon. It has an atomic mass of 14; the atomic mass of non-radioactive carbon is 12. As it has a half-life of only 5700 years, this process is especially useful for dating recent geological events.

November 1989, the oldest known rock was discovered in the Canadian Shield. It was formed 3.96 billion ybp, according to the Canadian Geologic Survey and other experts. However, meteorites found in ancient sediments are even older. Some have been dated at 4.6 billion ybp. Scientists believe these meteorites were formed at the same time as the earth. This would therefore put the earth's age at 4.6 billion ybp.

Things To Do

1. Using old geographic magazines, collect pictures to make a collage illustrating the lithosphere.

2. a) Obtain a piece of rock. Take it home and change it in some way.
   b) Show the rock to your group. Have them guess how you altered it.
   c) Explain how what you did to the rock could have occurred naturally.

3. a) Using a relative time scale, prepare a timetable of events that have occurred in your school over the past week.
   b) Revise the list in a) to make it an absolute time scale.
   c) How do geological and human time scales differ?

4. Geological change can occur rapidly or slowly.
   a) Give an example of a cataclysmic geological event that has occurred recently.
   b) Give an example of a slow geological event that is occurring in your community.

5. Prepare fully labelled diagrams explaining the principles of superposition, uniformitarianism, and fossil correlation.

6. Use the geological time scale to determine the era, period, epoch, and date in ybp for each of the following events:
   a) first animals
   b) first plants
   c) first dinosaurs
   d) formation of the Rocky Mountains
   e) continental glaciation
   f) formation of coal
   g) first humans
   h) first vertebrates
   i) extinction of dinosaurs.

7. Figure 19.4 shows relative time but there is not an accurate scale to show the length of each period.
   a) Using a long piece of paper, make a scaled timeline to show each era and period. Consider using 1 mm for every million years.
   b) Draw pictures to show major events for each era.

8. If the earth was created in twenty-four hours, estimate the length of each era of the geological time scale in hours and minutes. Copy the following organizer into your notebook. The Archaeozoic Era has been done for you.

Era	Start	End	Difference	Ratio to 24-hr Period
Cenozoic				
Mesozoic				
Palaeozoic				
Proterozoic				
Archaeozoic	4600	2500	2100	$\frac{2100}{4600} = \frac{\chi}{24} \therefore \chi = 10.96$ hr

Chapter 20

The Structure of the Lithosphere

INTRODUCTION

Little was known about the internal structure of the earth until recently. New technology has enabled scientists to probe deep beneath the earth's surface by studying vibrations in the different layers of the lithosphere.

Earthquakes and underground nuclear tests send shock waves through the many layers of the lithosphere. These **seismic waves** are measured using electronic sensors called **seismographs**. Vibrations move through the layers of

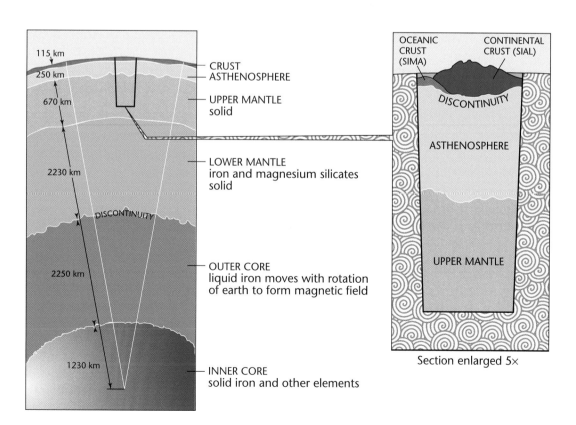

FIGURE 20.1 *The Earth's Interior*

the lithosphere differently, depending on temperature. Cooler layers are generally more rigid than hot layers. When these cool layers vibrate, they send out seismic waves at a higher frequency than the shock waves that travel through hot layers. Rock density also affects how shock waves travel. Sometimes the layer is under so much pressure that it is **plastic** and absorbs the shock waves. In geological terms, plastic does not mean it is made of the same material as toys. It means that it is soft and pliable. The substance is solid but it behaves like a liquid. It flows like a liquid and rebounds when pressed down. Vibrations through the layers of the earth are altered by differences in temperature and plasticity. Piecing together evidence from seismic readings, scientists have determined what the interior of the earth is like.

Why There Are Layers

When the solar system was created 4.6 billion years ago, the earth and the other planets formed from a cloud of dust, gas, and ice. Gravity pulled the different bits and pieces of the solar system into huge chunks of solid matter. One such chunk became earth. As the new planet spun, gravity pulled heavier elements to the centre while lighter objects floated to the surface. This is why the earth is made up of layers. Each layer has different properties. They are made of different elements and have different thermal characteristics. Seismic evidence supports theories that geologists have long had about the earth's structure. But more detailed information about transition zones between layers is increasing our understanding of geology.

The Core

At the centre of the earth is the core. It is made up of two parts, the inner core and the outer core. While the core appears quite small compared with the other layers, it actually makes up a third of the planet's total mass. When the planet was formed the heaviest elements were pulled to the centre of the planet. The inner core is believed to be made of iron, with some silicon and pockets of oxygen and sulphur. The density of the inner core is estimated to be 12.7 to 13 g/cm^3, about the same density as the element mercury.

Estimates of the temperature of the inner core vary from 4000°C to 6650°C. That is even hotter than some estimates of the sun's temperature! What makes the earth's core so hot? Scientists speculate that this incredible heat is the result of decaying radioactive material. Each time an alpha ray is emitted from radioactive matter, nuclear energy is given off. So the same radioactive isotopes that help geologists date rock layers create the heat of the earth's inner core. All of this heat is held in by the insulating effect of the rock layers that lie above. You might think that these temperatures would melt the core, but this is not the case. The incredible pressure of the rock pressing down makes the layer solid rather than liquid.

The earth's outer core is cooler and molten because the pressure is less. This fluid layer generates about 90 per cent of the earth's magnetic field. It also forms the magnetosphere, a force field that protects the planet from cosmic radiation and **solar wind**. Scientists think that electric currents deep in the outer core are caused by the earth's rotation. These currents produce the magnetic field that surrounds the earth.

The outer core is less dense than the inner core, about 10.7 g/cm^3. Like the inner core, it is made up primarily of iron. The transition zone between the inner and outer core is approximately 5200 km below the earth's surface.

The Mantle

Above the core is the mantle. This layer makes up about 80 per cent of the earth's total volume. It is less dense than the core, averaging 4.5 g/cm^3, but denser than the layers above it. Iron, silica, and magnesium are the mantle's main elements. Like the core, the mantle is divided into two parts: the lower mantle and the upper mantle. The lower mantle starts about 2900 km beneath the earth's surface and extends upwards almost 1200 km to the much narrower upper mantle. As you move towards the surface, the temperature of the mantle gradually cools.

Until recently, the discontinuity, or boundary, that separates the lower mantle from the outer core was believed to be a smooth one. Recent seismic studies have shown that the discontinuity is in fact uneven, with many peaks and valleys much like the surface of the earth.

The Asthenosphere

Technology has enabled scientists to discover a new layer in the upper mantle. Called the **asthenosphere**, this layer is thought to affect **plate tectonics** and mountain building. Unlike the rest of the mantle, the rock in this layer sometimes acts like a liquid. Convection cells, most likely caused by decaying radioactive elements, sometimes occur, making the rock less dense. Convection currents distribute this heat throughout the asthenosphere. As the material moves through the layers, continents sitting on the crust directly above the asthenosphere also move. This movement causes the crust to be **folded** or **faulted** to create mountains. When the hot spots reach the upper limit of the asthenosphere, volcanoes may erupt through the crust, forming new mountains. The discovery of the asthenosphere helps to explain many of the processes that shape the lithosphere.

The Crust

The outer layer, the earth's **crust**, is found near the surface. The crust forms the continents and underlies the oceans. This relatively thin layer (less than 0.1 per cent of the earth's volume) consists of two parts. The part that underlies the oceans is often called the **sima**. It is made up of a rock called basalt. Comprising iron, magnesium, and silica minerals, this rock has the same composition as the mantle. It is less dense, however, at 2.9 g/cm^3. This layer of the crust moves with convection currents in the asthenosphere. Where the sima collides with and plunges beneath flows under the continental crust, friction occurs. Volcanoes and earthquakes are often found along these continental margins.

The part of the crust that underlies the continents is mainly granitic. This rock is made up of many different elements, including silica, potassium, and aluminum. It is less dense than

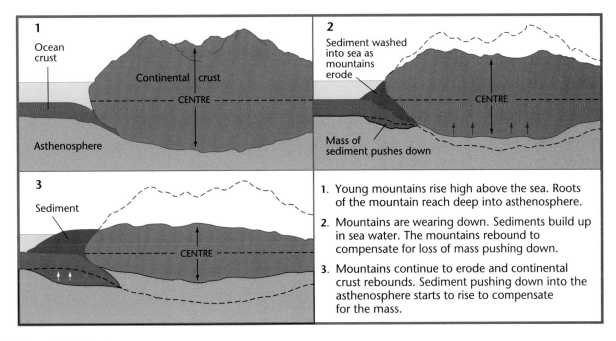

FIGURE 20.2 *Isostasy*

the asthenosphere at 2.8 g/cm^3 and floats on the denser plastic layer beneath it. The continental crust, often called the **sial**, varies from 20 to 80 km in thickness. Where there are mountain ranges it is thicker, sinking deep into the mantle. As the mountains erode over millions of years, the crust rises. It gradually becomes thinner until the mountains are nothing more than deeply eroded upland regions like those found in the Canadian Shield. This process in which the sial rebounds is called **isostasy**.

It is important to understand the structure of the earth because it explains many of the geological processes that occur in the lithosphere. As scientists learn more about the inner workings of the planet, our knowledge of the complex world in which we live will increase.

Things To Do

1. Prepare an organizer comparing the different layers of the lithosphere. Include the following criteria: name, depth, thickness, principle materials, whether it is solid or liquid, density, temperature, and special features.

2. a) Why is it important that the outer core is liquid?
   b) The magnetic field of the earth has reversed nine times in the past 4 million years. If this were to happen again, what problems could it create?

3. In what ways is the asthenosphere different from the other layers?

4. Explain *isostasy*.

Chapter 21

Elements, Minerals, and Rocks

Introduction

Elements, minerals, and rocks are the building blocks of the lithosphere. Because they are so closely connected these terms are often confused. **Elements**, the smallest building blocks, make up all minerals, which in turn make up rocks. Ninety-two elements are found in nature, but only eight are needed to form most of the earth's crust. Elements such as gold, silver, and copper are rare, which makes them particularly valuable. These precious elements are often confused with minerals but, as we shall see, they are not minerals.

Element	Formula	% of the Earth's Crust Mass	Volume
oxygen	O	46%	91%
silicon	Si	28%	1%
aluminum	Al	8%	1%
iron	Fe	5%	1%
calcium	Ca	4%	1%
sodium	Na	3%	2%
potassium	K	3%	2%
magnesium	Mg	2%	1%
carbon	C	<1%	<1%
sulphur	S	<1%	<1%
lead	Pb	<1%	<1%
hydrogen	H	<1%	<1%

Figure 21.1 *Major Elements of the Earth's Crust*

Figure 21.2 *Pink Granite*

FIGURE 21.3 *Rock Samples*
Top left *hematite (iron)*;
lower right *copper sulphate*;
centre *quartzite*;
lower right *galena (lead)*;
top right *malachite (copper)*.

The combination of two or more elements creates a **mineral**. For example, silicon (Si) and oxygen (O) are both elements. When they join together chemically, they form the mineral silicon dioxide (SiO_2), more commonly known as quartz. While minerals have common names, chemical formulas are better for identifying them. These formulas list all the elements contained in the mineral. Some may look complicated but they are not too hard to figure out. Take *forsterite*, for example. This common mineral has the formidable formula $(MgFe)_2SiO_4$. What this tells us is that forsterite is made of magnesium (Mg), iron (Fe), silicon (Si), and oxygen (O).

Pure minerals are rarely found in nature. Usually several minerals are cemented together to form rocks. Rocks contain all kinds of minerals mixed up together. For example, granite is a common rock often made up of quartz, mica, and orthoclase or plagioclase feldspar.

Things To Do

1. Study Figure 21.1 and determine what elements make up the following minerals:
   a) quartz: SiO_2
   b) calcite: $CaCO_3$
   c) pyrite: FeS_2
   d) magnetite: Fe_2O_3
   e) galena: PbS
   f) corundum: Al_2O_3
   g) gypsum: $CaSO_4 \cdot 2H_2O$
   h) dolomite: $CaMg(CO_3)_2$
   i) orthoclase: $K(AlSi_3O_8)$

2. Obtain a collection of ten to twenty different minerals from your teacher.
   a) Prepare an organizer comparing the different samples. Criteria could include colour, hardness, density, etc.
   b) Use a field guide to identify the minerals.
   c) What criteria were used in the field guide to identify minerals?

3. Obtain a collection of ten to twenty different rocks.
   a) Identify the minerals that make up each rock using the samples from activity 2 and the field guide.
   b) Use the field guide to identify the rocks and to check if your answers in part a) were correct.

Three Classes of Rock

Because the earth is constantly changing, rocks are being created and destroyed all the time. Rocks are classified according to how they are made. There are three rock types: **igneous rock**, which is formed from molten rock; **sedimentary rock**, which is made when layers of deposited material are cemented together; and **metamorphic rock**, which is created when other rocks are heated under great pressure.

Igneous Rocks

The minerals that form igneous rocks come from **molten rock** deep within the crust and upper mantle. Igneous rocks are often found on the earth's surface as well as deep within it. They reach the surface in one of two ways. **Extrusive igneous rocks** are formed when molten rock is carried to the surface by volcanic processes. When this **lava** cools new rock is formed. **Intrusive igneous rocks** originate deep in the earth as molten rock called **magma**. The magma also cools and hardens to form igneous rock. It reaches the surface as erosion gradually wears away the rock layers above it. Much of the igneous rock in Canada was formed deep in the earth millions and even billions of years ago. The only exceptions are Palaeozoic volcanoes in eastern Canada and more recent volcanoes in the Western Cordillera.

Igneous rocks are not only classified by the minerals they contain, but also by their texture. The size of the crystallized minerals in rocks is determined by how fast the magma cooled. If it cooled slowly the texture is coarse; individual crystals can be seen clearly. These coarse-grained igneous rocks are usually intrusive—that is, they were formed deep beneath the earth's surface where cooling is slow. Rocks with a fine texture result when the magma cooled so quickly there was no time for individual crystals to form. These extrusive rocks are usually formed when volcanic lava comes into contact with relatively cold air or water. Some igneous rocks that are formed under water as a result of lava flows have the same texture as glass. Obsidian, for example, is extremely hard and looks just like black glass. Sometimes an igneous rock contains both coarse and fine grains. This indicates that cooling occurred in two stages. Perhaps the outside cooled rapidly since it was in contact with the air but the inside cooled slowly because of the internal heat of the lava. Mineral composition and texture are the two ways that igneous rocks are classified.

Sedimentary Rocks

Sedimentary rocks differ from igneous rocks in that they are not formed from molten magma but are created from sediment, or deposits that build up in oceans and lakes. There are three types of sedimentary rocks: **clastic sedimentary rocks**, which comprise broken up pieces of other rocks; **chemical sedimentary rocks**, which are formed from chemicals dissolved in shallow seas or lakes; and **biogenic sedimentary rocks**, which are made from the remains of living creatures.

Sediment builds up when layers of material are deposited by natural processes. First rock is **weathered**, or broken up into little pieces. Then it is **transported** to another place by wind, glaciers, or water. Finally it is **deposited**. Layers are created as more and more weathered material accumulates. In time, thick deposits of sedimentary rock are formed.

For sediment to be transformed into rock, it must be compacted and cemented together.

FIGURE 21.4 *An Igneous Rock Formation*

FIGURE 21.5 *A Sedimentary Rock Formation*

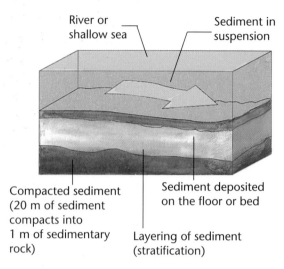

FIGURE 21.7 *How Sedimentary Rocks Are Laid Down*

Compaction occurs as new layers of sediment are deposited on top of older layers. The mass of the upper layers squeezes out any spaces between particles and presses the weathered material tightly together. This flattens the individual particles and causes them to interlock with one another. **Cementation** occurs when water filters

FIGURE 21.6 *Sedimentary Rocks: Conglomerate (top) and Cross-bedded Sandstone*

through the sediment. Minerals dissolve in the water. At a certain temperature and pressure, the minerals precipitate out of the water solution and remain in the spaces between the rock particles. They now act to cement the grains of **source material** together to form sedimentary rock. This is exactly the same process used in making concrete. Once it has been weathered, concrete looks a lot like sedimentary rock.

Sedimentary rocks are classified by their origin and texture. Clastic sedimentary rocks are made from other rocks that have been weathered over time and deposited in thick layers. Textures range from shale to sandstone to conglomerate. Made from clay, shale contains minute particles, usually deposited in a lake or other still body of water. Sandstone, of course, is made from sand that is laid down in moving water. Conglomerates are stones and particles as big as boulders that are cemented together and deposited in rapidly moving water. These three rocks—shale, sandstone, and conglomerate—are the main clastic sedimentary rocks.

Chemical sedimentary rocks are classified by texture and source material. Salt and gypsum are formed when the water in which these minerals were dissolved partially evaporates. Thick deposits of salt occur where inland seas once existed. Chert or flint is a hard chemical sedimentary rock. It forms from quartz crystals that are cemented together by a fine silica gel.

Limestone and dolomite are the most

common types of biogenic sedimentary rocks. The shell and skeletal remains of marine animals leave calcium deposits. Calcium carbonate cements these fragments together, creating limestone. Dolomite is formed when magnesium carbonate is the cement that binds the biogenic particles. Like other sedimentary rocks, biogenic rocks usually form where inland seas once occurred.

All three types of sedimentary rocks can be found mixed together. Limestone, for instance, may contain fragments of flint, and sandstone and dolomite may be found with beds of salt or gypsum.

Metamorphic Rocks

Meta means "change" and *morph* means "shape." Metamorphic rocks change more than just their shape, however. They change in mineral composition, structure, and texture as a result of the great pressure and heat that exist in the lithosphere. Unlike igneous rocks, in which change occurs during melting, changes in metamorphic rocks occur while the rock is solid. The degree of change depends on the amount of heat and pressure.

There are more types of metamorphic rocks than there are igneous or sedimentary rocks. In

FIGURE 21.8 *Metamorphic Rocks: Gneiss (top) and Schist*

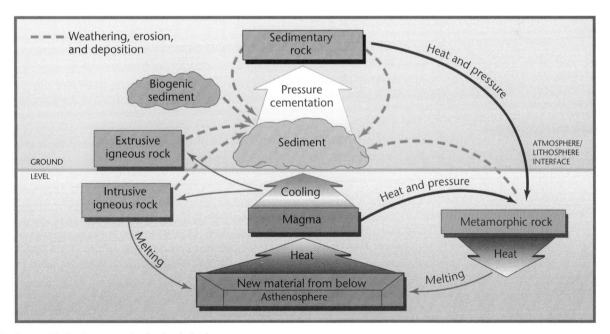

FIGURE 21.9 *Processes in the Rock Cycle*

fact, each sedimentary and igneous rock has at least one metamorphic equivalent. Some have several. For example, the metamorphosed form of sandstone is quartzite. Limestone changes to marble. Granite becomes gneiss (pronounced "nice.") Shale changes to slate. But as more heat and pressure occur, the slate changes again to schist. It has been estimated that 85 per cent of the upper crust is made up of metamorphic rock. Most of it is buried under sedimentary rock, so you don't see too much of it unless you go to places where erosion has removed the earth's upper layers.

Metamorphic rocks are formed wherever there is enormous pressure and temperatures reach over 300°C. Where these conditions exist, two types of metamorphic rock are formed. **Contact metamorphic** rocks are rocks that change as a result of the heat given off during the formation of intrusive igneous rocks. **Dynamic metamorphic** rocks are formed when rocks are compressed by the mass of rock layers above them or by tectonic forces.

Metamorphic rocks are sometimes hard to identify. They often look like igneous rocks. But there is one trait that sets them apart—a banded structure called **foliation**. When pressure is intense, minerals may align themselves in layers within the rock. A good example is gneiss. The minerals in the original granite from which gneiss is made—feldspar and quartz—are concentrated in alternating bands.

Things To Do

1. Obtain a collection of twenty to thirty rocks.
   a) Work as a group to sort the rocks into the three rock categories.
   b) Use a field guide to identify the different rock types and confirm that you placed them in the correct category.
   c) After discussing the reasons why the rocks were grouped as they were, make a list of characteristics of igneous, sedimentary, and metamorphic rocks.
   d) Prepare a display to summarize your results.
   e) Have another group evaluate how accurate your facts were.

2. Study Figure 21.9 on page 161 and reread this section of the text.
   a) Write a short essay explaining the flow chart. Don't forget introductory and concluding sentences for each paragraph. Include a detailed explanation of each important point.
   b) Self-evaluate your paragraph. Give one mark for each point you believe has been fully explained.
   c) Exchange your work with another student and evaluate each other's work.
   d) Discuss any discrepancies between your mark and the mark that your peer gave you.

FIELD STUDY

At your field study site, collect a representative sample of rocks and minerals.

1. Display your samples on a piece of plywood, in an egg carton, or in some other way.

2. If there is a lot of variety in your rock samples, identify each one and classify it as igneous, sedimentary, or metamorphic. Write a brief explanation to support your classification.

3. If all of your samples are of the same rock type, write an explanation of how you think the rock was formed. What generalizations can you make about the geology of your study site?

CHAPTER 22
Minerals in Our Lives

INTRODUCTION

Minerals are essential natural resources in our modern world. No matter where you look, minerals provide raw materials for many of today's conveniences. Look around your classroom. The graphite in your pencil, the slate chalkboard, and the chalk are all made from minerals. What other things can you find in the classroom that are made from minerals?

Often minerals are hidden, so they go unnoticed. We don't usually see the copper wiring and steel beams in the walls, so we may not appreciate the fact that these, too, are made from minerals. And we don't see the fuels that power our society, but we know that our industrial world could not function without minerals like coal, natural gas, and oil.

Valuable minerals usually occur in small quantities. Sometimes the extraction of these deposits can cost more than the actual mineral is worth. For example, it has been estimated that there is $1 billion worth of gold dissolved in the waters of Vancouver Harbour. But the cost of extraction is far greater than even this valuable mineral is worth. The key to mineral extraction is to find deposits that have been concentrated through natural processes. Only when the cost of extraction is less than the value of the mineral is the resource considered to be an **economic deposit**.

FIGURE 22.1 *A Rare Find*
This champagne-coloured diamond, discovered in southern Brazil, is estimated to be worth $1.5 million!

Career Profile: Geologist

Do you find the study of rocks and minerals fascinating? Do you like studying geography and other scientific subjects? Do you enjoy travelling and working outdoors? Have you enjoyed the field studies you have conducted in this course? If you answered "yes" to any of these questions, then you might be interested in becoming a geologist.

The Nature of the Work

There are many different types of geologists. Some specialize in oil drilling. Geological engineers design and build mines. Other geologists study earthquakes or volcanoes. Hydrogeologists study groundwater supplies. The list of related fields is endless.

Jim Tilsley is an economic geologist. He travels all over the world, studying ore samples and determining if mineral deposits can be mined profitably. Economic geologists are generalists who bring together specialized knowledge from many different related fields. Structural geology tells them where economic deposits may be found. Geochemistry indicates the minerals and elements that make up an ore body. Knowledge of mineralogy, engineering, geography and economics is essential to become a successful economic geologist.

Tilsley came from a mining family in Nova Scotia, where he received an undergraduate degree in science from Acadia University. He worked in the uranium mines of Elliot Lake as a sampler before moving on to north Africa. In Morocco and Mauritania, Tilsley was an underground geologist working in silver and gold mines. Upon his return to Canada, Tilsley enrolled at the University of Toronto, where he obtained a postgraduate degree in mineral economics. Today he works as a consultant, evaluating mineral deposits in such places as Guyana, Brazil, Zaïre, and Peru as well as radioactive gas in Ontario school sites.

The three stages in Tilsley's work are similar to those you have followed in your own field studies. First, a preliminary study of the area is conducted using existing topographic maps, air photos, and geophysical surveys. This enables Tilsley to determine whether geological structures associated with certain minerals are present. Geographic factors such as climate, transportation, government regulations, and the availability of skilled labour are also taken into account.

Next Tilsley goes into the field to survey and register the claim. At this stage a sizable sample must be dug out of the ground for analysis. Because sites are often in remote areas, equipment must be brought in by truck, plane, or even barge! Once the sample is collected, assayers determine the proportion of the mineral that makes up the ore. Using this information, Tilsley is ready for the final stage—his report and recommendations. The report analyses and synthesizes the results of the field work. The value of the ore body and the cost of extraction are estimated. If the ore body is economic, then the resource will likely be mined.

Qualifications

Geologists need skills from many fields, including chemistry, physics, geography, computer science, mechanics, and economics. Geologists must hold an undergraduate degree in science as a minimum requirement. Postgraduate degrees are usually required for all fields of specialization.

Related Fields

If you do not intend to go to university but are interested in a technical field related to geology, there are many possibilities. You can collect ore samples or prepare topographic maps from air photos and satellite images. Or you might want to become an assayer. Whatever your interest, your guidance counsellor can give you more information about the many fields associated with geology.

Igneous Ore Bodies

Many economic mineral deposits are found in the igneous rocks of the Canadian Shield and the Western Cordillera. **Ore** is the name given to rocks that contain metals or minerals that can be recovered and sold at a profit. These metals are divided into three categories.

Precious metals, including gold, silver, and platinum, have a great many industrial uses in addition to their use in jewellery. Gold and silver are used in medicine and dentistry because they are easy to work with and do not react with other chemicals. Platinum is used industrially for many things, including catalytic converters in the emission controls of all cars. These valuable minerals are found in minute quantities, often comprising only a tiny part of the rock in which they are found. Gold, for example, can be economically mined even if it makes up a mere 0.005 per cent of its ore.

Non-ferrous metals are less glamorous than precious metals, but they also have important industrial uses. Copper, lead, zinc, tin, and aluminum are used to make all kinds of products, from auto parts to food containers to construction materials. **Ferrous metals** include minerals that can be combined to make steel **alloys**. Iron, nickel, molybdenum, chromium, and manganese are the main minerals in this group.

All three classes of metals are the result of **igneous intrusions** that occurred millions of years ago when molten rock filled cracks in the earth's crust. The way in which magma cools determines how minerals are concentrated in veins. These veins are formed in one of three ways. Sometimes heavier metals sink in molten rock and solidify at the bottom of the igneous intrusion; the nickel in Thompson Mine, Manitoba, was concentrated in this way. The second method is the result of the way in which the minerals **precipitate**, or are deposited. As the magma cools in the intrusion, it shrinks. Hot water containing minerals flows into the spaces that the shrinking magma leaves behind. The water cools and the minerals precipitate at different temperatures. These minerals form **veins** containing precious deposits as well as **gangue**, or waste material, such as quartz. The third way in which mineral veins are formed is when water is driven through the spaces between the rock layers in front of igneous intrusions. Most often, however, mineral deposits are the result of a combination of these processes.

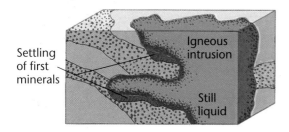
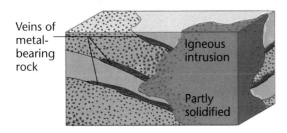
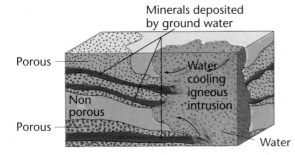

FIGURE 22.2 *Three Types of Igneous Intrusions*

Sedimentary Ore Bodies

Sometimes veins of economic minerals are exposed through erosion. These deposits are easy to mine and can be quite profitable. **Open-pit mines** are used to extract these ores. More often, however, the minerals are buried deep in the ground. Shafts and tunnels are required to extract these ore deposits.

Placer deposits occur when rock is eroded by running water. Minerals are released from the ore body. As the sediment being carried by a river is deposited, so too are the precious metals. These deposits are found where river currents are fast enough to remove lighter, worthless minerals but slow enough to allow heavy minerals to be deposited. The gold found in the famous Klondike River in the Yukon was concentrated in this way. Finding these rich deposits requires a good understanding of geology—and a little bit of luck!

The creation of natural gas and oil is a complex process that takes millions of years. Existing deposits were formed when minute floating plants and animals called **plankton** accumulated in shallow seas during the Palaeozoic and Mesozoic eras. (The Palaeozoic was the era when marine life dominated the earth; the Mesozoic Era was the period of the dinosaurs.) Ash from volcanoes, sediments from rivers, and wind-blown dust accumulated on top of the layers of dead plankton. Energy from the sun was converted into plant material by photosynthesis. Some of this may have been converted to animal tissue as microscopic animals fed on the tiny plants. This "fossilized sunshine" was trapped in the sedimentary rocks that formed around it for millions of years. It lay there waiting to be set on fire to free the stored solar energy.

This layer of organic ooze was unable to decompose because there was limited oxygen. However, bacteria transformed it into methane and a complex hydrocarbon called **kerogen**. Where the temperature rose to 100°C, the kerogen broke down to form oil. But if the temperature continued to rise above 150°C, the oil was transformed into carbon dioxide. Oil is really a freak of nature that occurs when

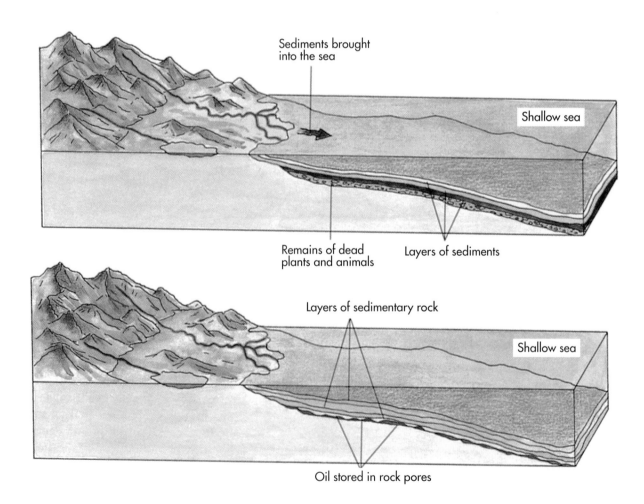

FIGURE 22.3 *Sedimentary Structures that Support Oil and Gas Deposits*

conditions are just right. It takes a great deal of scientific knowledge to find an economic oil deposit.

As with metallic minerals, natural gas and oil have to be concentrated before they are economic deposits. These two hydrocarbons are contained when they are squeezed into layers of porous rocks that are surrounded by non-porous layers. The precious fossil fuels accumulate in the spaces within the pores of the **reservoir rock**. The non-porous rock above holds them in. The **cap rock** prevents the natural gas and oil from floating up to the surface. Salt water under the reservoir rock usually prevents the oil from seeping down deeper into the crust. Because natural gas is lighter than oil, it usually rises above the oil deposit. When a hole is drilled in the cap rock, the natural gas explodes out of it. The oil follows soon after.

Coal is another valuable fossil fuel found in sedimentary layers. Like natural gas and oil, it formed from living things. In the Carboniferous Period, giant forests grew in the shallow tropical swamps. When the trees fell into the water they decomposed slowly. Eventually methane and peat, a flammable organic substance that can be used as a fuel, were formed.

If the layers of peat were buried under sediment and heated, water vapour was squeezed out of the deposit. Eventually the carbon concentration in these deposits rose and **bituminous coal** was formed. This concentrated fossilized vegetation burns hot and makes an excellent fuel. If further heat and pressure drive off more impurities and concentrate the carbon even more, **anthracite** is formed. This metamorphic form of coal burns hotter and with less pollution than bituminous coal.

Unlike oil, coal forms in thick seams within sedimentary structures. It does not need to be held in place by non-porous layers of rock. It can be mined easily if the layers of ore are exposed through erosion. Often shafts and tunnels follow the seams deep under the earth's surface.

Today, coal's popularity has declined because it does not burn as cleanly as natural gas or oil. Soot, sulphur dioxide, and other pollutants make it unacceptable to many people. Coal is used to make **coke** for **smelting** iron and provides the heat for many thermal power plants. A useful resource, coal fuelled the Industrial Revolution in Europe and North America. It may be just as important a resource for emerging industrial nations like India and China.

GEO-Fact

All the conditions were right for finding oil in the St. Lawrence Valley east of Montreal. The rock was the right age and geological structures indicated that oil may be present. Unfortunately, the degree of dynamic metamorphism was so great that the temperature exceeded 150°C, so carbon dioxide was produced instead of oil.

But even carbon dioxide is valuable. It is what makes the bubbles in soft drinks or sodas. The enterprising mining developer decided to mine the carbon dioxide. A pipeline was built from the deposit to a local bottling plant and the gas was used to make soft drinks. The next pop you drink may have bubbles in it that are millions of years old!

Other Economic Minerals

Many other minerals are used by people. Crushed limestone is used to make highways and concrete. Marble, limestone, and granite are used in the construction industry. Clays are used to make everything from bricks to fine china. Gypsum is used to make wallboard used in construction. Cement is made from a mixture of ground limestone and clay. In addition, many fertilizers are made from mineral deposits. All of these minerals are derived from rock formations found in abundance in Canada.

a. Swampy forest by the sea

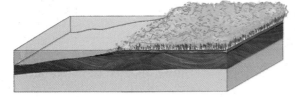

c. Sea level remains high

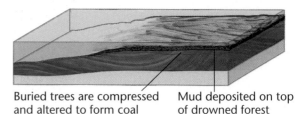

Buried trees are compressed and altered to form coal   Mud deposited on top of drowned forest

b. Sea level rises about 20 m

Trees and plants drowned by sea

d. Sea level falls again

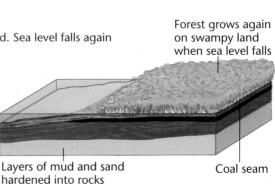

Forest grows again on swampy land when sea level falls

Layers of mud and sand hardened into rocks   Coal seam

FIGURE 22.4 *Coal Sedimentary Structures*

TECHNOLOGY UPDATE: FINDING DIAMONDS

Almost everyone loves diamonds! Not only are they valuable, but their everlasting beauty gives them a special magic. But there is no magic in finding diamonds. This is serious business requiring technology, knowledge, and skill.

Scientists believe that diamonds are found in abundance deep within the earth where the mantle and the crust interface. This is because the conditions 80 to 225 km below the earth's surface are right for diamond formation. The temperature is between 900°C and 1500°C and the pressure is immense! Ordinary carbon—the same substance that produces coal—becomes one of nature's most beautiful minerals. Unfortunately, it is impossible to mine these diamonds because they are too far beneath the earth. So geologists must look for the rare diamond deposits found near the earth's surface.

Finding these diamond deposits takes a lot of detective work. The first thing that geologists look for is a stable **Archaean crust**—the part of the earth's surface that is extremely old. Diamonds are 1.2 to 3 billion years old, so there is no point looking for them in relatively new rock! Geological surveys tell geologists where this ancient rock can be found. The Canadian Shield is a prime location for surface diamond deposits.

Once a prime location has been determined, geologists look for places where igneous intrusions may have carried diamonds from deep underground up to the surface. The molten rock bores

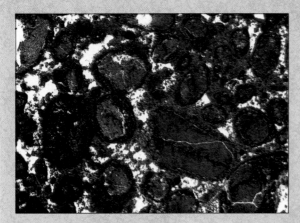

FIGURE 22.5 *Kimberlite Magnified 100 Times*

up through the crust at approximately 17 km/h. If it moves more slowly the diamonds may burn up, leaving common graphite—the stuff in pencils—instead of diamonds! Once the rock reaches the surface, it forms distinctive geological formations called **kimberlite pipes.** Where these structures exist, there is a *possibility* of finding diamonds. But only 3 to 4 per cent of all kimberlite contains enough diamonds to be economically viable. Diamond ore should contain at least 20 to 60 parts per billion (ppb)—in other words, 20 to 60 mg of diamonds in 1 t. It's somewhat like searching for a needle in a haystack!

To find kimberlite pipes, airplanes fly over a potential area with a **magnetometer** to record the magnetic fields of the surface rocks. Kimberlite has a special "signature" that makes it different from other magnetic bodies. Places where the magnetic field differs from the rest of the rock suggest possible igneous intrusions.

Once the kimberlite pipes have been identified, the next step is to stake a claim and obtain the mineral rights. Now a magnetic survey is conducted on land to locate the exact position of the deposit. Once this is established, geologists look for less valuable **indicator minerals** that are commonly associated with diamond deposits.

Back in the laboratories, geochemical studies determine the chemistry of the indicator minerals. If sufficiently high levels of chromium and low levels of calcium are detected, it is likely that diamonds are present. If there is no chromium, the search ends here. But if the geochemical studies are positive, a huge ore sample of several thousand tonnes is removed and analysed to determine the number of diamonds present. In this process, the rock is crushed, but diamonds are so hard they remain intact. The crushed ore is mixed with water and placed in a jig. This contraption pulses water through the ore slurry, allowing the heavier diamonds to separate from the lighter gangue minerals. Only then is it possible to determine if there are enough diamonds to make mining the deposit economically viable. After all this, perhaps only one or two diamonds may be found for every tonne of ore!

Things To Do

1. Prepare a comparison organizer detailing the similarities and differences between igneous and sedimentary mineral deposits.

2. Give examples to show how mineral deposits are concentrated in geological structures.

3. What geological structures would you expect for each of the following economic minerals:
   a) coal
   b) silver and gold
   c) oil and natural gas
   d) nickel
   e) diamonds?

4. Visit a local quarry or mine. Interview a worker and determine:
   a) the nature of the mineral
   b) how it was formed
   c) how the ore is mined
   d) how the minerals are removed from the ore.

CHAPTER 23
Tectonic Processes: Building Continents

INTRODUCTION

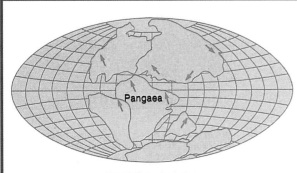

180 million years ago

(a) The supercontinent Pangaea had started to break up and drift apart

135 million years ago

(b) Both Gondwanaland and Laurasia continued to drift resulting in India and Antarctica–Australia becoming isolated

65 million years ago

(c) South America, completely separated from Africa, moved quickly north and westwards. Madagascar broke free from Africa. The Mediterranean sea is recognizable. In the south, Australia is still connected to Antarctica

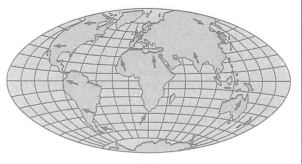

Today

(d) India has moved northwards and is colliding with Asia, crumpling up the sediments to form the folded mountain range of the Himalayas. South America has rotated and moved west to connect with North America. Australia has separated from Antarctica

FIGURE 23.1 *The Process of Continental Drift*

The ground beneath our feet is in constant motion. Old *terra firma* isn't so firm! People have difficulty appreciating how much the earth is changing because they think in terms of the human time scale rather than the geological time scale. The hill that your school was built on or the lake by your cottage has not always been there. The physical features of the earth are constantly changing. In many parts of Canada, the landscape is relatively recent. It was formed when the last ice age ended 10 000 to 14 000 years ago. The earth continues to change. In another 10 000 years, Canada will be different than it is today.

It is only recently that people have started to realize how much the earth changes. **Tectonic processes** are the result of forces operating deep beneath the earth's surface that cause the movement of oceanic and continental plates. When plates collide, the elevation of the ground changes. Some places rise and others sink.

Plate Tectonics

Have you ever noticed that the continents seem to fit together like pieces of a puzzle? Look at the east coast of South America. The coastline of Brazil looks like it fits together with the west coast of Africa. Greenland fits together with Labrador and Baffin Island. Madagascar looks like it has floated free from the east coast of southern Africa. These similarities in shape cannot all be coincidental. At one time, the continents must have been joined together.

German meteorologist and physicist Alfred Wegener was not taken seriously when he proposed the theory of **continental drift** in 1924 in his book *The Origin of Continents and Oceans*. He believed that the continents were once joined together in one single great land mass that he called Pangaea. About 200 million years ago, Pangaea split into two large continents, Laurasia and Gondwanaland. These further broke up to form even smaller continents as the land masses moved apart. Scientists did not accept the theory because Wegener could not explain how the continents moved. It was not until thirty to forty years ago that scientists began to uncover evidence that confirmed Wegener's theory. Further evidence being discovered today provides even more support for the idea that the continents are, in fact, still drifting.

Evidence to Support the Theory

1. The Shape of Coastlines:
   Maps of continental shelves beneath coastal oceans show that the edges of continental shelves fit together even more closely than shorelines. There is less erosion under water, so shapes are closer to what they were like when the continents split apart. Evidence suggests that the continents are resting on moving **plates**. This led to the new name for continental drift—**plate tectonics**.
2. Mid-oceanic Ridges:
   These are a series of ridges in the middle of each ocean. These ridges are like mountain ranges that run parallel to adjacent coastlines. Theorists believe these might be where the continental plates split apart.
3. Fossil Evidence:
   The fossil record shows that 300 million years ago the same carboniferous worm fossils were found on all of the continents. This indicates that the continents must have been joined in order to share this common species. Later sediments indicate that a reptile order called mesosaurus lived in present-day southern Africa and Brazil, but nowhere else. This suggests that the continents of Africa and South America were still joined 200 million years ago but the others had broken away. By 100 million years ago, according to the fossil record, each continent had its own distinct animals, except Australia and Antarctica. These two continents must have been joined for some time after the others had separated. Continental drift explains why the animals found on each continent today are unique to that continent—because they were separated from their ancestors and so evolved differently.
4. Geological Formations:
   Similar landforms are often found on adjacent continents. For example, the same

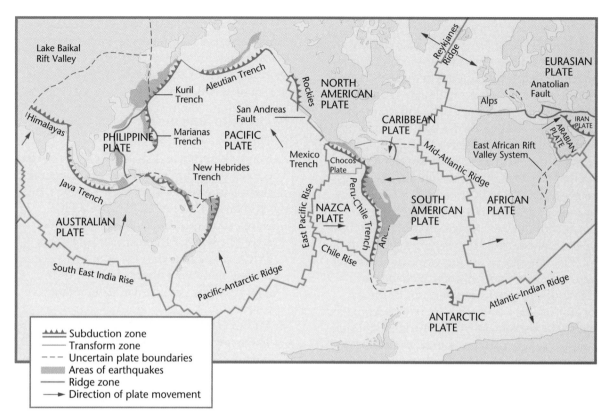

FIGURE 23.2 *The Earth's Major Plates*

Precambrian rock is found on the west coast of Africa as is found in Brazil on the east coast of South America. Such evidence strongly suggests that the continents must have been joined and supports the theory of continental drift.

How Continents Move

We know from seismic data that the upper mantle is 10 to 15 per cent liquid. This **plasticity** allows rock layers deep beneath the surface to flow like any other fluid. Some spots in the upper mantle are hotter than others due to friction, pressure, or radioactive decay. Convection currents cause the plastic rock to flow from regions with high temperatures to cooler regions.

The crust that rests on top of the upper mantle is made up of oceanic and continental plates. As the upper mantle flows from one place to another, these plates are dragged along with it. It is as if the plates are on giant conveyor belts. In regions where one plate collides with another, the continual pushing causes the plates to **subduct**, or sink back into the mantle. The pressure of the two plates grinding together is so great that the plates buckle, forming mountains.

Evidence that the plates actually move lies hidden in the mid-oceanic ridges. In the middle of each ridge is a deep canyon. As the plates move apart, magma oozes out to replace the crust that has moved away. Over the years, a series of tension ridges has been formed. Uranium dating shows that each ridge has a matching twin on the opposite side of the canyon. Each pair of ridges gets steadily older as you move away from the ridge until you reach the continental plate. Thus the sea floor must be moving away from mid-oceanic ridges. New crust is being created where the sea floor is spreading apart.

The theory of plate tectonics accounts for the movement of the continents. But it also explains other tectonic processes, such as why mountains occur on the colliding edges of plates and why active volcanoes are located where they are. It also explains why ocean trenches often occur alongside rugged fold mountains. Even earthquakes and fault patterns are related to plate tectonics. Thus we can see that plate tectonics is an important basis for further studies in geological processes.

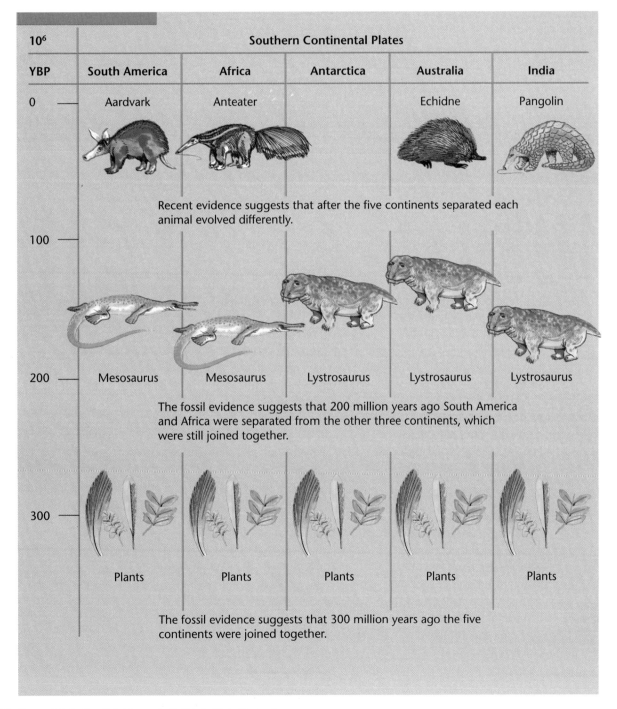

FIGURE 23.3 *Fossil Evidence to Validate Plate Tectonics*

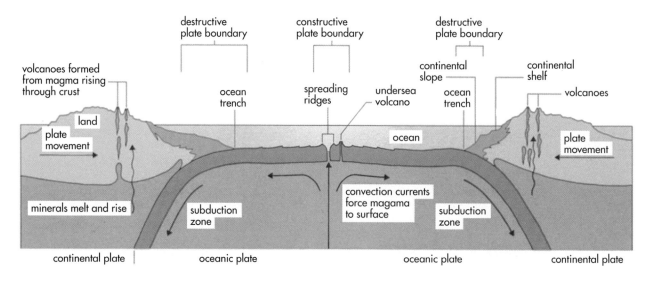

FIGURE 23.4 *Crustal Movement Under the Oceans*

Things To Do

1. Outline how the theory of plate tectonics is supported by the following geological evidence:
   a) fossils
   b) continental shelves
   c) mid-oceanic ridges.

2. Explain how the continents move from one place to another using the example of a conveyor belt.

3. a) Using a pair of scissors and photocopied maps of the world make cut-outs of each continent.
   b) Place the continents on a thick piece of paper so that they are all joined together to create Pangaea 300 million years ago. Trace each continent on the paper and colour them all one colour.
   c) Slide each cut-out continent into the position it would have had 200 million years ago. Trace each continent in its new position and colour it a new colour. Draw arrows to show how the continents moved.
   d) Continue the process described in part c) to show how each continent moved from 100 million years ago to the present. Add the mid-oceanic ridges and a symbol to show where collision has occurred.
   e) Measure the arrows to find out which continent has moved the furthest and which has moved the least.
   f) Complete your map display with a colour key and a title.

4. Speculate on how plate tectonics might explain each of the following:
   a) volcanic islands like Iceland on the Mid-Atlantic Ridge
   b) mountain ranges and volcanoes on the leading edges of moving continental plates
   c) unusual and unique animals evolving in Australia and Madagascar.

CHAPTER 24
Tectonic Processes: Building Mountains

INTRODUCTION

Mountains are formed in three ways: **folding, faulting,** and **volcanism**. Folding and faulting occur when pressures deep within the lithosphere cause the earth's surface to buckle, bend, and even split apart. These are called **diastrophic processes**.

Volcanism occurs when solid rock structures are created from molten rock. They may either be created below the surface or above ground when a volcano spews lava, ashes, and other material to form a new mountain.

FIGURE 24.1 *The Juras Mountains of the Swiss Alps*

Fold Mountains

Folding occurs when the earth's crust is pushed up from either or both sides. You can see how this works with a simple piece of paper. Lie the paper on a flat surface, then gently slide both edges towards the middle. The paper "folds" up and flops down on itself. This is exactly what happens when the earth's crust is pushed from both sides. **Fold mountains** occur where the crust is pushed up as plates collide and are often found near the leading edges of plates. The force of the two plates pressing against each other causes the crust to rise up in folds.

You can tell where folding has occurred in eroded rock layers because the different layers run up and down in regular parallel bands. Determining fold mountains is sometimes difficult for two reasons. First, erosion often wears away some of the folded layers. You have to imagine what the rock was like *before* the erosion occurred in order to see the parallel bands of rock strata. Second, folding often occurs along with faulting and even volcanism. This complicates the picture, making it difficult to determine how the mountains were formed.

The structure of fold mountains can be complex. The simplest fold mountains are made up of layers of rock that rise and fall like waves of water. The peaks or hills are called **anticlines**; the troughs or valleys are called **synclines**. The layers of rock often vary in hardness. If the harder cap rock erodes away, a valley may be carved in an anticline (hill). Broad synclinal valleys, narrow ridges, and narrow anticlinal valleys are thus created. Symmetrical ridges of resistant rock and valleys of softer rock result. Figure 24.3 shows this pattern in the Juras Mountains of northern Switzerland.

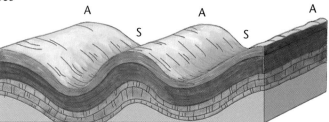

A = Anticline
S = Syncline
SM = Synclinal mountain
AM = Anticlinal mountain

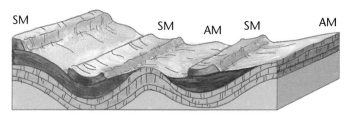

FIGURE 24.2 *A Simplified Model of a Fold Mountain*

Things To Do

1. a) Using an atlas, indicate the following mountain chains on a map of the world: Western Cordillera, Sierra Madre, Andes Mountains, Atlas Mountains, Alps, Zagros Mountains, Caucasus Mountains, Himalayas, Verkhoyansk Range.

   b) Complete an organizer like the one shown here for each of the mountain ranges in part a).

   The Western Cordillera has been completed as an example.

Mountain Range	Plate Boundary	Explanation
Western Cordillera	western edge of North American Plate	North American Plate is sliding past Pacific Plate

2. Develop a theory to explain why old fold mountains like the Appalachians, the Juras Mountains, and the Ural Mountains do not occur on present plate boundaries.

CHAPTER **24** TECTONIC PROCESSES: BUILDING MOUNTAINS *177*

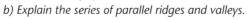

FIGURE 24.3 *The Juras Mountains, Switzerland*
   a) Describe the physical features of this area.
   b) Explain the series of parallel ridges and valleys.

178 PART 4 THE LITHOSPHERE: BUILDING UP THE LAND

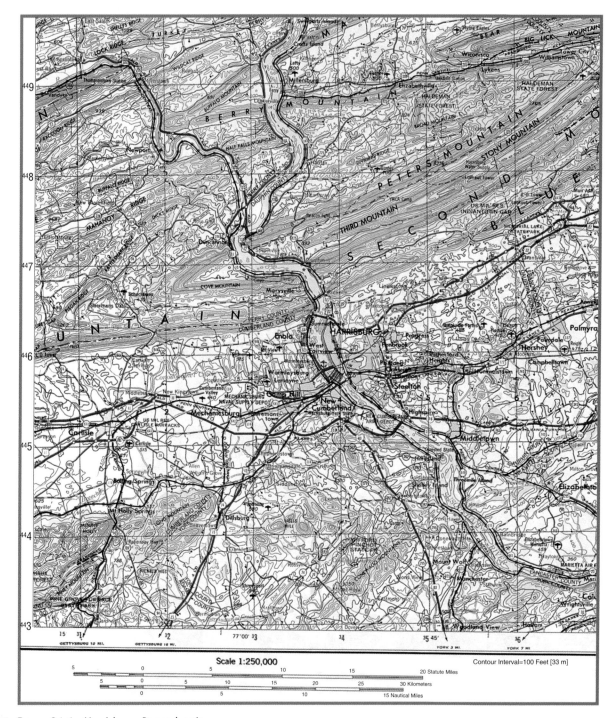

FIGURE 24.4 *Harrisburg, Pennsylvania*
  a) Draw a sketch of this map showing major physical features. Include rivers, mountains, valleys, and so on.
  b) Describe the relief of the region.
  c) Explain how Second Mountain, Cove Mountain, and Peters Mountain are really the same mountain.
  d) Make a profile to help you explain the geological history of the region.
  e) How have the different layers of folded strata affected drainage and land use patterns?

FIGURE 24.5 *Ridge and Valley Topography near Winston-Salem, in North Carolina*

Features Caused by Faulting

Faulting occurs when pressures are so great that blocks of rock fracture or break apart. The crust splits in a gigantic crack that may extend for hundreds of kilometres and deep into the crust. Unlike folding, faulting can occur rapidly. When two blocks of the earth's crust move, the ground shakes and vibrates in nature's most violent act—an **earthquake**.

Faults occur in any rock subjected to tension or compression. The many different landforms associated with faults are formed from the different ways in which plates move in relation to each other.

A normal or **gravity fault** occurs when rocks on either side of the fault move away from each other. The lower fault plate slides up relative to the other fault plate, creating a **fault plane**. If two major normal faults occur parallel to each other, a **rift valley** or **gräben** is sometimes formed. The most famous rift valley runs through East Africa from Lake Nyasa north to the Gulf of Aden. Many of the lakes and rivers in East Africa are found in this broad central valley. The Red Sea is a part of the same geological structure. Some geologists believe that the Great Rift Valley will one day become flooded as it sinks below sea level and water from the Red Sea rushes in. If this happens, East Africa will split away from the rest of the continent and slide slowly towards the east into the Indian Ocean. Of course, this process would take millions of years. Under the oceans, an enormous rift valley occurs where the plates are separating. The valley is being constantly filled with lava from deep in the crust.

If the land rises between parallel faults, then a block mountain or **horst** is formed. The Grand Tetons in Wyoming and the Sierra Nevadas in California are examples of block

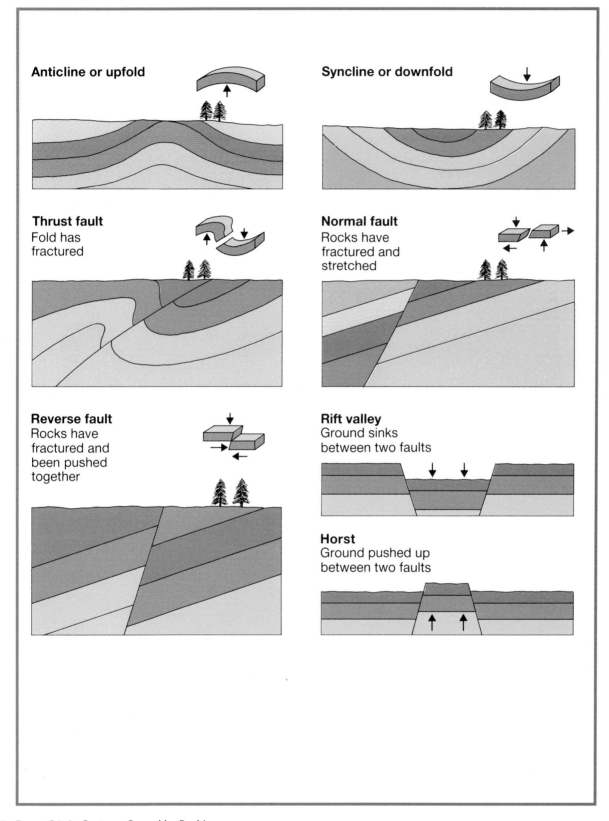

FIGURE 24.6 *Features Caused by Faulting*

CHAPTER 24 TECTONIC PROCESSES: BUILDING MOUNTAINS 181

FIGURE 24.7 *Study these photographs.*
a) *Make tracings of one or more of these images. Indicate where you think the faults occur.*
b) *Identify the type of fault, with supporting evidence from the photo.*
c) *Describe how the geological structure of the area affects drainage and land use patterns.*

Rocky Mountain Trench and the Columbia River

The San Andreas Fault

Grand Teton National Park

mountains that have been tilted as they were pushed up. Rift valleys and block mountains are two interesting features that result from parallel normal faults.

Reverse or **thrust faults** are the exact opposite of normal faults. They form when rocks are in compression, such as where fault plates are pushing together. One plate moves up while the other descends below it. This feature is common along plate boundaries where the two plates are pushing together. These are sometimes called **subduction zones**. The Alps of south-central Europe were formed from sedimentary layers thrust over five major faults caused by the collision of the African and the European plates. Much of the Western Cordillera was similarly formed from reverse faulting.

The **strike-slip** or **transform fault** in which two plates slide past each other does not result in mountain building. One of the most famous of this type is the San Andreas Fault in southern California. Here the Pacific Oceanic Plate is sliding northwest while the North American Continental Plate is moving southeast. In time, the western part of California will slide away from the rest of the continent and move northwest into the ocean. California is plagued with earthquakes because of the release of built-up energy related to the slipping that occurs along this fault. Although the movement itself is minimal, the suddenness of the movement causes the destruction associated with earthquakes.

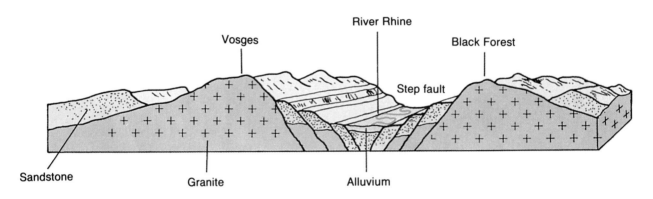

FIGURE 24.8 *The Rhine Rift Valley*

Things To Do

1. Use heavy cardboard, plaster, or modelling clay to make working models to illustrate different types of geological structures associated with faults. (See the section on model construction on page 190.)

2. Describe the different geological structures associated with faulting.

3. Explain why faulting often occurs along plate boundaries.

4. What effect would the following fault structures have on geographic patterns such as drainage, transportation routes, and settlements?
   a) gravity fault
   b) rift valley
   c) thrust fault
   d) horst
   e) transform fault.

Features Associated with Volcanism

Volcanoes are mountains formed from molten rock or **lava**. They are the result of **extrusive volcanism**. They form from magma lying beneath the crust. Friction and pressure melt the rock. It then expands into cracks and crevasses in surrounding rock, creating intrusive structures. When the pressure becomes so great that the magma reaches the surface, extrusive formations in the familiar form of volcanic mountains result.

Where would friction and pressure be so great that rock would melt? The answer is, along plate boundaries. The theory of plate tectonics helps to explain how volcanoes are formed and where they occur. When a plate is riding up over another plate in a subduction zone, friction and pressure melt the rock where the two plates grind together. The magma is often stored underground in a **magma chamber**. If the pressure is great enough, it flows to the surface through a **magma conduit**. Magma is often thin and mobile. It may ooze out of the ground and flow a long way before it solidifies, forming almost flat plateaus of basalt. Sometimes the magma is thick and viscous. It does not flow as well as basaltic magmas and solidifies rapidly when it is cooled, forming the familiar, cone-shaped mountains.

Often a **volcanic plug** forms in the magma conduit. As the pressure builds up, a dramatic explosion may occur. The lava solidifies into ash, rocks, and other volcanic debris, which rains down on the earth's surface for many kilometres. The volcanoes of the Western Cordillera and the Andes Mountains are examples of this type. Sometimes called **Vesuvian volcanoes**, the natural disasters these can cause can be devastating. When Mt. Vesuvius erupted in the year 79, the people were buried in searing ash. Many were instantly fossilized. On the Caribbean island of Martinique in 1902, a **nuée ardente** killed 30 000 people. They were instantly suffocated by this 600° to 700°C cloud of ash, water vapour, and glowing gases. When Mt. St Helens erupted in 1980, 2.75 km^3 of rock, ash, ice, and trapped water vapour roared down the mountain at 250 km/h, destroying everything in its path. Fortunately few people lived in the area, so the death toll was light. The people of Colombia were not as lucky in 1985 when Nevada del Ruiz erupted. Lava melted the icecap. The soil became lubricated with meltwater, creating a **lahar**. This wall of

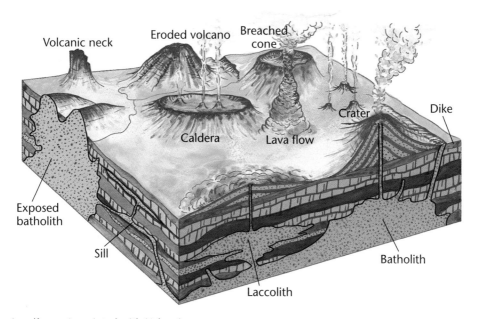

FIGURE 24.9 *Landforms Associated with Volcanism*

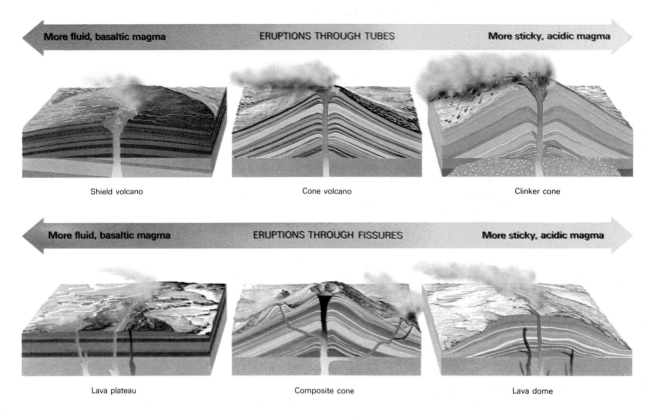

FIGURE 24.10 *Different Types of Volcanoes*
Types of volcanoes are created by the nature of the magma and the conduit through which the magma flows.

mud 40 m high moved down the mountain at 160 km/h. More than 23 000 people, some as far as 50 km away, were buried under the mud. What makes these Vesuvian volcanoes even more dangerous is their unpredictability. They are like sleeping giants. They lie dormant for hundreds, perhaps even thousands, of years. Over time, people forget how dangerous they are—until they erupt again.

Volcanoes also occur along mid-oceanic ridges where plates are pulling apart. The magma of these volcanoes is basaltic. It is thinner and runnier. It oozes out of the vents slowly and regularly without volcanic plugs forming. The classic conical mountain is often absent with this type of extrusive volcanism. The lava is so fluid that it spreads out to form a plateau rather than a mountain. Iceland is an example of this type of volcanism. It has been formed entirely from lava flowing out from the Mid-Atlantic Ridge as the North American and Eurasian plates pull apart.

Hawaiian volcanoes are named after the Pacific island. These volcanoes do not occur along plate boundaries. They are often located far from mid-oceanic ridges or subduction zones. Volcanologists believe these volcanoes occur over "hot spots" in the asthenosphere. As the plates move the earth's surface slides over the hot spot. Figure 24.12 shows the chain of islands that formed as the crust moved over the hot spot that currently lies under Hawaii. Scientists do not know why these hot spots occur.

Hawaiian volcanoes tend to be more gentle than Vesuvian volcanoes. The low viscosity lava from the regular eruptions forms massive dome-shaped mountains. Mauna Loa, the main volcano of the Hawaiian chain, is the largest mountain on earth, higher than Mt. Everest. We do not recognize its great height, however, because most of the mountain is under water.

Volcanoes are fascinating, but volcanism includes other structures as well as these familiar cone-shaped mountains. Intrusion occurs from the high pressure injection of magma into existing rock structures. The magma hardens, creating igneous intrusions of various sizes and shapes (see Figure 24.9). Smaller ones are called **dikes**, **sills**, and **laccoliths**. Larger ones, called **batholiths**, may be hundreds of kilometres across. When the surrounding rock erodes away, these harder igneous intrusions are left, often forming curious landforms that may seem out of place.

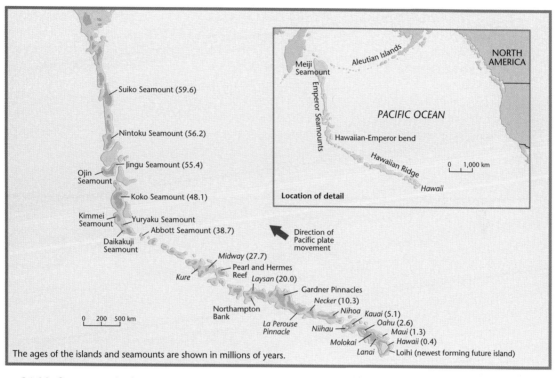

FIGURE 24.11 *Seamounts in the Hawaiian and Aleutian Archipelagos*

Things To Do

1. Compare the three main types of volcanoes in an organizer.

2. Use Figure 24.12 to make a map showing the locations of major volcanic eruptions. Use a colour key to identify where each type of volcano occurs.

3. Study Figure 24.13.
   a) Make a profile across Mauna Loa using the method described on page 190.
   b) Compare your profile with the profiles prepared by other members of your group. How do you know that Mauna Loa is a volcano?
   c) How do you know whether the volcano is still active?
   d) What other features indicate that Mauna Loa is volcanic?

4. Many technical terms are associated with volcanism. Make a crossword puzzle for the new words in this section. Use a computer if you have access to one.

5. Explain the relationship between plate tectonics and volcanoes.

6. List as many ways as you can think of in which volcanoes are dangerous. Give an example of each.

Things To Do (continued)

7. Research one volcanic eruption of your choice.
   a) Make a list of focus questions for your study. Narrow the list down to two or three questions.
   b) Use a periodicals index or computerized information retrieval system to find articles about this volcano.
   c) Make point-form notes of the information you find that is relevant to your focus questions. Be sure to include complete sources and bibliographical information.
   d) Synthesize and evaluate your research material.
   e) Draw conclusions about this volcano. For example, what type of volcano is it? What evidence supports this conclusion?
   f) Apply what you learned to predict what could happen in the future.
   g) Present your findings to your group or class. Decide on the method of presentation—a bulletin board display, a written paper, an oral presentation, or something else. What maps, diagrams, models, graphs, or charts will you include?

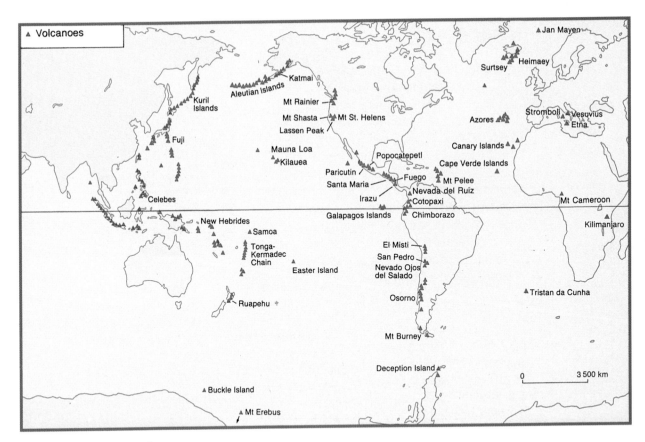

FIGURE 24.12 *Active Volcanoes*

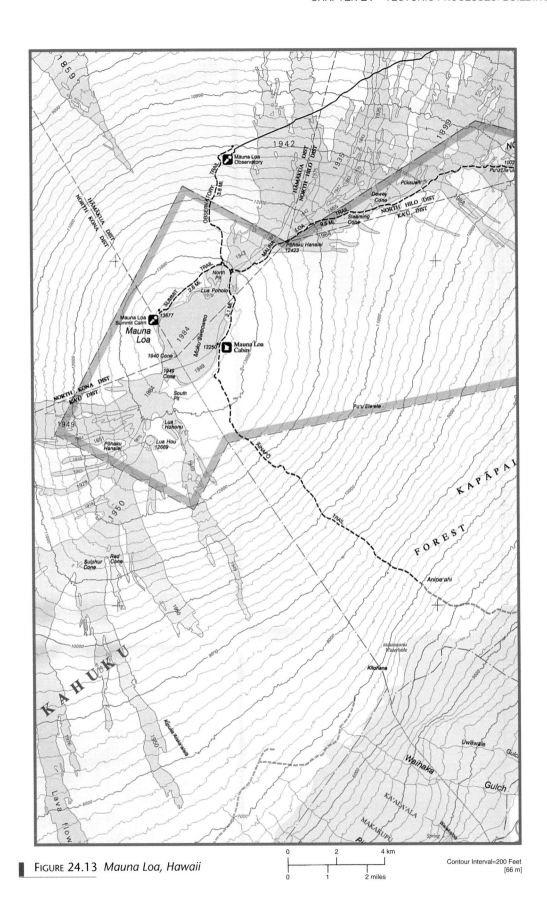

FIGURE 24.13 *Mauna Loa, Hawaii*

Chapter 25
Mapping Techniques: Contours, Profiles, and Models

INTRODUCTION

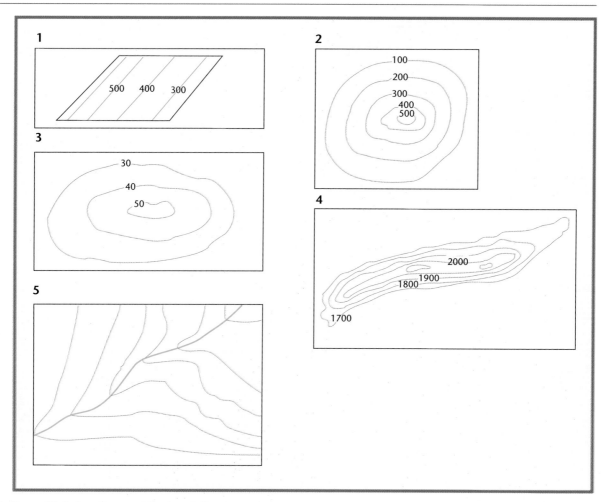

FIGURE 25.1 *Landform and Contour Patterns*
After reading about contours on page 189, match diagrams 1 through 5 with the following descriptions: a) a cone-shaped mountain; b) a flat plain sloping gently east; c) a river flowing northeast; d) a steep-sloped ridge; e) a gently sloped oval hill.

Geographers use many different mapping techniques to show physical features. The most difficult aspect to illustrate is **relief**, or the relative height and shape of landforms. Shading, cross hatching, colouring, and symbols are all used with varying degrees of success. These methods provide an aerial view but they are inaccurate because they do not represent relief in measurable terms. The method most commonly used today is **contour mapping**. This technique is relatively simple; it shows the exact height in measurable terms and enables geographers to create cross-sectional diagrams or profiles as well as accurate models to scale.

Contours

Contour mapping is the standard means of showing elevation. Contour lines connect all places of the same height. Lines that are far apart indicate the land is fairly flat. Lines that are close together indicate a slope; the closer the lines, the steeper the slope. With a little practice you can visualize what landforms look like by studying contour maps.

Here are some basic rules to keep in mind when working with contours:
1. The **contour interval** is the number of units between each contour line—for example, 10 m.
2. Contour lines never cross because they show places with the same elevation.
3. Contour lines never end. They join together or run to the end of the map.
4. When a contour line crosses a river it creates a v-shape upstream. The valley is lower than the surrounding land, so a valley is shown as contour lines pointing uphill.

Profiles

Profiles are two-dimensional, cross-sectional diagrams made from contours. When studying a map, they are helpful in visualizing what a landform looks like. (See activity 1 in Things To Do on page 190.)

Three-dimensional Diagrams

Three dimensional-diagrams also show what the land is like by showing depth and height on an exact scale. Once you have created a number of profiles, it is easy to draw a three dimensional diagram. (See activity 2 in Things To Do.)

Models

Creating a model is the best way to show what a place looks like. It is an exact representation of the landscape drawn to scale. The steps for creating a three-dimensional model are similar to those for making a three-dimensional diagram. (See activity 3 in Things To Do.)

FIELD STUDY

Mapping the Land

1. Obtain a topographic map of the field study area you selected earlier.
2. Prepare a tracing that identifies the significant geological features of the area.
3. Make several parallel profiles of the site.
4. Prepare either a three-dimensional drawing or a model of your field study area. Label the significant features and colour your work appropriately.

Things To Do

1. To prepare a profile follow these steps:
   a) Place the straight edge of a small piece of paper across the map where you want to show the profile. Carefully mark where each contour touches the piece of paper.
   b) Label the contour lines and prominent features such as roads, rivers, and lakes on the paper.
   c) Draw an x and y axis on a piece of graph paper. Select an appropriate scale for the x axis, for example, 1:10 000 (1 cm = 100 m). It is necessary to exaggerate the vertical scale to see the feature.
   d) Lay the piece of paper on which you recorded the contour lines along the y axis on the graph paper. Plot the elevation using the scale on the x axis.
   e) Join the dots together and label the features shown in the profile.
   f) Make profiles of the contour maps shown in Chapter 24.

2. Create a three-dimensional diagram following these steps:
   a) Make several profiles parallel to each other. Space them evenly across the map. The closer the profiles are, the more details your diagram will show.
   b) Trace each profile onto another piece of paper, making sure they are the same distance apart as they are on the map.
   c) Connect features across the profiles so that you have drawn a surface grid. Add details and colour.
   d) Use several profiles for one of the contour maps in Chapter 24 to make a three-dimensional diagram.

3. Create a model following these steps:
   a) Cut out each profile and attach it in a vertical position to a piece of light plywood. (Draw the profiles on a piece of cardboard so they are rigid.)
   b) Fill the spaces between each profile with crumpled newspaper.
   c) Soak gauze impregnated with plaster of Paris, available from craft stores and medical supply companies. Lay each piece of gauze over the profiles, overlapping them so that each layer is at a right angle to the layer it covers. Press the gauze down to fit the hollows and shape the material to follow the form of the profiles.
   d) Once the model has dried, paint and label the exhibit. You now have an exact scale model of the map.

4) Make a three-dimensional diagram or model of one of the contour maps shown in Chapter 24.

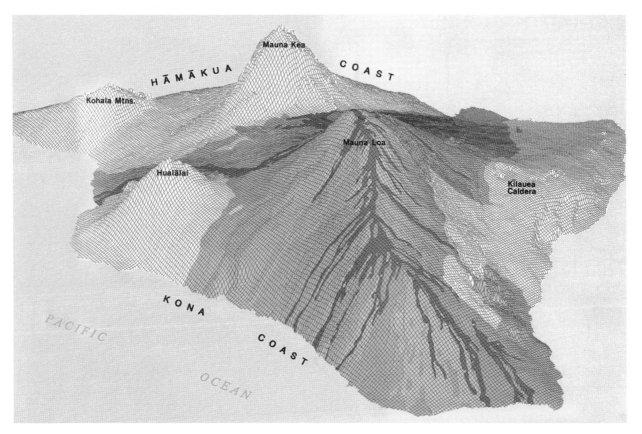

FIGURE 25.2 *A 3-D Model of Mauna Loa*

Chapter 26
Earthquakes: An Integrative Study

INTRODUCTION

The residents of Mexico City will never forget the 19th of September 1985. It was the day the earth moved! Thousands of people died and the world's most densely populated city lay smouldering in ruins. Although this earthquake created an unexpected disaster, earthquakes of various intensities are quite common. Thousands of tremors occur every year. Fortunately, most of these are small or happen in unpopulated areas. The earthquake in Mexico City was unusual because it hit a large metropolitan area in which many buildings were constructed before earthquake safety standards were established and it was intense enough to cause serious damage. By studying what happened in Mexico City, **seismologists** can improve their understanding of these devastating natural disasters. They need to determine what caused the earthquake, where similar earthquakes could occur, what precautions can be taken to minimize damage, and how earthquakes can be predicted.

FIGURE 26.1 *Devastation Caused by the Mexican Earthquake*

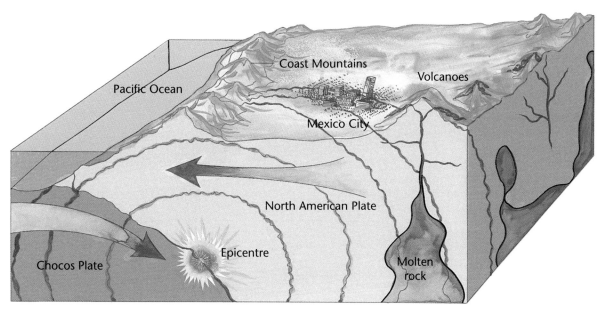

FIGURE 26.2 *Cross-section Showing the Cause of the Mexican Earthquake*

Earthquake Damage

The earthquake that shook Mexico City in 1985 caused more extensive damage than was expected because of the way in which the city was built. In ancient times, Mexico City, called Tenochtitlan, was the home of the legendary Aztecs. For defensive reasons, the city was originally built on an island in Lake Taxcoco. Over the years, the growing population consumed the water in the lake and eventually the lake dried up. Buildings were constructed on the lake bed as the city began to spread out in all directions. But this ground is not solid. It is made up of thick layers of sediment that formed beneath Lake Taxcoco. In fact, many of the buildings have been gradually sinking for years. When the earthquake sent shock waves under the city, the unstable soil gave way, causing buildings to collapse.

Another factor that contributed to the extensive damage was building construction. Much of Mexico City was built during the colonial period, using brick and cut stone. When the earth shakes these rigid structures crack. Some collapse, while others become structurally unsafe and thereby prime candidates to collapse during the next strong tremor. In addition, the many makeshift shantytowns across the city are too unstable to withstand the tremors. All of these factors contributed to the massive destruction Mexico City experienced in 1985. It has taken many years to rebuild the city. Today improved zoning regulations and construction methods appropriate for an earthquake zone have reduced the chances of such extensive destruction when Mexico City experiences another earthquake.

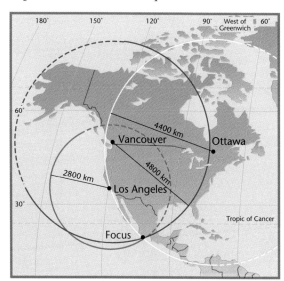

FIGURE 26.3 *Triangulation Map Showing the Mexican Earthquake*
An arc is drawn from each seismic station indicating the distance the quake is away. (i.e., Vancouver 4800 km, Ottawa 4400 km, and Los Angeles 2800 km). The focus is the point where the three arcs intersect.

Place	Lat./Long.	Size*	Date	Place	Lat./Long.	Size*	Date
Argentina	25°S 67°W	8.2	1977	Algeria	37°N 0°	7.3	1980
Indonesia	6°S 127°E	8.1	1979	Japan	35°N 135°E	7.2	1995
Mexico	19°N 99°W	8.1	1985	Italy	40°N 17°E	7.2	1980
China	39°N 188°E	8.0	1976	Aegean Sea	40°N 25°E	7.1	1983
Ecuador	0° 75°W	7.9	1979	Guatemala	15°N 91°W	7.1	1983
Philippines	16°N 121°E	7.7	1990	Turkey	40°N 43°E	7.1	1983
Iran	38°N 46°E	7.7	1978	Kuril Is.	45°N 150°E	7.0	1984
Indian Ocean	10°S 80°E	7.6	1983	Turkmenistan	39°N 57°E	7.0	1984
Solomon Is.	10°S 155°E	7.5	1984	Japan	36°N 137°E	7.0	1984
Chile	20°S 70°W	7.3	1983	Afghanistan	37°N 70°E	7.0	1983

*Size is based on the Richter scale. Anything over 7.0 is a major quake and can cause serious damage.

FIGURE 26.4 *Recent Strong Earthquakes*

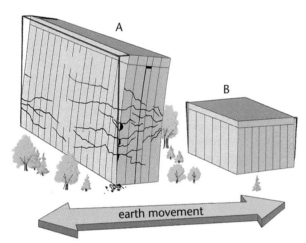

A and B are the same size. Why is B the safer building?

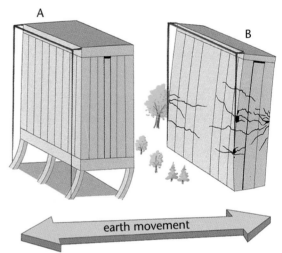

Which is the safer structure? Why?

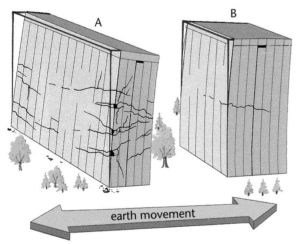

Both buildings are the same height. Why is A swaying? What advice would you give to builders in quake-prone cities?

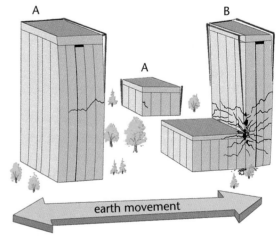

Why is the tall building in A less damaged than the tall building in B?

FIGURE 26.5 *Buildings Designed for Safety*

Twenty Seconds of Terror

The quake struck at 5:46 a.m. and lasted a mere 20 seconds—but virtually every part of Kōbe suffered major damage. The death toll soared toward 5000 and more than 25 000 persons were reported injured: city officials said about 300 000 people were homeless.

Fearful of aftershocks, wary survivors picked their way through a nightmare landscape of rubble and ruination.

Fires fed by ruptured gas mains raged out of control in the chill winter wind: in many neighborhoods, firefighters were unable to douse the flames because of damage to the city water system. Office buildings lay like crumpled cardboard boxes. Some 50 000 buildings were destroyed and early estimates of property damage ranged from $30 billion to $80 billion.

From *Newsweek*, 30 January 1995.

FIGURE 26.6 *An Earthquake in Kōbe, Japan, January 1995*

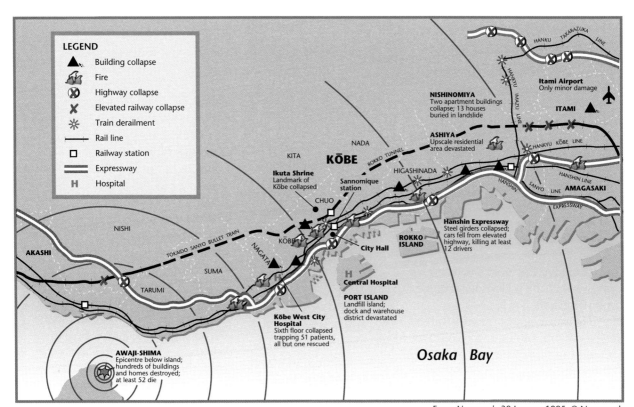

From *Newsweek*, 30 January 1995. © Newsweek Inc. All rights reserved. Reprinted by permission.

FIGURE 26.7 *The Path of Destruction of the Kōbe Earthquake, January 1995*
What similarities and differences are there between the earthquakes in Mexico City and Kōbe?

Things To Do

1. Study Figure 26.2.
    a) Explain what caused the earthquake in Mexico City. Include the terms *epicentre, Chocos Plate, North American Plate, shock waves,* and *subduction zone* in your explanation.
    b) Explain the relationship between earthquakes and plate tectonics.
    c) Develop a hypothesis of where you would expect earthquakes to occur.

2. a) How did building construction make the devastation even more disastrous in the Mexican earthquake of 1985?
    b) What other geological factors made the devastation worse?

3. a) Use the co-ordinates in Figure 26.4 to prepare a map showing the location of recent strong earthquakes.
    b) Does the map confirm the hypothesis you developed in activity 1?

    c) Label major cities located in this earthquake zone.
    d) Explain why earthquakes are more significant in populated areas than in unpopulated ones.

4. Study Figure 26.5.
    a) Which pair of buildings seems to be safest in an earthquake? Explain.
    b) What general principles of building design should planners consider in earthquake-prone regions?

Measuring and Predicting Earthquakes

Seismometers measure the intensity of earthquakes. A rod anchored deep in the ground vibrates when a quake occurs. Joined to the rod is a pendulum with a pen attached to it. As the ground vibrates, the rod, pendulum, and pen vibrate with the same frequency. The pen draws a graph (called a **seismograph**) on a paper drum that slowly rotates at a constant rate. The greater the earthquake, the more the pen moves and the wider the swings on the graph.

The seismograph records three types of vibrations. L waves travel along the earth's surface. Smaller P-waves (primary waves) are followed by more violent S-waves (secondary waves). By measuring the time between P and S waves, scientists can figure out how far away the quake was using a seismic wave graph (see Figure 26.10).

GEO-Fact

Some people believe that earthquakes can be predicted by studying nature. In China, people have predicted earthquakes for thousands of years by observing certain animal behaviour, such as chickens refusing to roost and snakes leaving their burrows. Unfortunately, this does little to determine the intensity of the tremor or provide adequate forewarning of an impending quake. Moreover, these animal behaviours also occur when there are impending storms and floods, so there is no way to be sure what natural disaster may be approaching.

Seismologists can also determine exactly where the earthquake occurred using seismic readings from three or more different stations and a mapping technique called **triangulation**. (See Figure 26.3.) Each station is located on a map. A circle is drawn, with its centre being the seismic station and its radius representing the distance from the quake. Circles are drawn for other stations that picked up the vibration on their equipment. The **epicentre** occurs where the circles intersect.

The **Richter scale** is the most popular scale used to measure earthquakes. It is based on intensity over a given distance. Scientists thousands of kilometres away are often able to detect an earthquake on the other side of the world and determine how serious it is. When the earthquake devastated Mexico City, seismologists in Hawaii, California, and British Columbia were able to monitor the event.

Seismographs do not predict

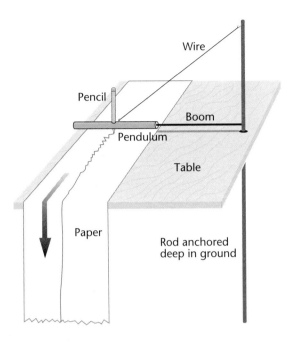

FIGURE 26.8 *Simplified Diagram Showing How a Seismograph Works*
Notice seismogram being drawn as the table moves.

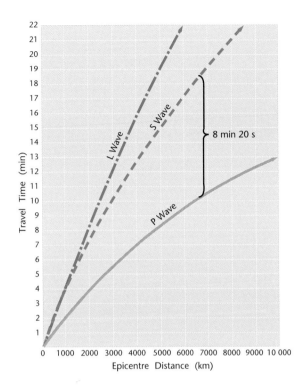

FIGURE 26.10 *A Seismic Wave Graph*
A seismic wave graph can be used to find the distance to an earthquake. For example, if the difference between the arrival times of the P wave and S wave is 8 min 20 s, the earthquake occurred 6800 km from the seismograph station.

earthquakes; they simply record them as they happen. Predictions are much more complex. Three things must be known: when it will happen, where it will happen, and how intense it will be.

Computers and an understanding of chaos theory help scientists to find complex patterns in seismic readings. For example, Parkfield, California, has had an earthquake measuring between 5 and 6 on the Richter scale on average every twenty-two years since 1857. The last one was in 1989. If the pattern holds, there should be another in 2011. Most earthquakes, however, follow much more chaotic patterns and are much harder to interpret.

More sophisticated techniques rely on statistics gathered from subtle changes in the shape and movement of geological features. In California, the San Andreas Fault is closely monitored. Lasers on one side of the fault are aligned with sensors on the other side. The beam of intense light gradually moves along the sensor as one plate slides past the other. If the rate of movement slows or stops altogether, it is likely an earthquake will occur as tension builds up along the fault plane. When the tension reaches the breaking point, a sudden jolt or earthquake occurs.

FIGURE 26.9 *Seismographs*

Size	Damage Expected	Number Each Year
10	felt worldwide	unrecorded
9	felt most places	unrecorded
8	most buildings collapse	less than 1
7	most buildings damaged	20
6	some buildings damaged	100
5	slight damage	500
4	felt by most	6000
3	felt by some	30 000
2	not felt but shown on a seismograph	more than 150 000

Note: Numbers are estimates.

FIGURE 26.11 *The Richter Scale*

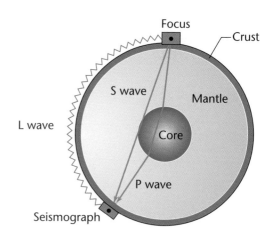

FIGURE 26.12 *Seismic Waves*

Things To Do

1. a) Explain how a seismometer works.
   b) What information does a seismometer provide?
   c) How can seismic readings be used to determine the location of the epicentre using triangulation?
   d) How can computers be used to study seismic data?
   e) What other technology aids in the study of earthquakes?

2. Study the following statements, then answer the accompanying questions:
   - Parkfield, California, will receive a tremor in the next two months.
   - Sichuan province in China will experience an earthquake measuring between 4 and 7 on the Richter scale within the next week.
   - South America will have a major earthquake in 2007.
   - Southern British Columbia will have an earthquake sometime in the next century.
   - Manila, the Philippines, will have a major earthquake within the next twenty-four hours.

   a) Which statements would help residents prepare for an earthquake?
   b) Which statements lack important information? What is missing in each of these?
   c) What three pieces of information are needed for an earthquake prediction to be useful?
   d) What precautions could people take if an earthquake was predicted for your area?
   e) What effect could inaccurate predictions ultimately have on a community?

3. Explain how planners could modify each of the following unsafe practices to make it less hazardous in earthquake areas.
   a) shantytowns sprawling up mountain slopes
   b) underground shelters
   c) skyscrapers with elevators only and no stairs
   d) elevated highways
   e) large buildings constructed on faults
   f) municipal water reservoirs in the mountains above the city
   g) long suspension bridges
   h) loud sirens notifying residents of an earthquake.

4. Use the inquiry model to research an earthquake of your choice. (Refer to page 116 for information about the inquiry model.)

PART 5

Gradational Processes:
Wearing Down the Earth

Chapter 27

Forces that Shape the Earth

Introduction

We have seen how tectonic processes—those forces operating *within* the earth—form the lithosphere. But the lithosphere is also shaped by forces operating *outside* the earth. These are called **gradational processes**. They include gravity, rivers, waves, glaciers, and winds.

Gradational forces are at work everywhere. Rivers cut deep valleys through mountain passes and create deltas far downstream. Waves and currents cut away coastlines and polish sandy beaches. Glaciers ebb and flow over vast continental regions and in remote mountain uplands, sometimes carving out landforms while at other times erecting massive features. Desert winds carve beautiful rock formations and carry sand into the hollows. All of these forces shape the planet on which we live.

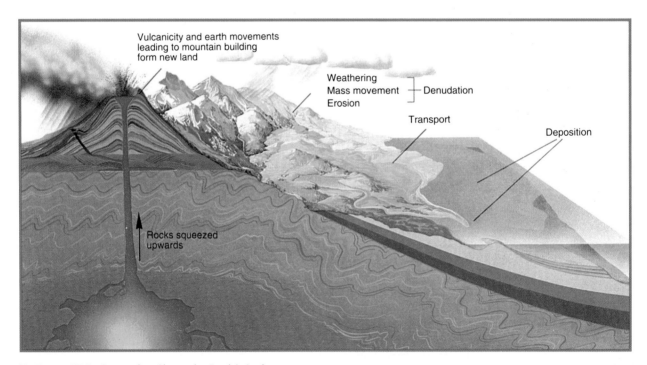

FIGURE 27.1 *Forces that Shape the Earth's Surface*

Base Level

Have you ever watched a mechanical grader operating on a construction site? The machine moves back and forth over the land, scraping away the high points and filling in the low spots. Nature acts in much the same way. Natural forces work the land to get it flat and smooth. Ultimately they are trying to achieve a hypothetical state called **base level**.

An American geologist named John Powell first formulated the idea of base level in 1875. He theorized that base level is the point at which a river no longer cuts down into the valley and instead diverts its energy to making the valley wider, not deeper. Base level can be considered a smooth curve running under the continents that gently rises from sea level on the coast inland towards the higher interior. The earth's gradational processes work towards reaching this base level. Powell's theory originally applied only to rivers. But today the concept has been expanded to include all processes because they all work to smooth out the earth's surface and bring it to base level.

If left undisturbed, gradational processes would wear down the mountains and fill in the oceans so that eventually the planet would end up as smooth as a billiard ball! Of course, we know this will never happen. Tectonic forces within the earth are also constantly changing the landscape. Mountains rise, rift valleys form, and coastlines sink. It is the interplay between tectonic and gradational forces that creates the many interesting and varied landscapes that form the surface of our dynamic planet.

Principles of Gradation

Even though each gradational process is different and shapes the land in its own distinct way, they all operate under the same basic principles of weathering, transportation, erosion, and deposition.

Weathering

In order for the rough, high spots of the earth's surface to be worn away, they must first be broken down into pieces that can be easily removed by gradational forces. This process is called weathering. As the name implies, forces in the atmosphere work to break down rock to form **weathered material**,

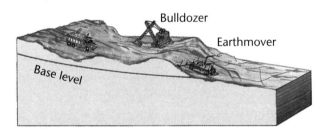

Mechanical graders work to smooth the land

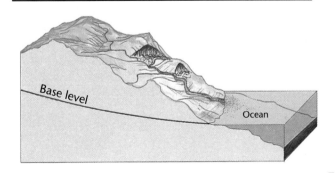

FIGURE 27.2 *The Concept of Base Level*

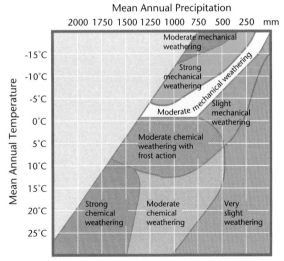

FIGURE 27.3 *The Relative Importance of Different Types of Weathering Under Different Temperature and Precipitation Conditions*

often referred to as the **regolith**. Rain, snow, rapid temperature change, and other processes all work to wear down the rock and create the regolith. Weathering can be a mechanical or chemical process. Usually rocks are weathered by both, but climatic conditions often determine which type is dominant.

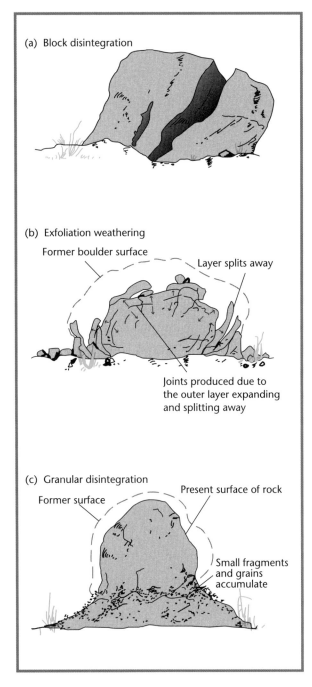

FIGURE 27.4 *Types of Mechanical Weathering*

Mechanical Weathering

Mechanical weathering—the physical wearing down of rock—is most active in cool, dry environments. **Frost shattering** occurs in polar regions and temperate climates. Moisture seeps into cracks in rock. If the moisture freezes, the ice expands and widens the crack. Over time the crack becomes so wide that the rock breaks apart. Frost shattering occurs more rapidly in regions where the temperature rises above freezing during the day but drops below freezing at night. It is easy to understand how this would be a significant form of weathering in Canada.

Mechanical weathering also occurs as a result of plant growth. Roots often push into cracks in the rock. As the roots expand, they force two pieces of the same rock to break apart.

Thermal expansion is another form of mechanical weathering. If you were to place a cold glass straight out of the freezer into a hot oven, it would shatter. Thermal expansion causes this to happen. Theoretically, the same thing happens to rocks in desert regions where night temperatures sometimes drop below freezing. When the hot desert sun rises, a rock that has been frozen by the evening chill may heat up so rapidly that it shatters. The importance of thermal expansion is challenged by some geologists whose field studies show that temperature changes are seldom extreme enough to cause this type of weathering.

FIGURE 27.5 *Mechanical Weathering*
Exfoliation has broken up this block of granite.

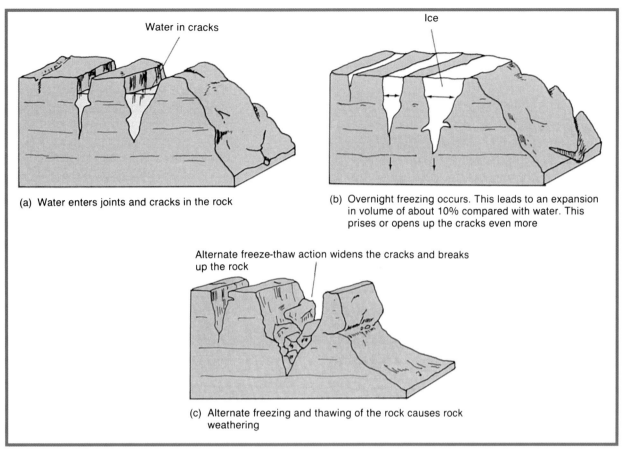

FIGURE 27.6 *Freeze-Thaw Action*

Exfoliation or **sheeting** is a form of mechanical weathering common in Canada. Rocks that were formed under great pressure deep within the earth often end up on the surface. The pressure that held the rock together is gone, so it literally falls apart. Layers of rock "peel" off, just like layers of an onion. You often see this in the Canadian Shield where great slabs of rock litter the ground. Of course, other forms of mechanical weathering are also at work here.

Chemical Weathering

Chemical weathering is more common in warm, humid climates than in temperate or polar regions. When carbon dioxide is dissolved in water, it forms an acid that acts upon the rock to chemically change it. Calcite, for example, is changed into calcium bicarbonate when it is exposed to carbonic acid. The new soluble compound is carried away by water. What is left are broken up pieces of rock. This is weathering by **solution**.

Chemical weathering also occurs when carbonic acid acts upon minerals called silicates. This process is called **hydrolysis**. Ions of water replace ions of the silicate so the rock falls apart. Water soluble minerals are carried away, but clay minerals are left behind. Soils of the tropics are often very deep and clayey because hydrolysis is so common here. Even when the rock is buried deep beneath layers of earth, hydrolysis works on the bedrock to break it down.

You know what happens to iron when it is exposed to water. It rusts. Oxygen dissolved in water transforms iron to iron oxide. Like the calcium bicarbonate, it is dissolved in water and carried away. Since many rocks contain iron, **oxidation** is a common form of chemical weathering. You can tell when this process has occurred because the rocks are stained rust red. These three chemical processes—solution, hydrolysis, and oxidation—usually operate together.

Erosion and Transportation

People often confuse erosion with weathering. Erosion includes the processes of weathering and transportation. Transportation occurs when weathered material is carried away. Gravity, rivers, waves, glaciers, and winds are all powerful erosive forces that transport weathered material.

It takes energy for material to be carried away. The greater the energy, the more material can be carried. For example, if there has been a drought and a river has little water in it, it will not have much energy. Therefore only tiny grains of sand can be carried by the water. Under normal conditions, the river may have been able to transport pebbles and small rocks. If there is a flood, however, the ability of the river to carry weathered material is much greater. There is more erosive energy in

FIGURE 27.7 *Granite Boulders Eroded Chemically, Malawi*

the raging waters. Large rocks and even boulders may be washed downstream. Geological engineers are able to determine how powerful an ancient flood was by the size of the rocks it carried.

The same basic principles apply to other gradational processes. Strong waves cause

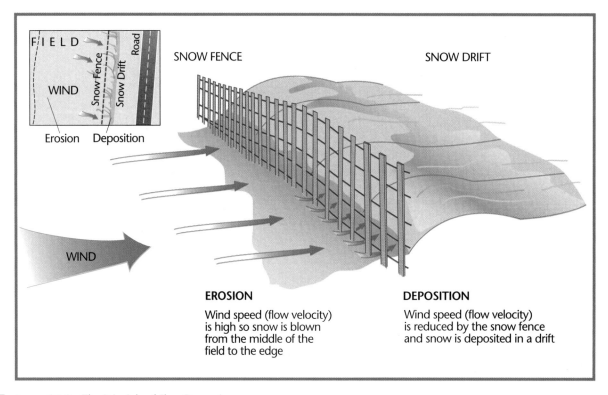

FIGURE 27.8 *The Principle of Flow Dynamics*

more erosion than calm seas. Strong winds whip around more weathered material than gentle breezes. Steep slopes enable gravity to pull more weathered material down them than gentle slopes. It all comes down to energy.

Deposition

All material carried by winds, rivers, and other agents eventually must come to rest. The laying down of weathered material is called deposition. As we know, it requires energy for erosion to take place. The greater the energy, the more weathered material that can be transported. When this energy, or **flow velocity**, is reduced, it cannot carry the same amount of weathered material. So the material is deposited.

You can see deposition in operation during a snowstorm. Winds are strong over an open field, so snow is picked up and carried away (erosion). If the flow velocity of the wind is reduced, it drops its load and forms a snow drift (deposition). Snow fences are designed to keep the snow off highways. They are erected every fall, 10 to 20 m from the highway on the windward side. As the winds blow through the fences, they lose energy and so they slow down. The snow is then deposited on the side of the road instead of on the highway.

The same principle applies to gradational processes. When water in a river slows down, a deposit of sand is made, forming a **delta** or sand bar. **Talus slopes** are masses of coarse rock fragments that accumulate where landslides come to rest when they reach a flat surface. Glaciers form hills of rock debris called **moraines** when they no longer have enough energy to keep transporting this material forward. These are just a few examples of depositional features that are created when the flow velocity drops.

Things To Do

1. Explain the term *base level*.

2. Give definitions of each of the following, along with an example of each: weathering, transportation, erosion, deposition.

3. In an organizer, summarize the different ways in which rocks weather.

4. On a nature walk, take photographs of different examples of weathering.

5. Explain flow velocity and how it relates to the processes of transportation and deposition.

6. Experiment with a stream table.
   a) Set the table perfectly flat. Smooth the sand so that it is flat and add water at one end. What do you observe? What does this simulate?
   b) Increase the slope by raising one end. Add a steady flow of water at the high end, and observe what happens to material on the table. Where does transportation occur? Where does deposition occur? Make a sketch to show your simulation.
   c) Gradually reduce the slope. Note changes in flow velocity, transportation, and depositional features.

7. How can an understanding of flow dynamics help the following people?
   a) engineers building a dam
   b) highway workers in a landslide zone
   c) farmers irrigating their fields
   d) glaciologists
   e) skiers in avalanche areas
   f) farmers on the edge of a desert oasis
   g) seaside resort operators.

Chapter 28

The Work of Gravity

INTRODUCTION

Gravity is such an obvious gradational force that it is sometimes overlooked! Landslides, rock falls, avalanches, and soil creep are all examples of gravity moving **unconsolidated material** from one place to another. Collectively, this process is called **mass wasting**. The factors that influence the amount of mass wasting include degree of slope, particle shape and size, depth of the unconsolidated material, vegetative cover, tectonic stability, and lubrication. Each of these factors can combine with other factors to create many different types of mass wasting.

FIGURE 28.1 *Rock Slide*
This massive rock slide buried large parts of the town of Frank, Alberta, in less than 100 s back in 1903.

Slope

The steeper the slope, the greater the chance that weathered material will move. When a slope is increased for some reason, mass wasting is likely to occur. A change in slope is usually the result of another gradational process in operation. For example, a river may undercut its bank or wave action may erode part of a shoreline. The resulting change in slope could cause the soil to collapse and move.

Unconsolidated Material

The shapes and sizes of rocks in unconsolidated material also affect mass wasting. Particles that are smooth roll more easily than rocks with jagged edges. The sharp corners of jagged rocks fit together like a jigsaw puzzle, so they are not as likely to roll downhill. They have a steeper **angle of repose**—that is, the angle at which unconsolidated material can remain stable on a slope—than round, smooth rocks. Similarly, smaller particles are likely to have a steeper angle of repose than larger rocks, which may balance precariously on top of one another and may set off a chain reaction if one started to roll. Still another factor in the stability of weathered material on a slope is the depth of the unconsolidated material. The greater the depth, the less stable the material. In such cases it may take just one small rock rolling downhill to start a devastating rock slide.

Natural Vegetation

Vegetation also affects mass wasting. Roots hold unconsolidated material in place. This means that erosion is less likely in areas where there are lots of plants. You may have noticed that grass is often planted on the sloping ground adjacent to highways. This is because the roots of the grass hold the soil in place and reduce the likelihood of mud flows onto the road.

GEO-Fact

Avalanches occur when there is a build up of snow over a valley. The snow has a very steep angle of repose but can become unstable due to melting or vibrations. In many alpine ski areas, snow conditions are monitored so that avalanches can be forecast or controlled. Often explosives are used to break up overhanging accumulations of snow before they become dangerous. In Switzerland elaborate barricades have been built above alpine villages to reduce damage caused from these deadly hazards.

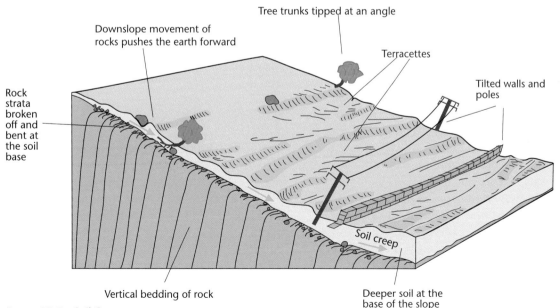

FIGURE 28.2 *Soil Creep*

FIGURE 28.3   *Mass Wasting: The Kotewall Road Landslide in Hong Kong, 1972*
This hillside became extremely unstable following heavy rains; a mud and rock slide resulted. The landslide toppled a thirteen-storey apartment building, killing sixty-seven people.

In Switzerland, the problems created by mass wasting have become increasingly serious in recent years. Acid deposition from automobile exhausts has reduced the size of the forests on the mountain slopes. The reduction in vegetation has resulted in numerous avalanches and rock slides in this mountainous country.

Vegetation does not always eliminate mass wasting, however. In some situations the soil slides away beneath the vegetation. This is because the holding power of the vegetation is counteracted by other forces. **Soil creep** is the slow and gradual movement of soil down slope, usually under intense pressure. (See Figure 28.2.) This is particularly noticeable in pastures where the grass is kept short by grazing animals. This causes the ground to subside, and the vegetation sinks along with it.

Tectonic Stability

Tectonic stability simply refers to how much the earth moves due to tectonic processes. If there is an earthquake or volcanic eruption, the vibration this creates often accelerates mass wasting. In addition to natural vibrations, explosions, loud noises, and even heavy traffic can trigger mass wasting. In some alpine ski areas, explosions are purposely set off to start avalanches in order to remove dangerous accumulations of snow above ski slopes.

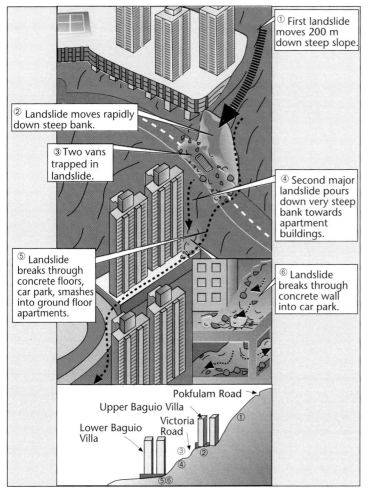

FIGURE 28.4 *The Path of a Landslide*
In May 1992, heavy rains triggered several landslides in Hong Kong. This diagram illustrates the movement of the Baguio Villa landslide.

Moisture Content

Water is the lubricant that helps unconsolidated material to slide better. Water fills the spaces between particles, which reduces friction and the angle of repose. In California, for example, heavy winter rains often result in mud slides. The combination of rain, sparse natural vegetation, steep mountain slopes, and frequent earthquake tremors provides all the ingredients for devastating mud slides.

Solifluction is an unusual form of mass wasting that is common in Canada, especially where permafrost occurs in the far north. In this cold climate, the upper part of the earth thaws each summer, but the temperature of the soil 1 to 2 m below the surface never gets above freezing. Any moisture that is in the upper soil has nowhere to go. As a result, the soil becomes saturated. Weathered material can creep as much as 5 cm per year down slopes as gentle as 1 or 2°. Solifluction can occur outside of the permafrost zone, but it is not as dramatic in more temperate regions.

Things To Do

1. Use an organizer to differentiate between rock slides, mud slides, soil creep, solifluction, and avalanches.

2. Use a periodicals index or computerized data retrieval system to find articles about natural disasters caused by mass wasting.
   a) Prepare a map showing where each disaster occurred. Use a colour key to identify each type of mass wasting.
   b) Research one of the natural disasters using the inquiry model and present your findings to the class.
   c) Suggest how the disaster you researched in part b) could have been avoided.

3. Identify the factors that influence mass wasting and write an explanation for each.

4. Find an area in your community that is prone to mass wasting. Determine what steps could be taken to reduce the possible damage. (Be sure to take appropriate safety precautions when dealing with potentially dangerous situations.)

Chapter 29

The Work of Rivers

Introduction

Rivers have played an important role throughout human history. Our first great civilizations were developed alongside rivers. Later, trade centres and industrial sites established themselves next to rivers. Rivers have provided us with drinking water, transportation routes, recreation, sanitation, and even power. There is little doubt that people throughout time have had a special interest in rivers.

Have you ever noticed that many cities are built on rivers? New York, Toronto, Montreal, and Vancouver were all built on trade routes where people travelled down rivers to trading posts. St. Louis was built on the interfluve between two rivers. Many European towns and cities were built where the river was narrow enough to ford or cross. (Why do you think Oxford was so named?) London, England; Paris, France; and Budapest, Hungary, were built near river crossings. Still other cities such as Alexandria, Egypt; Calcutta, India; and Bangkok, Thailand, were built on the rich farmland of deltas. Rivers were truly good places for civilizations to develop.

FIGURE 29.1 *A Magnificent Waterfall in the Rainforest of Ecuador.*

River Systems

Rivers seem to last forever, at least in terms of the human time scale. The Nile River was an integral part of ancient Egyptian civilization. The Huang He in China—the "river of sorrow"—was as important to early dynasties as it is to the country today. Literature reveals the role of the River Thames in medieval Britain. These rivers and many others are glorified in song and story. But how are these life-sustaining rivers formed?

River systems evolve over many years. At first, **drainage patterns** are poorly organized. Rainwater washes over the new landscape, finding low areas through which to flow and hollows in which to settle. In time, channels form and the hollows are enlarged to become ponds or wetlands. Rivulets of water flow across the landscape. These tiny channels eventually join together to form streams. Eventually, a well developed river system evolves and the land is efficiently drained of water.

Drainage Patterns

River systems develop intricate patterns over the earth's surface. But it is not until we look at a map or satellite image that we can see the patterns created by flowing water. From the air, river systems like the Amazon and its many tributaries often look much like trees. The main river is the trunk, the tributaries are the branches, and the distributaries in the delta are the roots. Where the tributaries join the main river acute angles are formed. This **dendritic drainage pattern** occurs when there is little variation in the bedrock over which the river flows.

This is not the only possible drainage pattern, however. In areas where there are fold

> ### GEO-Fact
>
> Did you know rivers can actually steal water from other drainage basins? Two conditions promote **river capture**: there must be two rivers systems flowing close to each other and there must be a shallow divide between them. It is quite common for the river with the greatest discharge rate to undercut the bank of the weaker river. The **pirated river** dries up as its upstream flow is diverted into the stronger river. Called **wind gaps**, the dry valleys that are left seem oddly out of place in humid landscapes.

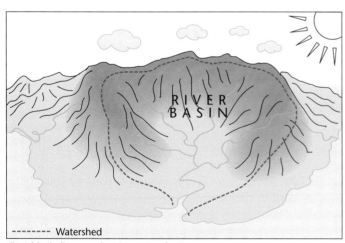

(i) A block diagram showing a river basin

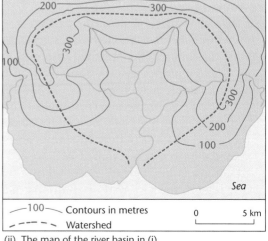

(ii) The map of the river basin in (i)

FIGURE 29.2 *A River Basin and Its Watershed*

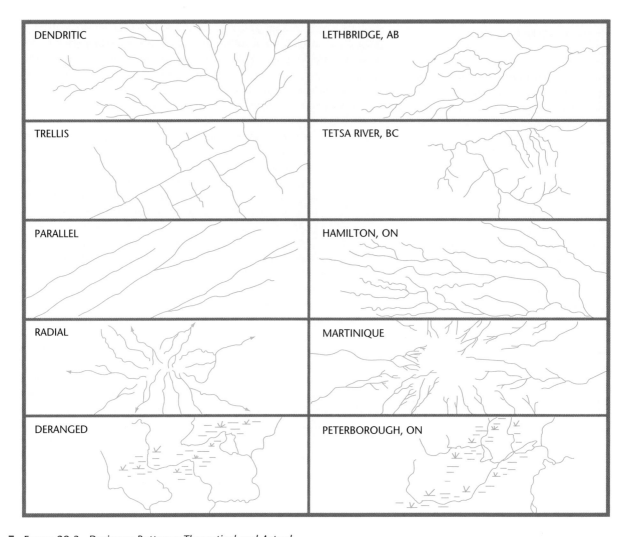

FIGURE 29.3 *Drainage Patterns: Theoretical and Actual*

mountains, like the Appalachians and the Western Cordillera in North America, rivers follow the valleys. A dendritic pattern is impossible because the mountains get in the way. Instead, a **trellis drainage pattern** is created. The tributaries join the main river at approximately right angles (90°), creating a formation resembling a garden trellis.

Drainage patterns can reveal a lot about the rock over which they flow. In many parts of Canada, bare rock is exposed because of recent glaciation. Surface water follows fracture lines in the rock that often run parallel to each other. This **parallel drainage pattern** is found in many parts of Canada.

Another distinct drainage pattern created by topography occurs where there are volcanoes or similar cone-shaped hills. This **radial drainage pattern** is found on many of the volcanic islands of the Caribbean region. As the name suggests, rivers flow out from a height of land in the centre.

Sometimes there does not seem to be any drainage pattern at all. In much of Canada, drainage patterns are relatively recent, having been developed in the last 10 000 to 20 000 years after the last ice age. This system is called a **deranged pattern**. There are often many wetlands, bogs, and lakes in these drainage systems. What drainage pattern is found where you live?

Things To Do

1. a) Use an atlas map or topographic sheet to make a tracing of a river system. Label the main river, significant tributaries, the mouths and sources of each tributary, and other significant drainage features such as lakes and wetlands.
   b) Establish where the divide exists between this drainage basin and adjacent ones by drawing a line equidistant between the sources of each tributary.
   c) Estimate the area of the drainage basin using a grid placed over your tracing. If each square is a standard unit of measure (say 100 km^2), you can count up the squares to get the area.
   d) Determine the elevation of the source of each major tributary. Subtract the elevation at its mouth from the elevation at the source to get the **rise** of the river.
   e) Measure the length of the main river and major tributaries to find the **run** of each.
   f) Calculate the **gradient** of each tributary and the main river by dividing the rise by the run.
   g) Based on your findings in f), list the tributaries in order from fastest flowing to slowest.
   h) Determine the drainage pattern for your river system.
   i) Summarize your information in a display.

2. Based on your findings in activity 1, write a descriptive paragraph telling what the drainage basin is like. Evaluate if the system does an efficient job of draining the land.

3. a) Use a collection of topographic maps to find examples of dendritic, trellis, parallel, radial, and deranged drainage systems like those shown in Figure 29.3.
   b) Make a tracing of each drainage system and mount it on a piece of cardboard.
   c) For each drainage pattern, make assumptions about the rocks over which they flow.
   d) Study the maps to determine how drainage patterns have affected the ways in which people use the land.

River Systems and Flow Dynamics

The dynamics that make rivers seem so alive are the same as for all gradational processes. Water that falls on the earth's landforms either evaporates or flows under the force of gravity to the lowest possible point. The water is working to reach **permanent base level** (sea level). But its passage to the ocean is often blocked by mountains and other landforms. When this happens, water accumulates in lakes and inland seas above sea level, creating a **temporary base level**. The water stays here until an outlet enables it to seek a lower base level.

Erosion occurs when water flows over land above base level. As we have already discovered, weathered material is transported because flowing water has energy. The material the water carries is deposited in hollows that are below base level. Flowing water slows down and even stops altogether in these hollows. When it does so, some of its **load** is dropped. In the St. Lawrence-Great Lakes Basin, for example, erosion occurs where water flows rapidly—in places like Niagara Falls and La Chine Rapids. Deposits of sediment occur in the Great Lakes because the water slows down in these giant basins. (But you can imagine how many years it will take for these lakes to be completely filled with sediment!)

Flow velocity is not the only factor that determines a river's erosive power. The volume of water flowing through a river, called the **discharge rate**, is also important. A tiny stream flowing rapidly down a steep hill will obviously not have the erosive power of a mighty river with an enormous volume of water but a much slower flow velocity. The third factor that affects the ability of water to erode is the nature of the river bed. If the rock is hard and resistant, relatively little erosion will occur. Erosion is greater in a stream bed that is made up of soft rock. So flow velocity, discharge rate, and the nature of the stream bed are the factors that determine the erosive power of a river.

Rivers transport material in one of three ways: **solution, suspension,** and **saltation**. In solution, water flows over rocks, picking up minerals which it then dissolves. The minerals are removed only when the water evaporates and the solution becomes more concentrated. Eventually a point is reached where the water cannot hold any more of the mineral. The mineral then **precipitates**, or separates from the solution.

Unconsolidated material can also be held in suspension. The size of the particles the river is able to carry depends on the energy the water has. Where the energy is great, everything from tiny silt and grains of sand to small stones can be carried. Rivers with less energy cannot transport large particles; the slowest rivers may only be able to carry fine particles of silt and clay. This material settles only when the water slows down or stops flowing altogether. Often the material that is deposited is of fine texture and forms rich **lacustrine** soils when a lake is eventually filled in. Lake Winnipeg—the remnant of the much larger Lake Agassiz—lies in the middle of such a lacustrine plain.

Larger particles are swept along the river by saltation. Some rocks bounce along, landing on the river bed and then jumping up again or causing other particles to in turn bounce up. Large rocks may even roll along the stream bed, occasionally coming to rest in places where the flow velocity is reduced. No matter how material is carried in a river, however, it is all part of the river's load.

There is a positive correlation between flow velocity and the size of the particles a river can carry. The faster a river flows, the larger the particles it can carry. The opposite also applies. If flow velocity decreases, the largest particles in the load are deposited first. The size of the deposited material gradually decreases until the finest material and dissolved minerals settle when the water stops moving.

If there is a steady flow from one direction for a long period of time, the particles sort themselves by size. The largest particles separate first. The size of the particles that separate gradually decreases, leaving the smallest silts and clays to be laid down last. Known as **sorted sediment**, these layers of different-sized particles are often found where ice age deltas once formed into ancient lakes or seas. Today these deposits are mined for the various grades of gravel and sand they provide for the construction industry.

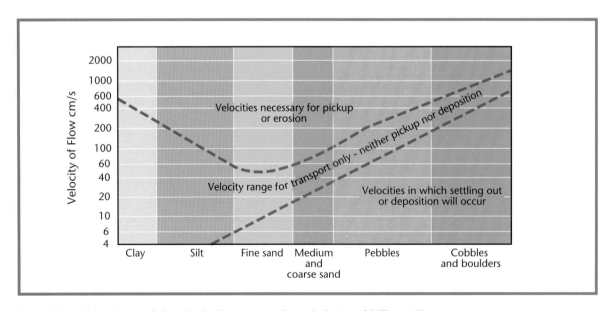

FIGURE 29.4 *Velocities Needed to Erode, Transport, or Deposit Grains of Different Sizes*

Things To Do

1. Explain the relationship of base level to erosion and deposition in flowing water.

2. Explain why most lakes are temporary landforms.

3. List and explain the factors that determine the erosive power of flowing water.

4. Study Figure 29.4.
   a) What do each of the two line graphs represent?
   b) What velocity is needed for water to erode fine sand?
   c) What velocity is needed for water to erode pebbles?
   d) What velocity is needed for water to erode cobbles and boulders?
   e) What correlation—either positive or negative—is there between flow velocity and erosion of particles larger than fine sand? Explain why this occurs.
   f) What velocity is needed for water to erode clay and silt?
   g) Why do you think fine particles require a greater flow velocity than larger particles?
   h) Prepare a chart to show the velocities where deposition occurs for different sized particles.
   i) When would clay and fine silts settle out? Why?
   j) Is it possible for a river to be neither eroding nor depositing material? When would this occur?

The Geomorphic Cycle

William Davis, an American geographer, formulated the **geomorphic cycle** in the early twentieth century. This theoretical model sought to explain how landforms develop. Today his theory is not as popular as it once was because it implies that all landforms were created by rivers. We know, of course, that other agents also play a role. Still, the geomorphic cycle enables us to understand the unique role of rivers in creating the land.

Davis contended that landscapes go through periods of development, much like people do. Rivers start out young, then mature, and eventually become old. But unlike people, rivers can experience a rebirth called **rejuvenation**.

In the young stage, an initial period of uplift is followed by a long period in which the drainage system organizes itself. The base level is very low, so the water has a lot of potential energy as it tries to find a way to the sea. Channels and ponds are formed until an outlet to a lower elevation is found. Then the river flows rapidly through its new channel, cutting down into the bedrock. Young rivers are characteristically straight and flow rapidly down steep gradients. Valleys are v-shaped, narrow, and straight like the rivers they contain. Because these rivers flow rapidly, they can transport huge amounts of unconsolidated material. The stream beds of young rivers are lined with rocks, cobbles, and boulders. Smaller particles are washed away. Between the river valleys are broad, poorly drained upland regions. Young rivers are found in many mountainous and highland regions of Canada.

In time the river changes. The main stream is close to reaching base level, but the river still has a lot of energy because of all the water flowing into it from tributaries upstream. Instead of cutting downwards, the river now begins to cut sideways into its banks, creating **meanders**, or winding curves. The valley begins to grow wider. The stream bed now migrates across the land like a snake. The meandering river deposits sediment in old channels and cuts through its banks to build new ones. Unlike the valleys of young rivers, mature river valleys can be extremely wide. Characteristically, a mature river valley is u-shaped because of all the side cutting. During periods of high water, the river may overflow its banks and flood the surrounding

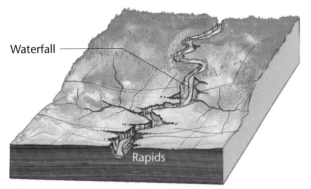
(a) Youth

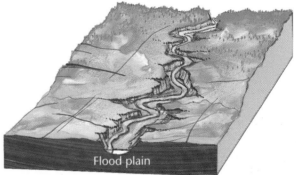
(b) Maturity

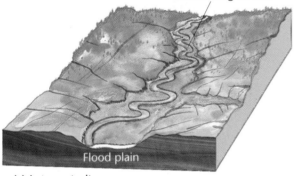
(c) Late maturity

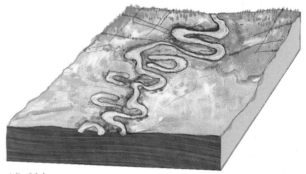
(d) Old age

FIGURE 29.5 *The Geomorphic Cycle of a River*

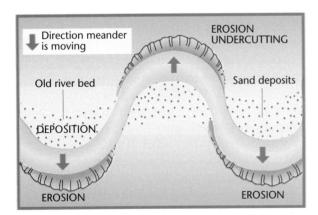

FIGURE 29.6 *How Meanders Are Formed*

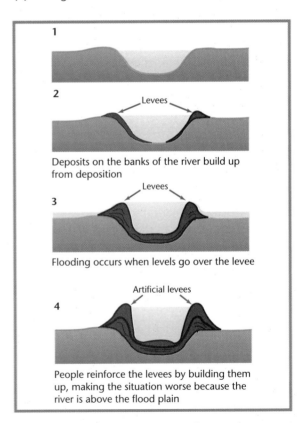

FIGURE 29.7 *How Levees Are Made*

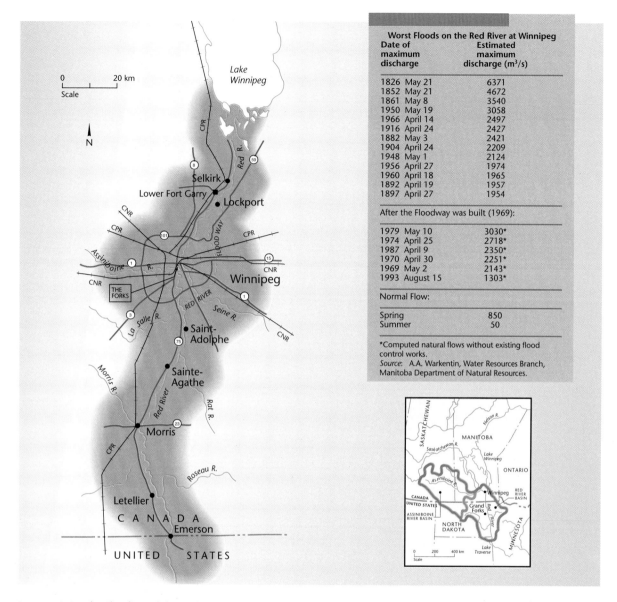

FIGURE 29.8 *The Flooding of the Red River*

land. For this reason, mature valleys are often called **flood plains**. The **interfluve**, or land between rivers, is better drained here than in young rivers and covers a much smaller area. Mature rivers flow into coastal plains and other flat landforms all across Canada. The Saint John River between Edmundston and Fredericton, New Brunswick, is one such river.

A river's old age begins when it starts to deposit more sediment than it transports. Old rivers are slow moving and capable of carrying only silt-sized particles and minerals contained in solution. Unlike more youthful rivers, the water is muddy and sluggish. Often the river actually rises above the flood plain. Sediments deposited in the stream channel build **levees** that contain the water. (See Figure 29.7.) When heavy rains increase the discharge rate, the water breaks through the river's natural banks and fills the flood plain with silty brown water. It will remain there until it either evaporates or reenters the river. The nutrients contained in the river water cover the flood plain with a rich natural fertilizer. Much of world's best farmland has been the result of the flooding of old rivers. The Nile, the Indus, the Ganga, the Chang Jiang, and the Mekong are just a few rivers whose rich sediments have created great agricultural societies.

Even more elaborate meanders develop in old rivers than in mature ones. To compensate for the shallow gradient, the river flows from side to side in increasingly greater meanders. The length of the river actually increases, even though the straight-line distance from the source to the mouth remains the same. Several features are commonly found along old rivers. **Oxbow lakes** are formed when a river cuts through its levee and truncates a meander. (See Figure 29.9.) When these curved wetlands dry up, they leave behind **meander scars**. Sometimes swampy areas called **bayous** develop where the river once flowed. When tributaries have difficulty cutting through the levees to join the main stream, a **yazoo stream** may form. This will flow parallel to the main river for some distance until it can join it. The wetlands formed from oxbow lakes, meander scars, bayous, and yazoo streams make the flood plains of old rivers a haven for waterfowl and other wildlife. Unfortunately, the land is difficult for people unless artificial structures like levees contain the river's waters.

GEO-Fact

The very floods that create rich agricultural land can also wreak havoc. In the summer of 1993, the Mississippi-Missouri basin flooded. Artificial levees were built to contain the flood waters, but these were largely unsuccessful. Fifty people were killed. Damage caused by the flood waters totalled over $50 billion.

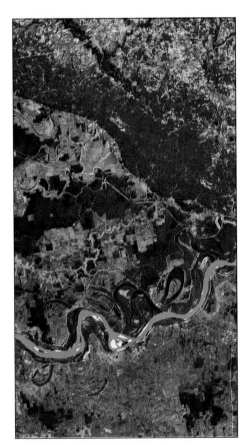

FIGURE 29.9 *An Oxbow Lake in Mississippi*

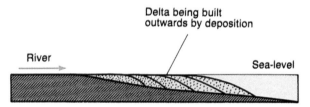

a) A cross-section of a typical delta

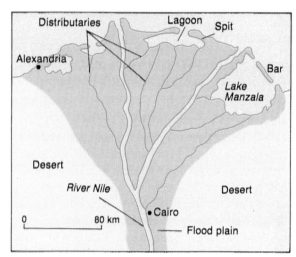

b) The formation of the Nile delta

FIGURE 29.10 *Delta Formation*

When the mouth of the river is reached, the enormous amount of sediment the water has carried finally comes to rest. Where the water enters the ocean, deltas are sometimes created by this deposition.

Old rivers are found all over the world. The Mississippi, the Missouri, the Nile, the Amazon, the Zaïre, and the lower Mackenzie are all old rivers.

According to the geomorphic cycle, if the base level remains the same, a river will pass through the three stages described here. In reality, however, this seldom happens.

Rejuvenation occurs when a river gains new energy as the result of the land being uplifted by tectonic forces. So this could cause an old river to become youthful once again. Perhaps the best example of rejuvenation is the Colorado River as it flows through Arizona. As tectonic forces caused the land to continue to uplift, the river began to cut more deeply into the land. The incredible gorge known as the Grand Canyon was the result!

Rejuvenation can also occur as the result of a drop in the base level. The river now begins to cut down into the flood plain once again. When the base level rises, the flood plain becomes flooded with sea water, forming an **estuary**. Many North American rivers flow into the Atlantic through flooded estuaries. The Delaware, the Susquehanna, and the Saint John rivers were all flooded when the last ice age ended.

FIGURE 29.11 *River Systems*
Identify each map or photograph shown here and on page 220 as a young, mature, or old river system.

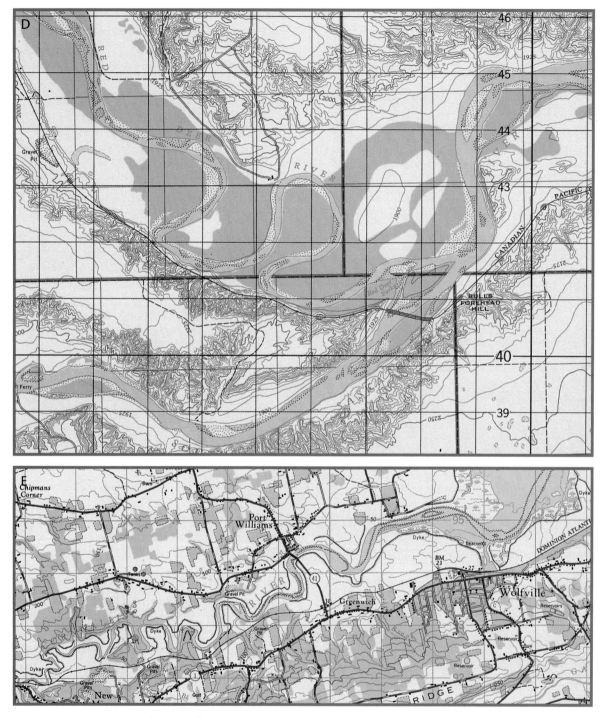

FIGURE 29.11 *River Systems (continued)*

Things To Do

1. Prepare an organizer comparing the different stages of the geomorphic cycle. Include the following criteria: flow velocity, valley shape, sediment size, gradient features, interfluve characteristics, and flood plain.

2. Explain how each of the following features are created: meanders, yazoo streams, flood plains, oxbow lakes, deltas, levees, estuaries.

3. Explain two ways in which rejuvenation can occur.

4. Which stage of river development do you think is best suited for human occupation? Explain your answer.

5. Experiment with a stream table to make simulations of rivers in the following stages of development: young, mature, old, rejuvenated.

6. Refer to Figure 29.8
    a) Explain why it was necessary to provide flood control for Winnipeg.
    b) How was flooding reduced?
    c) List the dates of high discharge rates in chronological order since 1969. How many years would flooding have been a problem?

Karst Topography: Rivers Underground

Many of us are fascinated by underground caves. There is something mysterious and exotic about them. Named after the Karst Mountain region in Slovenia where early studies were first conducted, **karst topography** has many interesting features, including caves, caverns, and sinkholes or **dolines**. About 15 per cent of the earth's surface contains rock formations where karst topography could form. Few regions, however, have the spectacular caves found in parts of the United States. In Canada, recent glaciation has limited the development of karst topography.

Drainage patterns in karst regions are bizarre. Rivers flow on the surface for a while, then go underground, only to re-emerge some distance away. The surface is pockmarked with circular hollows and the occasional small, round lake. Drainage seems disorganized and remarkably limited. The water just seems to disappear! What could explain this pattern?

The underlying rock in karst regions is limestone with at least 80 per cent calcium carbonate. When it rains, this mineral is dissolved in the water and the rock erodes. Since the rock is **impermeable,** water does not seep through it. So where does the water go? For a karst region to develop, there must be cracks and joints in the rock. The water flows through these, gradually increasing their width until caverns and caves are created beneath the surface. The water is constantly trying to find its base level. When it does, underground lakes or pools are formed. When a cave collapses, the classic sinkhole is created on

FIGURE 29.12 *Karst Formations in Carlsbad Canyon, New Mexico*

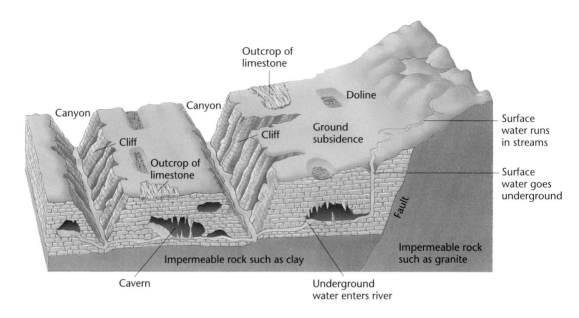

FIGURE 29.13 *Karst Formations*

the surface. In Florida, a farmer lost most of his farm in a giant sinkhole overnight! So while water is rare on the surface, the landscape is eroded by running water underground.

Karst topography created by deposition can be even more interesting. As water drips into underground hollows, it evaporates, leaving calcium carbonate behind. **Stalactites**, strange icicle-shaped rocks, hang from cave ceilings.

On the floor of the cave, upside-down rock icicles called **stalagmites** form where water drips. Where stalactites and stalagmites join, a **column** is created. Other odd creations, such as rock curtains and waterfalls frozen in stone, abound wherever there is dripping water. These features combine to make this underworld extremely beautiful—and just a little eerie!

Things To Do

1. What geographic characteristics are necessary for karst topography to develop?

2. a) Classify the following karst features as being caused by erosion or deposition: doline, stalactite, rock curtain, dry valley, cave, column, underground lake.

   b) Draw a sketch of each feature and explain how it was formed.

3. What challenges does karst topography pose for people?

Chapter 30

Measuring Rivers

INTRODUCTION

Valuable information is derived from the study of rivers. **Discharge rates** determine the amount of water available for human consumption and when flood conditions exist. Flow velocity measures when and where rising waters will reach the danger point. But how do we measure rivers?

It is fairly easy to measure a river's width, average depth, cross-sectional area, flow velocity, and rate of discharge, providing you observe proper safety precautions. In this chapter, you will be asked to work with your classmates to gather this data first-hand at a field study site.

FIGURE 30.1 *Measuring a River*
  Be sure to review the procedures and the safety tips before beginning any field activity.

Measuring Width and Depth

Measuring the width and depth of a river is fairly easy using simple equipment as long as you don't mind getting wet. Two or three people are needed to take these measurements.

To measure the width, have one person hold a tape measure at one end. Another person then wades to the opposite shore, extending the tape measure across the width of the river. The second person then reads the measurement from water's edge to water's edge. Taking three or four measurements and finding the average leads to greater accuracy.

Finding the depth is a little harder. Tie a weight to the end of a long piece of string. Lower the string and the weight into the river until the weight hits the bottom. Pull the string taut and mark where the surface of the water intersects with the string. Haul the weight out of the water and measure the length of the string that is wet. The depth of the river varies from place to place, so it is necessary to get readings at regular intervals. (Try taking measurements 1 m apart if the river is fairly narrow; increase the intervals if the river is wide.) Now it is possible to draw a profile of the river bed using these measurements. (See Figure 30.2.)

The **cross-sectional area** can be calculated using the average depth and width of the river. Simply multiply the two numbers together. For example, if the stream was 12 m wide and the average depth was 1.8 m, the cross-sectional area would be 21.6 m² (12 x 1.8 = 21.6). This measurement is needed later when the discharge rate is calculated.

SAFETY TIPS

For the sake of safety, bear these tips in mind.

1. Do not work in a river where the water is over your head.
2. Work under your teacher's supervision.
3. Always work with at least one other person.
4. Avoid rivers with strong currents.

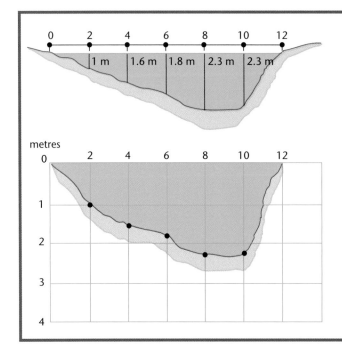

1. Measuring average depth:
$$\frac{\Sigma x}{n} = \frac{1 + 1.6 + 1.8 + 2.3 + 2.3}{5}$$
$$= 1.8 \text{ m}$$

2. Calculating cross-sectional area:
A = depth × width
A = 1.8 × 12 = 21.6 m²

3. Graphing the bank profile using depths:
 • plot each depth
 • join the dots

FIGURE 30.2 *How to Calculate the Cross-sectional Area and Depth of a River*

Measuring Flow Velocity

The flow velocity, or speed the river is flowing, can also be measured using a float, a tape measure, and a stop watch or a wrist watch with a second hand. Three people are needed to take this measurement: a timer/recorder, a starter, and someone at the finish line.

Measure 10 m along the river parallel to it. Mark a start line and a finish line at the beginning and end of the measured section. Have the starter place a floating object, such as an orange, in the water and signal to the timer to start a stop watch. When the object passes the finish line, the person at this point signals the timer to stop the watch and record the time it took the object to move 10 m. To ensure accuracy, repeat this procedure four or five times and calculate the average. You can now calculate how far the float would travel in 1 s using the formula t/d, where t is the time and d is the distance. For example, if the float moved 10 m downstream in an average time of 15.2 s, the speed would be 0.66 m/s (10/15.2). Multiply this by 60 to get the distance in 1 min. In this case, the flow velocity would be 39.6 m/min (10/15.2 × 60). To get the speed in km/h, multiply by 60 once again to change minutes to hours, and divide by 1000 to convert m to km. The answer would be 2.38 km/h (10/15.2 × 60 × 60/1000). The only problem with this measurement is that it shows only the flow velocity on the surface. To get the speed the water is flowing at greater depths you need more sophisticated measuring devices.

Now that you know the cross-sectional area and the flow velocity, it is possible to find the discharge rate in cubic metres per second (m^3/s). Simply multiply the flow velocity in metres per second by the cross-sectional area in metres squared (m^2). In the example, the discharge rate would be 14.26 m^3/s (21.6 × 0.66). This means that every second 14.26 m^3 (or tonnes) of water passes any given point.

FIELD STUDY

Measuring Rivers
Use your field study site (or another site if there are no rivers at your site), and prepare a detailed field study using the methods described in this chapter.

Getting Started: In the Classroom
1 Trace the river you are studying from a topographic map. Include interesting features such as waterfalls, rapids, meanders, and oxbows for later investigation.

2 Determine the direction of flow using contour lines and indicate this with an arrow on your sketch map.

3 Determine the gradient of the river for the section you are studying. To find the gradient of a river:
   a) Calculate the length of the river using the scale.
   b) Calculate the altitude at both ends of the river section.
   c) Subtract the altitude at one end from the altitude at the other end.
   d) Express the difference in c) as a ratio to its length in a) to determine the m/km figure. For example, a river drops from 1000 m to 500 m above mean sea level over a distance of 100 km:

   $$\frac{1000 \text{ m} - 500 \text{ m}}{100 \text{ km}} = \frac{500 \text{ m}}{100 \text{ km}} = 5 \text{ m/km}.$$

4 Prepare several profiles of the valley showing prominent features. (See page 190 for how to make a profile.)

5 Make several photocopies of your map for use in the field and back in the classroom.

6 Prepare a field manual from folded paper stapled in the middle. This can be used to make notes in the field.

In the Field
1. What are the physical characteristics of the river and valley? Sketch or photograph interesting features and indicate their locations on a sketch map.

2. Indicate on the map places where erosion, deposition, and flooding have occurred. What patterns of erosion and deposition do you notice? Explain why each occurs.

3. What patterns of vegetation are there? How does the river influence these patterns?

4. Measure the width, average depth, cross-sectional area, flow velocity, and discharge rate for three different sites along the river.

5. Draw cross-sections of the river at the three study sites.

6. Determine the river's stage in the geomorphic cycle and give evidence to support your decision.

7. How do people make use of the river and the valley?

8. What evidence of pollution is there?

9. What improvements could be made to the river?

Putting It All Together: Back in the Classroom
1. Use one of the map tracings you made of the site to make a display.
   a) Add the scale, a north arrow, and a grid.
   b) Label prominent features, such as evidence of flooding, meanders, oxbows, sandbars, etc.
   c) Use a legend to indicate areas of erosion and deposition.
   d) Use a legend to indicate natural vegetation in the flood plain.
   e) Draw diagrams or use photographs to illustrate prominent features. Mount these beside the map and use arrows to indicate their locations.
   f) Draw cross-sectional diagrams of the river bed where you measured the depth and refer these to the map.

2. Provide written explanations of each of the features shown in the display.

3. Prepare an environmental study to evaluate the impact of people on the river using a **needs assessment** organizer similar to the one shown below.

Present Situation	Action Plan	Vision
evidence of flooding	build dams and reservoirs upstream	reduced flooding

Chapter 31

The Work of Wind

Introduction

Another gradational process that shapes the land is wind. It is not as strong an agent as running water or flowing ice because air has a much lower density than either of these elements. However, there are elements of the world that have been shaped, at least in part, by the movement of air. We will look at some of these in this chapter.

FIGURE 31.1 *An Aeolian Landscape: The Sahara Desert in Libya*

The Aeolian Landscape

Landscapes shaped by winds are commonly called **aeolian landscapes**, so named after the Greek god of wind. Aeolian processes are most common in desert and steppe regions where conditions are extremely dry and there is little natural vegetation. In an arid environment there is little moisture to hold the soil particles together, so they are free to be carried away by wind. The lack of natural vegetation also exposes the ground to gradational processes because there are few plants to hold the soil in place.

Running Water in Arid Landscapes

Ironically, the main gradational process in arid land is running water. Even though there is little precipitation, the erosive force of the water that is received is much greater than in humid regions. The lack of natural vegetation and the fact that desert soils are often hard-baked prevents water from filtering down through the soil. Instead, it runs along the surface, carving deep gullies into the earth. **Wadis**, or intermittent streams, flow occasionally when there is a freak storm. These streams share the features of other rivers. Meanders, undercut banks, and **alluvial fans** (delta-like formations) emerge on the desert pavement. The alluvial fans form as the rivers empty onto flat land. The flow velocity drops and a load of **graded particles** is deposited. Flow velocity also drops when the water enters a basin or hollow. Once again, its load is deposited. A **playa**, or salt pan, is formed once the water evaporates and the dissolved load precipitates. In the Kalahari Desert of southern Africa, the Okavango Basin and the Etosha Pan are transformed into bountiful wetlands for birds and other wildlife when the summer rains come. Animals flock to the area to regenerate after months of enduring the challenges of living in a harsh, dry land.

The infrequent rains, which break up the hard surface of the desert, expose the soil to the erosive force of the wind. But people, too, can be responsible for the erosion of desert sands. During the Gulf War in 1991, for example, there was so much shell bombardment and tank movement that the hard surface of the earth was broken. After the war, strong desert winds carried sand from the desert to coastal farms and cities.

Wind erosion occurs when particles are carried away by flowing air. As with running water, the size of the particles the wind can carry is determined by its flow velocity. The higher the speed, the greater the wind's ability to transport particles. Wind does not have the power of running water, so it can act only on relatively small particles like dust, silt, and sand. Rocks and boulders obviously cannot be blown by winds. As with running water, when the flow velocity decreases the wind drops its load.

Wind-blown Sand

Despite what you may think, deserts are not often sandy. Most of the sand has been blown away. The larger particles that are left make a desert pavement sometimes called **reg**. Whether it's the Great Australian Desert, the Sahara Desert, or the American Desert, the predominant land surface is rock. So where does all the sand go?

Huge volumes of sand, called **erg**, accumulate in certain parts of some deserts. As much as 20 per cent of all deserts are ergs. The Grand Erg Oriental of the central Sahara is enormous, covering an area as big as the Maritime provinces with a layer of sand up to 1200 m thick. That's a lot of sand! Some of the fine particles are carried thousands of kilometres, to be deposited in oceans or even on other continents. Recent studies have shown that millions of tonnes of dust from the Sahara end up in the Amazon region of South America! Much of the prime agricultural land of northeast China is covered with layers of fine desert silt that has blown in from central Asia.

Sand dunes take many different shapes. **Barchans**, the classic crescent-shaped dunes, are formed in areas where there is limited sand and the winds consistently blow from one direction. At first the sand accumulates around a small bush or stone. The flow velocity is reduced and the sand is dropped. Increasingly

FIGURE 31.2 *Deserts on Three Continents:* Top left *the Great Australian Desert;* top right *the Sahara Desert in Niger;* above *the American Desert in Arizona*

more and more sand builds up the dune. When it reaches its maximum height, it starts to travel forward. With its horns leading the way, the dune marches across the desert. Sand erodes from the windward slope and accumulates on the downwind side or **slipface**.

When there is a lot of sand and winds are irregular, **transverse dunes** resembling giant ripples form. Other dunes include **parabolic dunes**, created when vegetation prevents erosion at each end of the dune; **longitudinal dunes**, which are aligned parallel to the prevailing winds; and **star dunes**, great pyramids of sand formed when the winds flow from many different directions.

Wind action not only carries away particles, it also contributes to weathering. When sand is blown against an object it wears it away. Known as **abrasion**, this action can be compared with the sandblasting techniques used to clean buildings. A stream of air shoots particles of sand against the surface, wearing away the dirt and soot and leaving a fresh, clean façade. (Of course, this technique can't be used too often or it will destroy the object being cleaned.)

Winds often wear away the rough edges of rocks, giving them a streamlined appearance. These **ventifacts** are aligned with the prevailing winds. Some people have speculated that the famous sphinx of ancient Egypt was originally a natural rock formation carved by the wind. Finishing touches created by a forgotten artisan transformed it into the familiar monument.

Because of gravity, the larger, more abrasive wind particles remain close to the ground. Consequently, rock outcrops experience more erosion at their base than higher up. This creates the crazy-looking pedestals, balancing rocks, arches, and other strangely beautiful landforms frequently found in the desert.

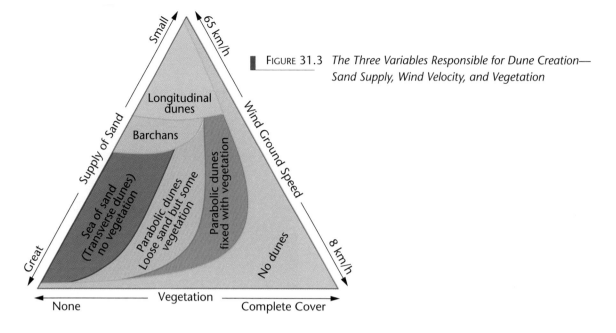

FIGURE 31.3 *The Three Variables Responsible for Dune Creation—Sand Supply, Wind Velocity, and Vegetation*

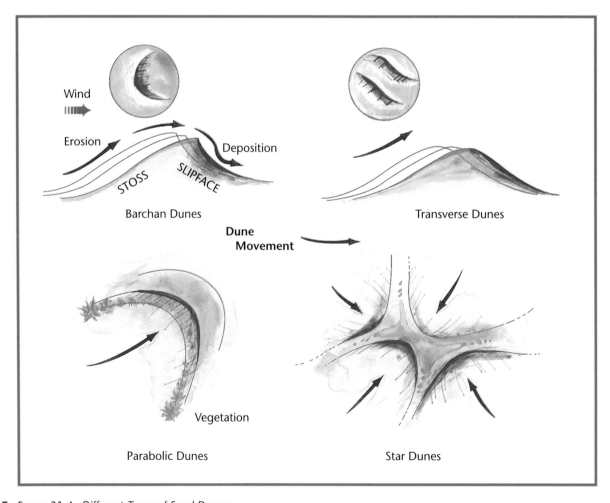

FIGURE 31.4 *Different Types of Sand Dunes*

Stages of Aeolian Development

The geomorphic cycle suggests that rivers experience three stages of development: youth, maturity, and old age. (See page 215.) Similarly, aeolian landscapes experience different stages of development. In its young stage, a period of uplift is accompanied by a period of gully erosion formed by running water. The gullies become increasingly wide and eventually form **canyons**. The land unaffected by erosion remains as table-like hills or **mesas**.

Enormous quantities of material are removed by the actions of water and, to a lesser extent, wind. Only the larger particles are left, forming a desert pavement. Alluvial fans form from gullies flowing between mesas. Sometimes the fans join together to form a **pediment**, or gentle slope, composed of graded sediments. In time this mature landscape ages. The mesas become increasingly smaller until all that is left are a few cone-shaped hills.

As with running water, the landscape is constantly dropping in elevation as the wind continues to erode the small particles and weather the softer rock. Eventually, in old age, all that is left are a few resistant boulders in a sea of worn-down gravel. Called **inselbergs**, or rock islands, these are the last remains of a landscape that was once much higher. Unlike rivers, winds do not erode to a base level. Because of this, aeolian landscapes tend to be much flatter and uniform than those that are formed primarily from flowing water.

Things To Do

1. List all the gradational features highlighted in bold in this section. Classify each as erosional, depositional, or other and explain how it is formed.

2. Describe the role of running water in aeolian landscapes.

3. Why are desert landscapes more susceptible to erosion than humid regions?

4. Explain how the flow dynamics of running water and winds are similar to and different from one another.

5. Study Figure 31.3.
   a) What does it explain?
   b) What three variables are used?
   c) Describe each variable for each type of dune.

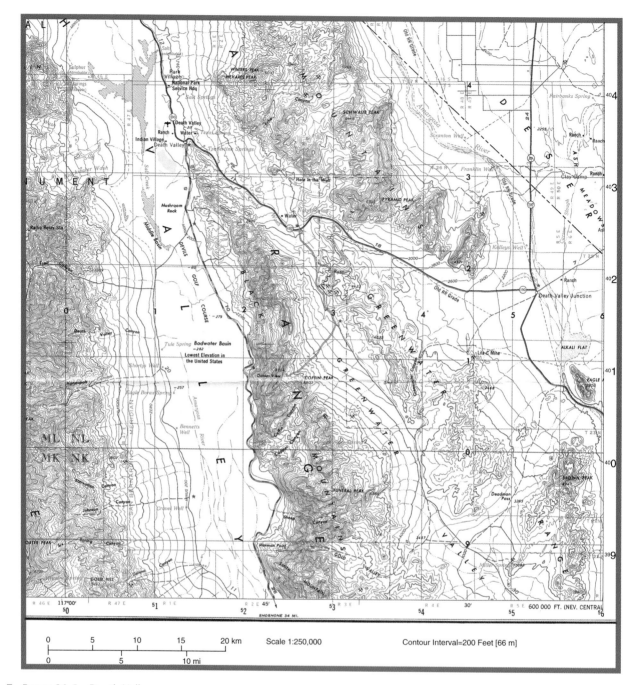

FIGURE 31.5 *Death Valley*
Study this topographic map.
  a) Find the lowest elevation in Death Valley. What is unusual about this location? How is it indicative of a desert region?
  b) Describe the drainage pattern in detail. Explain how it is characteristic of a desert region.
  c) List physical features that indicate this is an aeolian landscape. Use grid references to show locations.
  d) What stage of aeolian development would you assign to this landscape? Explain.
  e) What evidence of human activity is shown on the map?
  f) Use place names to establish some of the challenges of this place.

CHAPTER 32

The Work of Ice

INTRODUCTION

Of all the gradational processes, glaciers have had the most obvious impact on the Canadian landscape. As recently as 20 000 years ago, glaciers covered most of what is today Canada. That may seem like a long time ago, but it is very recent in a geological time scale that covers over 4 billion years.

Enormous sheets of ice, up to 2 km thick, flowed out of the north. Imagine ice the thickness of four CN Towers stacked on top of each other! The impact they had on the land was incredible. Even though most of the glaciers are gone now, glacial features are predominant throughout Canada.

FIGURE 32.1 *The Athabasca Glacier in the Columbia Icefield*

Types of Glaciers

There are essentially two types of glaciers, **continental** and **alpine**. Continental glaciation is responsible for Canada's transformation. A period when the climate cools and glaciers cover large parts of the earth's surface is called an **ice age**. Until recently, it was thought that there had been four ice ages. But recent analysis of deep sea sediments indicate that there could have been as many as eighteen ice ages over the past 900 000 years!

This period of glacial formation is known as the Pleistocene Epoch in the geological time scale. Within this epoch there were **interglacial periods**—that is, periods when there were no glaciers. These occur when the average world temperature rises, causing the glaciers to melt and returning the landscape to its earlier state. During these warm periods, continental glaciers shrink back to polar regions where temperatures remain below freezing year round. We are currently living in an interglacial period that is about half over.

Continental glaciers still exist in Antarctica and Greenland. The Antarctic ice sheet alone accounts for over 90 per cent of all glacial ice on the planet. Interestingly, the human species has never lived in a period when the earth was without glaciers. Compared with other periods, the earth is now experiencing a cold snap 1.5 million years long!

Of course, it is not enough just to have cold temperatures. There must be significant accumulations of snow each year to create a glacier. In the last ice age parts of northern Alaska, Siberia, and the western edge of Yukon were not glaciated. These regions were arctic deserts that did not receive enough precipitation to be glaciated. Many west coast glaciers are presently active because they receive enormous amounts of orographic precipitation in the form of snow.

Unlike continental glaciation, alpine glaciation is common in many parts of Canada today. The northeast coastal mountains of Baffin Island, Ellesmere Island, and many mountainous regions in British Columbia have glaciers extending down into the valleys. Although the term originated in the Alps, alpine glaciation is found wherever it is cold enough for snow to remain all year.

The point at which year-round snow occurs is called the **snow line**. In southern Greenland the snow line is about 500 m above sea level (asl). In the middle latitudes, it is about 2700 m asl because of the more temperate climate. Near the Equator, it is even higher still. In the Andes Mountains, year-round snow does not occur until 5000 m asl. During ice ages, these snow lines moved lower until they joined with the continental glaciers to create a total blanket of ice.

Glacial Movement

Continental and alpine glaciers shape the land in different ways. Yet the way in which they move is essentially the same. Glaciers are formed from water in its solid state. Deposits of snow are the basis for glacial ice. In the summer, this snow changes to ice pellets, called **firn**, as periods of freezing and thawing alter its crystalline structure. If you are a skier, you may know firn as "corn snow" since it resembles the shape and size of corn kernels.

Like sedimentary rock, glacial ice develops in layers as fresh accumulations of snow are added each winter. Deep beneath the surface the layers of ice are changing. Air pockets between the ice crystals are squeezed out. The ice crystals then grow and join together to form a solid sheet of ice. Essentially, they change because of the great pressure of the ice above. This metamorphosis can take a long time. In Antarctica, where there is little precipitation and temperatures are extremely cold, it may take 1000 years for surface snow to change into glacial ice.

The most significant property of glacial ice is that it flows like a thick liquid. The immense pressure of its own mass causes the ice to flow outward from its source. Most glaciers move so slowly—only a few metres a year—that change is unnoticeable from day to day. It is only when measurements are taken from year to year that movement is detected. Yet there are incidents of what is called **glacial surge**.

The Bering Glacier, an Alaskan ice floe two-thirds the size of Prince Edward Island and 225 km long, has moved faster than any glacier in modern times—10 km during a nine-month period. Each surge occurs every twenty-five to thirty years. The reason for these surges is uncertain, but it could be the result of **basal slippage**. Heat and friction between the glacier and the ground cause some of the ice to melt. The meltwater lubricates the surface, allowing the ice to slide more easily and rapidly.

The movement of alpine glaciers is directed by the shape of the landforms through which they flow. They move out of the mountains into the valleys and sometimes end up flowing into the oceans. When either continental or alpine glaciers flow into regions where summer temperatures are high enough to melt the ice, the glacier starts to become thinner. This area is called the **zone of ablation**. The glacier is in equilibrium if the same amount of ice is being added at the source as is melting in the zone of ablation. This usually never happens, however. Either the glacier is advancing or retreating. When it advances, it is building. When it retreats, it is melting. The term "retreat" is somewhat inaccurate because it suggests the glacier is moving back up into its source area. But this is not the case. The ice is continuing to flow out of the source, but it is melting faster than it can build up.

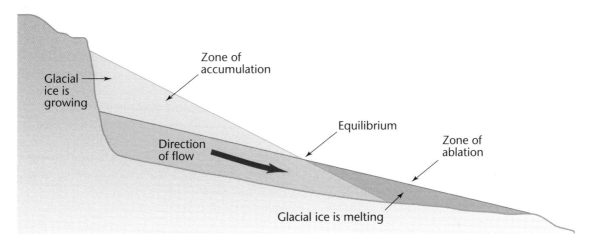

FIGURE 32.2 *Mass Balance for a Glacial System*
This mass balance diagram shows the glacier in equilibrium. It is neither moving forward or retreating. Ablation and accumulation are balanced.

Things To Do

1. Outline the conditions needed for glaciers to advance.

2. In what way can ice be compared with sedimentary rock?

3. Why do alpine glaciers still exist in most of the world's continents, but continental glaciation is restricted to Antarctica and Greenland?

4. Study Figure 32.2.
   a) Redraw the diagram showing what it would look like in a period of glacial advance.
   b) Redraw the diagram showing what it would look like in a period of glacial retreat.

5. Explain the dynamics that cause a glacier to move.

Continental Glaciation

Continental glaciation was just an unproven theory in the early 1800s. Scientists believed that the earth's surface features were created by running water and flooding. Glacial deposits were thought to have been laid down by icebergs that had drifted over the area—hence the term **drift**, meaning glacial deposit. Scientists acknowledged that glaciation occurred in mountains because they could see them at work. They did not make the connection between alpine and continental features, however. How could there be glaciers that covered whole continents?

In 1836, Louis Agassiz, a Swiss scientist, began to compare rock deposits in Swiss valleys with other deposits in northern Europe. He discovered that huge boulders in Switzerland were of the same igneous rock as the bedrock in Scandinavia hundreds of kilometres away. Using the theory of uniformitarianism (see page 148), Agassiz theorized that glacial movement was the only way these huge rocks could have been carried such great distances. Like most visionaries, Agassiz was scorned in his day. His theories were considered credible only much later. Today continental glaciation is widely accepted as a major gradational process.

Erosion

The erosive power of glaciers is incredible! There is practically nothing a glacier cannot pick up. In the last ice age, boulders as large as houses were moved hundreds of metres. Loose material on the surface was swallowed up as part of the glacier's load. Frozen into the ice, these particles scraped, sanded, and polished whatever bedrock the glacier could not move. This action is called abrasion.

The material the glaciers collected was not limited to loose particles. When a weak section of the bedrock was encountered, a large chunk of rock was sometimes plucked right out of the ground! Friction created by the glacier's advance would cause the ice to melt. The water would seep into cracks in the rock, where it would refreeze. When the glacier began to advance again the frozen chunk of bedrock was wrenched along with it.

Some people describe glaciers as being like giant bulldozers. This is not entirely correct, however. A bulldozer moves material with its front only. But most of the erosive power of a glacier is not in its front, or **snout**. Material is picked up along the length of the glacier's entire base. In this way glaciers are similar to rivers.

Glacial erosion is evident throughout the Canadian Shield. The region was cleared of most of its soil during the last ice age. The soil it has today either was deposited by the glacier or has formed since the ice age ended. Bedrock is exposed over most of the Shield. Rock surfaces are smooth and often polished. Stripped of most of its soil, the Shield has little agricultural value. But it offers rich mineral resources that are easy to find because of the absence of overburden. These deposits include precious minerals like gold and silver, metallic minerals like nickel and iron, and non-ferrous metals like copper, lead, and zinc.

There are millions of depressions formed mainly as the result of abrasion in the Canadian Shield. When the glacier left, many hollows were filled with water, forming long, thin **finger lakes**. The most famous of these glacial lakes are the Great Lakes. There are several theories about the formation of these lakes. Some believe they existed before the Pleistocene Epoch and so were not formed by glaciers. Others contend that the area they now cover was originally a river valley, and the deep sedimentary basin was excavated over several ice ages until the lakes, as we know them, were created. There is no question, however, that the Great Lakes were enlarged and shaped, at least in part, by glaciation.

No matter what their shape or size, Canada's lakes are a tremendous resource, although one that is sometimes taken for granted. In the south, close to where most Canadians live, the lakes provide many recreational opportunities. In the north, the lakes are an integral part of the traditional lifestyles of many Native peoples. Moreover, the water stored in Canada's vast wilderness areas is becoming increasingly important in a thirsty urban continent. Water resources were included in the North American Free Trade Agreement. Many Canadians were concerned that our water resources could be depleted if they were diverted to meet the needs of American cities. In addition, the hydroelectric potential of the Shield's water resources has caused controversy. The James Bay Project has been widely debated over the disruption it caused Native peoples and the environment. Many Canadians don't want their wilderness areas plundered to provide electricity to our neighbours south of the border. (See pages 311-313 for more on the James Bay Project.)

Historically, Canada's lakes have served as our highways. In colonial times, the fur trade prospered because traders and trappers could easily travel great distances on the many lakes. Of course, the Great Lakes have always been and continue to be important transportation routes. Coal, iron, and other raw materials are transported to the many industrial cities in Canada and the United States. In addition, goods are loaded onto ocean-going vessels hundreds of kilometres from the Atlantic and transported around the world. There is no doubt about the enormous contribution the Great Lakes have made to the prosperity of this region.

Isostasy

Another way in which glaciers affect the earth's surface is through **isostasy**. The surface of the continental crust is pressed down by the mass of glacial ice. (It is similar to erosion because the surface is lowered, although it cannot be considered erosion because no material is actually removed.) When the glaciers are gone, the crust rebounds.

Deposition

The mechanics of glacial erosion and deposition are complex. Much of the load is frozen into the ice and is only deposited when the ice melts. You can imagine the enormous amount of drift deposited when a continental glacier finally melts. Some of the material is deposited directly by the ice and some is carried away and eventually deposited by streams of meltwater.

Many depositional features in the last ice age occurred at the southern edge of the glacier. There were many periods of advance and retreat. Each time the glacier moved, the drift was moved around and reshaped, so across much of southern Canada and the northern United States there are many different glacial deposits.

Where the glaciers melted, glacial deposits often formed into hills. These **moraines**, as they are called, contain a variety of unsorted drift material. Tiny particles may lie alongside huge **erratics**. Sometimes blocks of ice were buried inside a moraine. After the glaciers

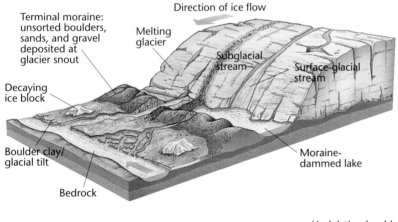

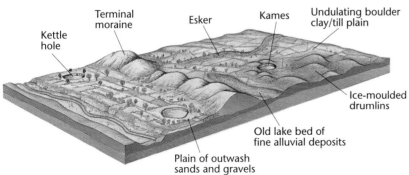

FIGURE 32.3 *Features of Glaciated Landscapes*
Top *At the ice front;* bottom *after the ice has melted*

238 PART 5 GRADATIONAL PROCESSES: WEARING DOWN THE EARTH

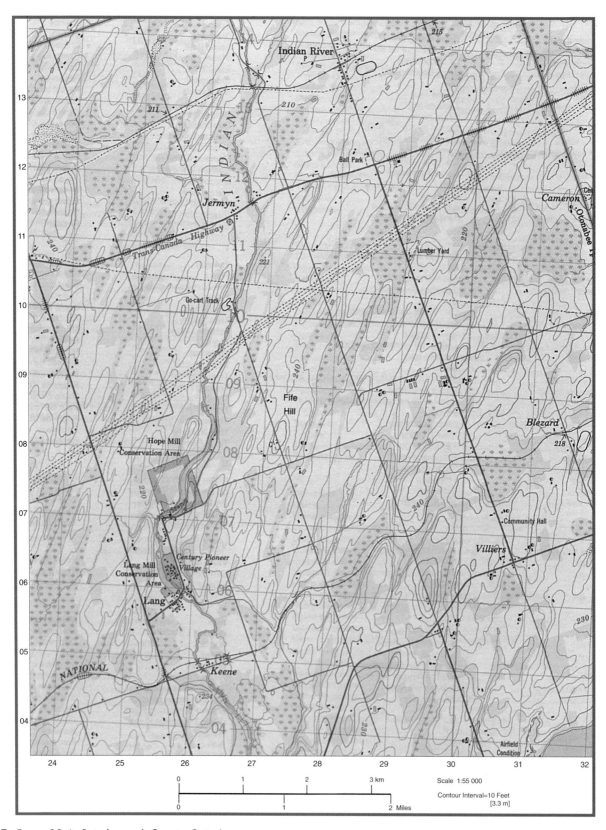

FIGURE 32.4 *Peterborough County, Ontario*

retreated, these ice chunks melted, creating hollows in the moraine. The hills in these formations are called **kames** and the hollows are called **kettles**. Often water fills the kettles, creating round kettle lakes. This hummocky landscape is common in many areas where glacial material was deposited.

One of the most famous moraines is the Oak Ridges Moraine in southern Ontario. Formed between two lobes of the glacier, this **interlobate moraine** forms the divide between Lake Ontario and Lake Simcoe. Since runoff flowing into Lake Ontario originates in the moraine, it is important that the area be preserved in its natural state to ensure the watershed will remain healthy. Unfortunately, this environmentally sensitive area already houses numerous landfill and development sites.

Drumlins are curious hills that frequently appear in groups. What is curious about them is their strange symmetry. They are all more or less the same shape, with one end wide and rounded and the other end thin and tapered. The thin end or tail of the hill points in the direction in which the glacier moved. These formations are created when a glacier encounters an obstacle of bedrock it cannot move. The moving ice reworks the till, creating the streamlined shape. The area around Peterborough, Ontario, is a classic drumlin field.

Stratified glacial deposits can be found in amazing variety. When a glacier melts it leaves an incredible amount of water and till. Rapidly flowing glacial streams exhibit the same characteristics as any other river. Outwash plains of stratified and sorted sediment form at the glacier's snout. Pre-existing drainage systems expand into enormous rivers known as **spillways**. It is obvious that in the past huge rivers cut these u-shaped valleys. The tiny streams in these spillways today are indicators of these massive post-glacial water flows. Sometimes the sediment is deposited in a temporary lake that may have been dammed by melting glaciers or moraines. Fine particles slowly sink to the bottom, leaving a rich lacustrine deposit. One such **ponding** occurred around present-day Schomberg, Ontario. The rich clay soils that were left behind make the area prime agricultural land.

Eskers look like upside-down river beds. These thin, meandering hills of layered material cross glaciated landscapes. They are formed when water flows through a stagnant glacier. This may occur on the surface of the ice, but it is more likely that the water flows at ground level or within the ice itself. Like other running water, this stream eventually drops its load. When the ice melts, the unusual esker landform is left. Eskers and **outwash fans**—deltas that form on land—provide important deposits of sand and gravel used in the construction industry.

Canada is not the only country to experience continental glaciation. Parts of the northern United States as well as Scandinavia and much of northern Russia share many of the same gradational features that are found in Canada.

Things To Do

1. Make a list of all the highlighted words in this section. Classify each feature as erosional or depositional, and explain how each was created.

2. How has glaciation shaped the history and the economy of the Canadian people?

3. Study the topographic map of part of the Canadian Shield in Figure 32.4.
   a) Which gradational process, erosion or deposition, is more noticeable?
   b) Describe the drainage pattern. How has it been affected by glaciation?
   c) What glacial features are obvious from the map?
   d) Draw profiles of glacial features.
   e) How could this map be used by people in the area?

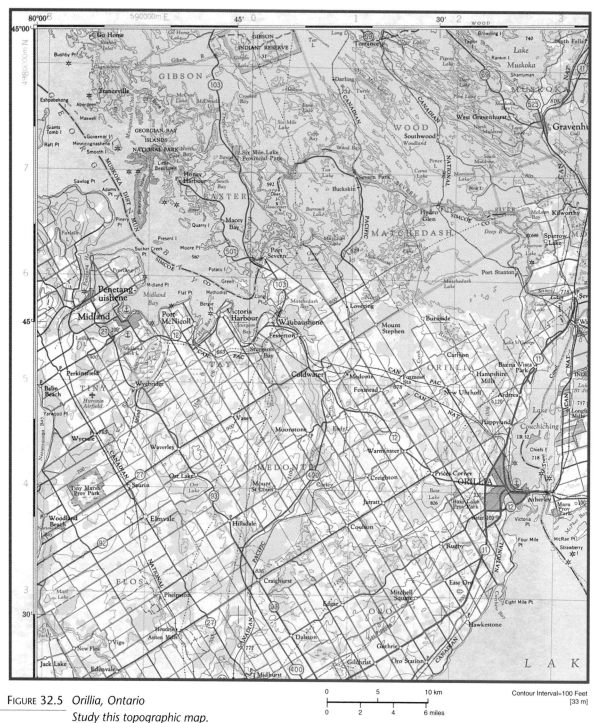

FIGURE 32.5 *Orillia, Ontario*
Study this topographic map.
  a) Draw a sketch map or trace the region showing major landform features.
  b) How is the northern half of the map different from the southern half? Include physiological as well as land use patterns.
  c) How has the physiography affected land use patterns?
  d) Use arrows on your sketch map to indicate the direction of glacial movement. Give two pieces of evidence to support this decision.
  e) Find examples of glacial features. Use grid references to indicate their locations.

Alpine Glaciation

Although mountain glaciers occur on a much smaller scale than continental glaciers, most middle and upper latitude mountain ranges have been shaped by them. The source area for these glaciers is high in the mountains where snowfall is heavy and temperatures are low enough for snow to last all year.

Erosion

Freeze-thaw action in the base of the glacier allows water to enter cracks in the rock. When the water freezes and the glacier moves away, surface rock is carried away with it. Over many years, horseshoe-shaped depressions called **cirques** are formed by this action. These eroded hollows are not visible until the glacier has retreated.

When two or more cirques develop close to one another, picturesque alpine features are formed. Two cirques against each other make a chiselled ridge, known as an **arête**. If the two cirques break through the arête, a **col** or pass is formed. Three or more cirques around a mountain peak result in a **horn**, that sharp-toothed feature so loved by mountain climbers! The Matterhorn of the Swiss Alps is the most famous example. The rugged beauty of alpine mountains is caused to a large extent by these cirque formations.

From the source region in the mountains, glaciers flow into mountain valleys. The force of gravity allows the ice flow to move steadily down and out of the mountains. Everywhere it goes, its erosive power modifies the shape of the land. Young v-shaped river valleys are carved into wide u-shaped valleys. **Hanging valleys** are carved where tributary glaciers flow into main glaciers. When glaciers eventually reach the ocean, they continue to flow out to sea until they are broken up by ocean currents and waves. **Fjords** are u-shaped valleys that are flooded by sea water when the glaciers melt. Unlike continental glaciation, the erosive power of alpine glaciers is concentrated by the relief of the land.

(a) Area before glaciation

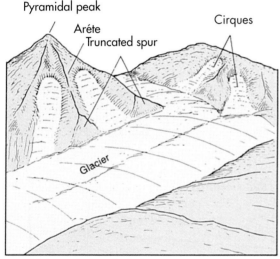
(b) Area during glaciation

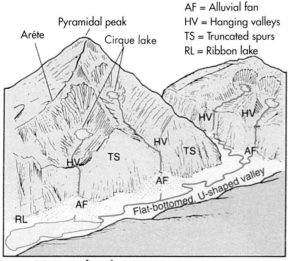
(c) Area after glaciation

FIGURE 32.6 *Effects of Alpine Glaciation*

Deposition

The depositional features of alpine glaciation are more or less the same as those of continental glaciation. Moraines are common. But instead of interlobate moraines there are **lateral moraines** and **medial moraines**. Lateral moraines form along the sides of a glacier as it flows through a mountain valley. When two glaciers come together, they remain separate, flowing along side by side. The ridge of till formed between them is a medial moraine. This moraine is made up of the two lateral moraines, one from each glacier. It can be a significant landform feature because there is twice as much till accumulating in the same place. Lateral and medial moraines are recognizable as dark streaks of sediment in the glacial ice. (See Figure 32.7)

Sometimes a moraine blocks the flow of a river. When this happens a lake is formed. Picturesque Lake Louise in Banff National Park was formed in this way.

Terminal moraines that form in fjords are called **skerries**. Most often these hills are submerged under water, but sometimes they emerge to form islands at the mouths of the fjords. In northern climates skerries have proven to be beneficial landforms as they provide protection against violent winter storms for some harbours.

Alpine Glaciation and Land Use Patterns

Alpine glaciers influence the way people live. Cirques, for example, provide upland basins. For centuries Swiss herders have pastured their animals in lush mountain meadows during the summer. Many ski lifts operate from the valley to a cirque. The flat ground gives ample space for restaurants and other tourist attractions, right in the middle of the best skiing. On a more practical note, **cols** are important landform features for travellers. These saddle-like dips in otherwise impassable terrain form mountain passes. In the Swiss Alps, most of the traffic between Switzerland and its southern neighbours travels through cols such as St. Gotthard Pass and Simplon Pass. In Canada, the famous Crowsnest Pass and Kicking Horse Pass provide routes through the Rocky Mountains between Alberta and British Columbia.

Down below in the valleys, glaciers have also worked to improve the landscape for human settlement. U-shaped valleys are much easier to settle than v-shaped ones. The wider valley floor allows for the development of agriculture, transportation routes, and urban development. Mountain towns in Switzerland, such as Interlaken, are prime examples. However, we have our own examples in Canada. Kelowna, in the Okanagan Valley in British Columbia, flourishes as the centre of orchard farming in western Canada. The broad, flat valley floor provides ample room for farming, transportation, and other land uses.

FIGURE 32.7 *Mt. Waddington, BC*
What glacial features can you identify in this photograph?

CHAPTER **32** THE WORK OF ICE **243**

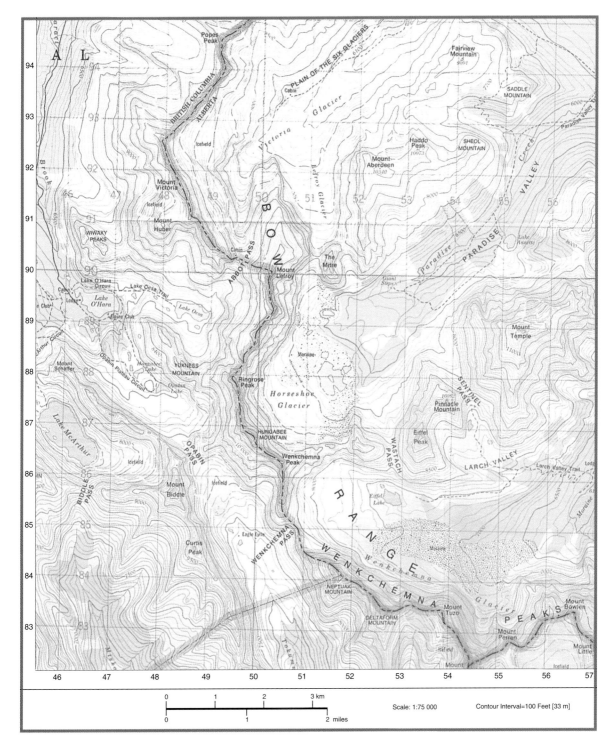

FIGURE 32.8 *Lake Louise, Alberta*
   a) Trace this topographic map of the Lake Louise area.
   b) Label the following features: cirques, arêtes, horns, u-shaped valleys, cols, hanging valleys, and moraines.
   c) Draw profiles for any three features.

Technology Update: Landsat Images

In 1965, the United States initiated the Earth Resources Survey Program. Satellites were put into orbit to survey the planet. Orbiting the earth at 919 km, these satellites cover every part of the planet every day. Despite what you may think, these satellites do not actually photograph the earth. If they did, the pictures would be up in space in the satellite, not on earth where they are needed!

What the satellite actually does is gather information about the earth using Multispectral Scanners. These devices scan a 35 000 km^2 area of the earth recording the variations in surface colour. The image is broken down into many tiny pictals. The colours are converted to numerical values for each pictal. So each satellite image is really a series of computer codes!

The information the satellite obtains is relayed to earth, where computers convert the codes into images. The information is displayed on a monitor or printed on paper. The process of creating satellite images is called **remote sensing**.

Can you imagine the incredible number of applications for this technology? So many things can be monitored, from glacial activity to plankton growth to troop movements. (In the Gulf War in 1991, the coalition forces used satellite images to target Iraqi military sites.) Satellite images are used to monitor pack ice in the high arctic and to identify the formation of icebergs where glaciers flow into oceans. This information is vital for ocean navigation and for monitoring changes in climate patterns. Infrared images are particularly good for showing plant growth and thereby revealing the health of remote forests and wilderness areas. For example, the destruction of the rainforest became a global issue in part because of the dramatic images of Amazon rainforests burning. In Canada, satellite images help us to survey the damaging effects of acid precipitation on forests in regions downwind of industrial and urban areas. Water pollution can also be observed. Black and white satellite images show oil slicks and help clean-up operations in determining where the worst messes can be found; the ecological damage caused by the Exxon *Valdez* disaster in 1989 was mapped in this way. Tankers and land-based industries can also be monitored, providing regulatory authorities with evidence of pollution, which can then be used to force polluters to clean up their activities.

Using satellite imagery to monitor soil moisture can be important in forecasting devastating floods, such as those that plagued the Mississippi-Missouri river basin in 1993. Hydrologists were able to assess the affect soil moisture would have on discharge rates and so could predict the floods. The ability to monitor soil erosion is also beneficial for agricultural regions.

Satellite images are especially good for observing geological formations. Mineral prospecting has improved significantly now that geologists are able to determine the location of geological structures that correlate to mineral deposits. In so many ways it is clear that these "eyes in the skies" are helping geographers to understand more about our dynamic planet.

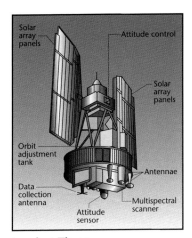
Landsat Three

The National Aeronautics and Space Administration (NASA) of the United States established its Earth Resources Survey Program in 1965. It has developed the space programs of *Gemini, Apollo, Skylab,* and *Landsat.* The first Landsat was launched in 1972 and the fourth in 1982. The satellite is put into orbit and gathers its information by remote sensing—that is, viewing the earth from a great height and signalling what it sees back to earth.

Multispectral scanner

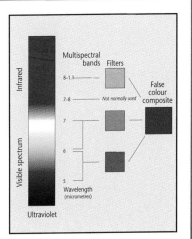
Landsat false colour

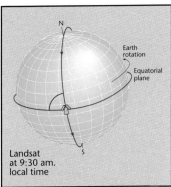
Landsat orbit

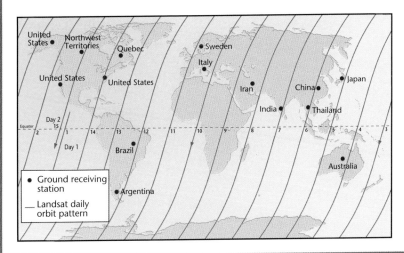

Some applications of Landsat

Geology — recognition of rock types, landforms, and specific minerals

Water resources — area of surface water, water availability: extent of snow cover or ice. Glaciers, floods, and irrigation: sediment and mapping of shallow waters: shoreline changes

Environment — effects of natural disasters: air pollution: mapping of remote areas

Agriculture — crop, timber, vegetation— surveys and health: soil conditions: crop areas and yield: land use surveys

FIGURE 32.9 *Landsat Satellites and How They Sense Images*

Things To Do

1. Prepare an organizer comparing alpine and continental glaciation.

2. How does alpine glaciation allow for human activity?

3. Study Figure 32.6.
   a) Draw a sketch or tracing of each feature.
   b) Identify the glacial feature and label it accurately.
   c) Explain how it was created.
   d) Determine how this feature has made life better for people in the area.

4. Assume that you were an entrepreneur about to develop the area shown in Figure 32.8. Choose one of the following activities:
   - a ski resort, including chalet, lifts, and runs
   - a quarry mining gravel for road construction
   - a mountaineering and hang-gliding school
   - an east-west highway showing bridges, tunnels, and gas stations
   - a white water rafting enterprise

   a) Show the location of your activity.
   b) Produce a map to a scale of 1:5000
   c) Draw several profiles of the site. Explain why the terrain here is more suitable for your project than other sites on the map.
   d) What other information would you need to know about the site before you proceeded further with your development?

5. Use a variety of reference materials, including atlases, satellite images, and air photos, to find global patterns for each of the following: fjords, continental icecaps, alpine glaciers, u-shaped valleys, glacial drift, and finger lakes.
   a) Using a legend, show the location of each of these features on a world map.
   b) Give a generalized statement about where each feature is found in terms of **absolute location** and **relative location**. Absolute location includes latitude, altitude, relief, etc. Relative location includes the relationships between each feature and the other features. For example, fjords have a lower elevation than alpine glaciers; they are extensions of u-shaped valleys; they have glacial till at their mouths; and so on.

6. Read the Technology Update on Landsat Images and explain how remote sensing aids the study of physical geography.

Chapter 33

The Work of Waves

INTRODUCTION

The part of the earth's surface where ocean meets land is affected by wave action. Many of the principles that operate in other gradational processes also work to create coastal landforms. As with other processes, erosion occurs when water is moving rapidly; deposition results as water slows down, loses its energy, and is unable to carry its load. Rivers, glaciers, and wind work to reduce elevation, often to base level. Wave action does not, however. Since all the action occurs at sea level, features are worn back from the shoreline, not down. This is the main difference between coastal and other gradational features.

FIGURE 33.1 *The Work of Waves*
Waves have carved out the coastal features of the Maritime province of New Brunswick.

Wave Size

Wave size is dependent on three variables: **wind velocity, wind duration,** and **fetch.** Wind velocity is the strength of the wind; the stronger the wind the higher the waves. Duration is the length of time the wind blows from one direction. The longer the duration the better able the waves are to organize themselves into regular **crests** and **troughs** called **wave trains.** If the wind shifts direction wave trains are disorganized; the water is choppy but the wave size is small. For really big waves to form, the wind must blow steadily from one direction. The third variable, fetch, is the distance the wind blows over the ocean. The longer the distance, the greater the opportunity for wind to transfer some of its energy to the waves through friction. So wherever strong winds blow steadily from the same direction over vast distances, the waves will contain enormous amounts of energy. Consider, for example, the steady northeast trade winds that blow across the vast Pacific to the beaches of Oahu in the Hawaiian islands. Here all of the factors that contribute to great wave energy are at work. The results are huge waves that make the island a surfer's paradise!

Some waves reach immense proportions. **Tsunami** is the Japanese word for "big wave," and that is exactly what these waves are! For years tsunami were erroneously called tidal waves. This is an incorrect term because these giant waves have nothing to do with tides. Submarine landslides, earthquakes, and volcanic eruptions are the main causes of tsunami. When the earth beneath the water moves, the resulting tremors create giant waves. Often the wave length of a tsunami may exceed 100 km! Initially, the wave height is usually not great, perhaps only a metre or so. Travelling at speeds of up to 800 km/h, these killer waves race across the ocean. But it is only once they reach shallow water near the coast that tsunami become really dangerous. The wave length shortens and the energy becomes more concentrated. Now the wave height grows rapidly, sometimes to 15 m or more. When this wall of water hits the shore, devastation is inevitable. People are swept out to sea. Towns are destroyed. The erosive power of these freak waves is greater than that of any other natural wave.

Hurricanes also generate giant waves capable of causing massive destruction. The powerful winds whip up the waves to such an extent that coastal areas are bombarded by abnormally large waves. Shorelines can be altered forever by just one catastrophic storm. Sandbars may be breached. Whole islands may even be washed away! The actions of one brief storm may be greater than the erosive forces of the ocean over many years. The Toronto Islands were created when a violent storm cut a channel through the sand bar that protected the city's harbour. This so-called Eastern Gap is now a shipping route for the busy port city.

Erosion and Wave Action

Contrary to popular understanding, water does not actually flow with the waves. Energy moves *through* the water, just as sound waves move through air. A cork floating in sea water bobs up and down with the waves but it does not move appreciably. The length of each wave is measured from crest to crest. Waves tend to extend down into the water about one-half their wavelength. Below this active layer, there is little movement.

Figure 33.2 on pages 250-51 shows what happens when a wave meets the shore. When a wave moves up a beach, it drags along the bottom. The top portion continues to travel forward, however. It topples over, resulting in the familiar breakers. The combination of momentum and gravity concentrates wave energy at the point where it breaks. If a rock or other formation is in the way, erosion occurs. The pounding hydraulic action, abrasion from particles car-

> **GEO-Fact**
>
> Surfers know where the really big waves can be found. In the United States, for example, the coastal islands of North Carolina have huge waves—much bigger than the waves found along coastal beaches that do not jut so far out into the ocean.

ried by the water, and corrosion from the salt in the water all work to erode the landscape.

Not every part of the coastline erodes at the same rate. **Headlands**—shorelines that project into the ocean—are subject to greater erosion than sheltered bays. This is the result of **wave refraction** (that is, waves bending around the headland), which concentrates wave action on these landforms. Notches are formed when rock is eroded at the point at which waves meet the cliff face. Overhanging rock above the wave action is not affected, so a rock shelf juts out into the sea. Eventually, the mass of the projecting rock becomes too great for the underlying rock to support and the overhanging rock collapses into the sea. Arches, caves, and sea stacks are formed in similar ways.

While all this action is happening on headlands, sand is being deposited in the bays where wave action is less rigorous. Waves seem to be working to smooth out the coast to a straight line. An old coastline is often much more regular than one that has only recently been glaciated or affected by some form of tectonic activity.

Wave Action and Deposition

Erosion and deposition are constantly at work along ocean beaches. Often sand dunes created by blowing wind form inland from the beach. Natural grasses hold the dunes in place. If the grasses are removed, the sand blows away and wave action may penetrate inland. This is why many tourist areas warn people to protect the grasses and stay off the dunes.

Beaches are formed by wave action. Waves carry particles of sand along the shore. These particles are deposited when the flow velocity of the water decreases. Sandbars and spits are formed wherever waves move sand along the beach. **Longshore drift** is responsible for many of the intricate coastal features found along the Atlantic coastline. Most of the eastern seaboard of the United States and parts of the Maritime provinces have depositional features formed from longshore drift.

Emergent Coastlines

Emergent coastlines are so called because the land is emerging out of the ocean. Long, thin sandbars form barrier islands just off the coast. Between the coast and the islands, fresh water **sounds** are fed from the water of continental drainage systems. Outlets between islands allow the fresh water to flow into the ocean. Sediment from runoff eventually fills in these sounds, creating swamps, saltwater marshes, bogs, and lagoons. Eventually, the area between the barrier islands and the mainland is filled in. Then a new series of barrier islands moves further out into the ocean.

An interesting ecosystem has evolved around these barrier islands. The mineral-rich fresh water from rivers flowing into the sounds is partially trapped behind the islands, providing excellent habitats for fish, waterfowl, and other sea life. People enjoy the islands because of the variety of recreational activities they offer, from surfing on the waves of the ocean side to sailing on the calm waters of the sound side. The sea life provides great opportunities for fishing, clamming, and shrimping.

Barrier waters do harbour their dangers, too, however. First among these is the risk of violent storms. In 1989, for example, Hurricane Hugo devastated the barrier islands along parts of the US east coast, destroying homes and businesses. But while the barrier islands take the brunt of these storms, they serve to protect the mainland coastal communities from violent storm surges.

Emergent coastlines also form when the land rebounds after a period of glaciation. Sea cliffs and raised **terraces** indicate that these land formations were once eroded by the sea. The raised beaches of many arctic islands are good examples of this type of coastline. The cliffs are constantly being eroded by the ocean, but beaches are also forming from the weathered material of the cliffs.

Emergent coastlines are common during glacial periods. Enormous amounts of water are frozen in continental glaciers. This results in a lowering of sea level. Large sections of continental shelves are exposed. Because we live in an interglacial period and sea level is relatively high, emergent coastlines created from reduced sea level are uncommon today.

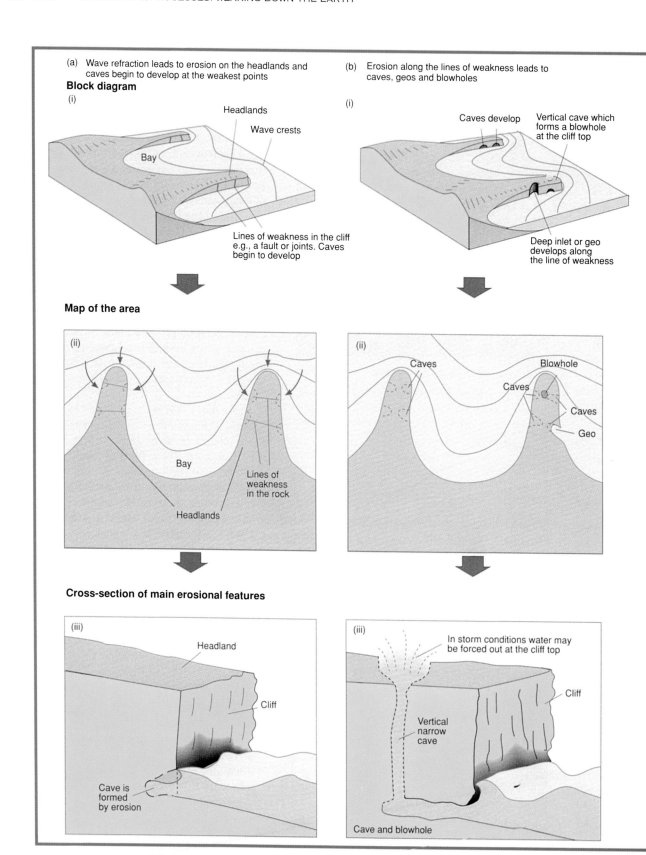

FIGURE 33.2 *Landforms Produced by Wave Erosion*

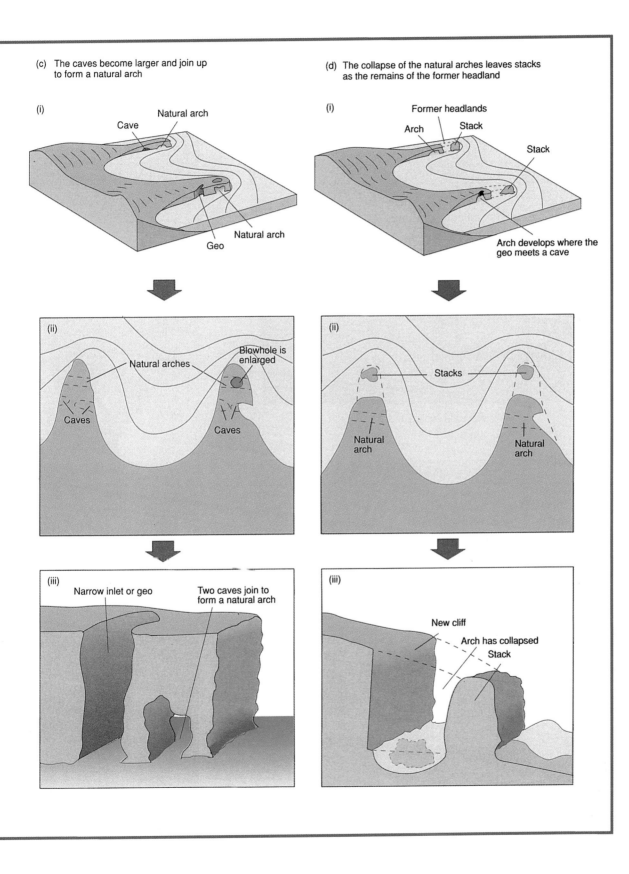

FIGURE 33.3 *An Emergent Coastline in South Carolina*

FIGURE 33.4 *An Aerial View of an Island and Lagoon in Bora Bora, Polynesia*

Submergent Coastlines

The opposite of an emergent coastline is a **submergent coastline**. These are formed when sea water invades river valleys and low-lying areas when sea level rises or when the shoreline sinks as a result of tectonic processes. Much of the Canadian coastline is submergent. Alpine glaciers in British Columbia, Newfoundland, Labrador, and the Arctic islands carved fjords out of river valleys. Sea level was much lower during the ice age. Once the ice melted and sea level rose, these u-shaped valleys flooded with sea water. **Ria coastlines** are formed when entire river valleys are flooded. The estuaries around Charlottetown, PEI, form an excellent example of a submergent coastline.

Coral Formations

Corals form much of the beach material in the tropics. Corals are the shells of tiny animals that live in colonies in shallow tropical water. As older animals die, new ones form on the outer surface, so the coral structure, or **reef**, is constantly growing.

Three different shoreline features result from coral formations: barrier reefs, fringing reefs, and **atolls**. The Great Barrier Reef off the northeast coast of Australia is the world's largest coral formation. Plants and animals thrive in the innumerable nooks and crannies of the reef. If sea level drops, as it did in the last ice age, the reef is exposed and forms an island. If, on the other hand, sea level rises, the coral continues to grow up towards the surface. Many islands in the Caribbean are also fringed by coral reefs.

When corals are weathered by wave action white sand is formed. Coral atolls are curious features found in much of the tropical Pacific. The corals form around a **seamount**, which is the crater of an extinct volcano. What results is a doughnut-shaped island, with the centre occupied by a shallow lagoon.

Things To Do

1. Identify the factors that cause erosion and deposition in coastal features.

2. Explain how coral formations could be an important tourist element for some Caribbean countries. Consider beaches, interesting structures, ecosystems, building materials, etc.

3. How is wave action similar to and different from other gradational processes?

4. Study Figure 33.5 on page 254.
   a) Give examples to show that both submergent and emergent coastlines exist around Charlottetown.
   b) Explain why most of the tourist industry in Prince Edward Island would be centred on the north shore.
   c) Explain how both the north shore and the south shore of the island were formed.

5. Study Figure 33.6 on page 255.
   a) Identify examples to show that this coastline has been flooded.
   b) Draw a sketch map showing what the island would look like if sea level were to rise significantly.
   c) How has the topography affected land use patterns in Nootka-Nanaimo?

6. Study Figure 33.7 on page 256.
   a) Draw a sketch map and indicate the probable direction of longshore drift.
   b) Explain how the sand hook was formed.
   c) What is the importance of the forested interior of the peninsula?
   d) How have human land use patterns in Provincetown been affected by this structure?

254 PART 5 GRADATIONAL PROCESSES: WEARING DOWN THE EARTH

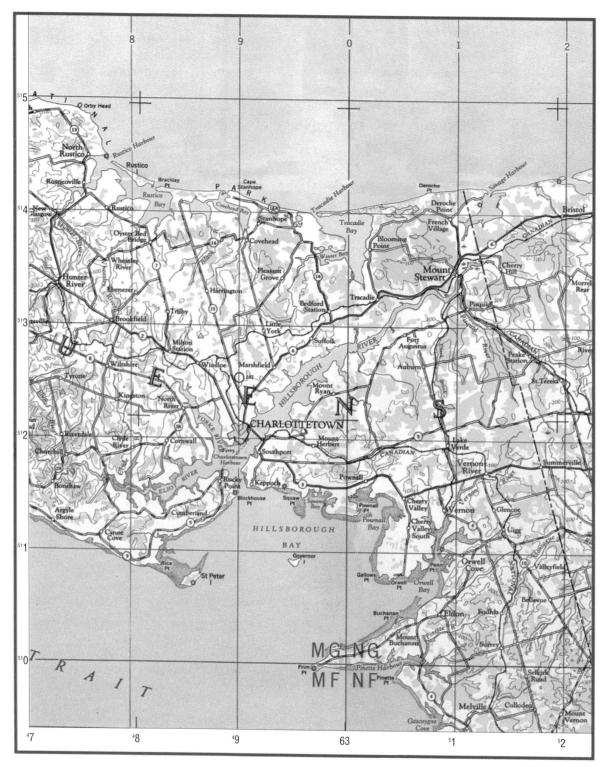

FIGURE 33.5 *Charlottetown, Prince Edward Island*

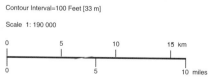

CHAPTER 33 THE WORK OF WAVES 255

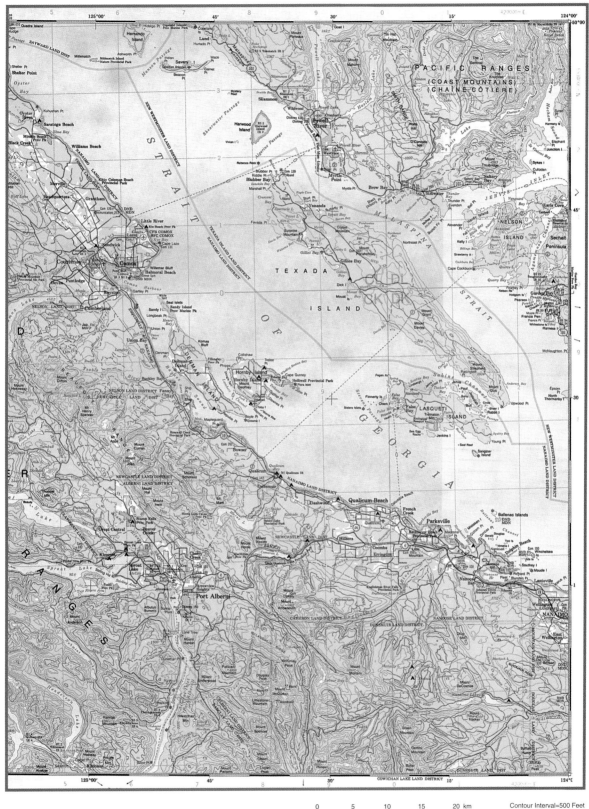

FIGURE 33.6 *Nootka-Nanaimo, British Columbia*

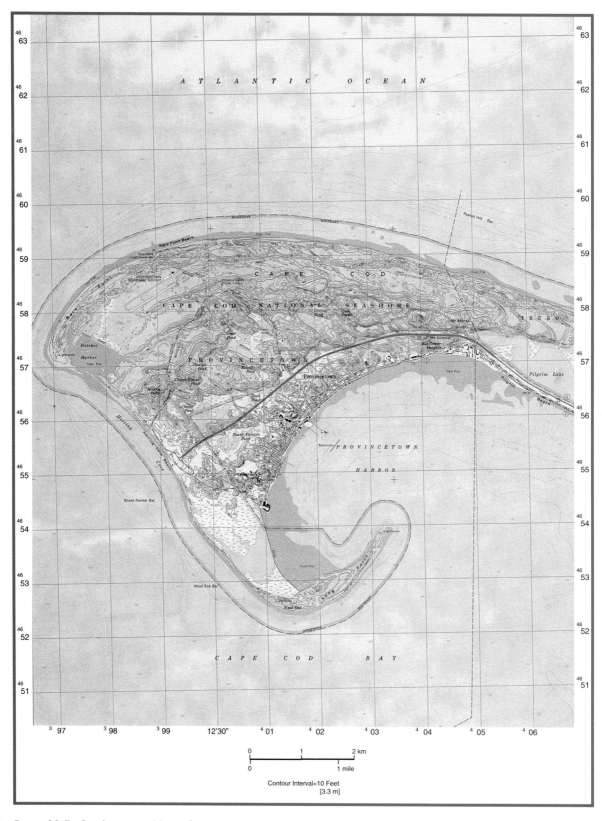

FIGURE 33.7 *Provincetown, Massachusetts*

PART 6

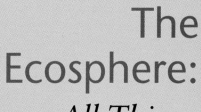

The Ecosphere:
All Things Living Under the Sun

Chapter 34

The Ecosphere and Our Place in It

INTRODUCTION

The world is made up of four interlocking spheres. The lithosphere is the ground on which we stand. The atmosphere includes the air we breathe, the weather, and climate. The hydrosphere provides the water needed for life to flourish. The ecosphere is the living part of the planet. It includes all the elements that allow life on earth to exist and it is interdependent with the other three spheres. The ecosphere is a vibrant and constantly changing part of our planet.

It is important to study physical geography in order to understand our place in the ecosphere. We have only recently realized the effects our actions have on natural systems. Rivers and lakes have been polluted, forests have been cleared, and swamps have been drained. In the process, countless ecosystems have been damaged. An understanding of our place in nature is essential if we are to protect the earth from environmental destruction.

Yet it is unrealistic to expect nature to stay constant and never change. Our planet is a dynamic one. Change is essential for its healthy evolution. As a species, we need to utilize the environment to obtain the raw materials we need to flourish. Our responsibility is to ensure that we do not change ecosystems to such an extent that plants and animals are unable to cope. The balancing act between resource development and conservation is a difficult one. The remainder of this book focuses on the challenges we face. Part 6 outlines patterns and relationships in the ecosphere. Part 7 deals with the ways in which people use the ecosphere to provide essential raw materials.

FIGURE 34.1 *Enjoying the Outdoors*
What is our responsibility to ensure that our actions do not harm the environment?

Ecosystems

Over millions of years, many complex ecosystems have evolved within the ecosphere. Each climate region develops ecosystems made up of distinctive soils, plants, and animals. Those plants and animals that adapt to their environment live and flourish. Those that do not die out. This process is called **natural selection**.

When there is a sudden major change in the environment, many plants and animals are often unable to cope. Geologists have discovered that several mass extinctions have occurred in the history of the planet. This is what happened to the dinosaurs. The climate changed so much, so quickly, that the giant beasts were unable to cope. Many people believe that we are currently living in a period of mass extinction because human domination of the earth is making it difficult for many species to survive.

Adaptation

Plants and animals that are well adapted to a particular environment are successful and survive. Adaptations can be either physical or behavioural. Physical adaptations occur when the structure of the plant or animal accommodates specific environmental conditions. In desert regions, for example, plants produce small, waxy leaves in order to reduce water loss through transpiration. Rainforest plants, on the other hand, often have large moist leaves that are well suited for releasing large quantities of water from the plant.

Animals also have adaptations to allow their bodies to cope with specific environments. Giraffes have long necks so they do not have to compete with other browsing animals for food. They are the only animals able to reach the leaves found high in the treetops.

Camouflage is another form of physical adaptation. Animals have markings or colouring that allow them to blend in with their surroundings. Herbivores are able to remain inconspicuous to their prey. On the other hand, predators are also frequently camouflaged so they can stalk an unsuspecting victim unnoticed.

Animals have many physical adaptations. Teeth, mandibles, and jaws are constructed for the food each species eats. Feet, claws, and legs are similarly designed to meet specific needs. Fur, hair, scales, and feathers have evolved to suit the physical needs of each kind of animal. Even eyes, antennae, ears, and other sense organs are specially adapted. What other examples can you think of in which animals have physically adapted to their environment?

Animals are also capable of behavioural adaptations—that is, they behave in certain ways to make the most of their environment. Many desert animals are nocturnal. They sleep during the day but come out at night when it is cooler. Birds migrate before winter sets in. Squirrels collect nuts in the fall. Bears hibernate for the winter. These are all examples of behavioural adaptations. Do you think plants have behavioural adaptations, too?

The Web of Life

Nature's variety is incredible! Not only is there a seemingly infinite variety of plant and animal species, but there are also innumerable ecosystems. Every ecosystem is an exceedingly complex community of plants and animals. Every living thing within an ecosystem is dependent on every other living thing. What happens to one species is sure to affect the whole community.

Plants and animals interact in intricate patterns and are usually mutually dependent on each other. For example, the mice that live in a meadow eat grass and small insects. These mice are in turn consumed by hawks, foxes, and other predators. If the mouse population declines, the number of insects could increase. Furthermore, the hawk and fox populations could decrease since there are fewer mice to eat. These changes occur naturally and are common. For example, if the climate suddenly became wetter, the soil could erode and lose essential minerals. Drought-resistant plants may not survive these new conditions, but new plant species may take over. The change in plant life could lead to a corresponding change in animal life. Some animals could leave the region or die out. Other species better suited to the altered environment may emerge. Changes like these indicate that the ecosphere is vibrant and alive. If it failed to change, the earth would be a dead planet.

The Effect of People on the Ecosphere

People interact with the environment in two ways: they adapt to it and they modify it to better serve themselves.

Human Adaptations

Our behaviour, lifestyle, clothing, and even our moods are influenced by the environment around us. Like other animals, humans have adapted to their environment by changing their physical structure or their behaviour patterns.

Physical adaptations occur when the human body changes in order to cope with a specific environment. Consider aboriginal peoples who evolved in tropical climates. The dark skin colour commonly found in these cultures is a physical adaptation designed to protect the body from strong sunlight. Longer limbs make it easier for the body to lose heat since there is a larger surface area exposed to the air. These are appropriate adaptations for people living in a hot climate. At the other climatic extreme, people indigenous to subarctic climates often have light skin because they have adapted to weaker sunlight. Their shorter bodies and limbs allow them to conserve heat, a necessary adaptation in a cold climate. Of course, people do not physically adapt to their environment in a few generations. It takes thousands, perhaps even millions, of years. Today with migration a common fact of life, people often live far from where their ancestors evolved.

Behavioural adaptations occur when people adapt their lifestyle to accommodate the environment. We have seen how desert animals are often nocturnal to avoid the daytime heat. Similarly, many people in Mediterranean Europe, for example, take long midday breaks, commonly known as *siestas*, to avoid the heat during this part of the day. During siesta, people go home for the main meal of the day and return to work later in the afternoon. This behavioural adaptation is unnecessary in a northern country like Canada. But we have our own behavioural adaptations. How do you think Canadians have adapted to our northern climate? In what ways have you personally adapted your behaviour to suit the environment where you live?

Modifying the Environment

People are unlike most animal species in that they can modify the environment to suit themselves. For example, instead of relying on food that grows naturally, humans have developed **hybrid** plants and animals that are produced on farms.

There are countless other ways in which people modify the environment, including mineral extraction, urban development, and forest management. Often these changes are for the good. Our lifestyles are enriched by the raw materials we harvest and the cities we build.

Sometimes, however, these changes are harmful, either as a result of thoughtlessness or incomplete knowledge of natural systems. Examples of the harm we have inflicted are numerous. The sinking of the Exxon *Valdez* oil tanker in Prince William Sound in 1989

FIGURE 34.2 *Living at High Altitudes*
People who evolved living at high altitudes have adapted physically to the lower concentration of oxygen. For example, their lungs are larger than those of people living at lower altitudes.

caused ecological damage to the Alaska and northern BC coastlines. It took several years for naturally occurring bacteria in the soil, combined with other natural processes, to destroy the oil slick that had spread along much of the coast. Another example of the negative impact of human activity can be found in agriculture. Farmers sometimes over-fertilize and over-irrigate their fields to produce our abundant food supply. Sometimes these chemicals contaminate water supplies. So when people modify the environment to meet our needs, we must be careful that we do not cause irreparable damage to the environment and indirectly to the quality of our lives.

Human overpopulation and mismanagement of resources are the root causes of our ecological problems. We need to understand the patterns and processes of physical geography so that we can avoid making mistakes that could destroy the ecosystems in which we live.

Humankind is the most successful species the world has ever known. *Or is it?* How do we measure success? By the level of our technological development? Or by the way in which we protect and develop the environment? Will we be considered the most successful terrestrial species in a thousand, a million, or a billion years? We will never know the answer to this question!

Things To Do

1. Select the atmosphere, the hydrosphere, or the lithosphere, and list the components that it provides to the ecosphere.

2. Prepare an organizer to compare how plants, animals, and humans cope with changing environments.

3. a) Provide examples of a food web (see page 58) for *one* of the following ecosystems: a marsh, a forest, or a meadow.
   b) Explain how the food web you prepared in a) could be affected by a change in one species.

4. a) Explain why a period of extinction occurs.
   b) Why do some people believe a period of mass extinction is occurring now?

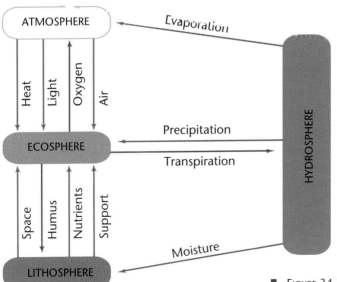

FIGURE 34.3 *How Different Systems Affect the Ecosphere*

Chapter 35

Natural Vegetation

INTRODUCTION

If you look out your classroom window, it is unlikely you will see any **natural vegetation**. Natural vegetation is plant life that grows naturally in a region without any human interference. Most of the plants we see every day have been put there by people.

Climate and soil determine the type of natural vegetation that develops in a region. Because there are many possible climate and soil combinations, there are many varieties of natural vegetation. Plants found thousands of kilometres apart often share characteristics because they develop under similar conditions. For example, cacti are found only in desert regions of North and South America. But similar plants are found in the arid climates of Africa and Australia. The high temperatures and evaporation rates require desert plants to retain as much moisture as possible. Small, thorny leaves with thick, waxy coatings and extensive root systems make the most of the water that is available. This characteristic is shared by desert plants in all regions of the world.

FIGURE 35.1 *Natural Vegetation in Desert Regions*
In the desert, the natural vegetation has adapted to the lack of moisture. These Joshua Trees are similar to plants found in deserts hundreds of kilometres apart.

Plant Needs

Plants have specific needs that must be met by the environment or they will not survive. These are air (carbon dioxide), water, sunlight, heat, nutrients, support, oxygen, and space. Without these essentials, plant life cannot exist.

The first four needs are provided by the atmosphere. Air is a **permanent resource**. As such, it is abundant in virtually every ecosystem. Unlike air, the availability of water varies from climate to climate. While different plant species require different amounts of water, all plants need some water to carry nutrients from the soil to plant cells. Without water, plants cannot survive.

The amount of sunlight also varies from place to place. Sunlight is essential for photosynthesis. Where there is no light green plants cannot survive. There is a negative correlation between the amount of direct sunlight at the earth's surface and rainfall. When there are few clouds and intense sunlight, rainfall is low. Climates that experience heavy precipitation often have gloomy skies. The lush natural vegetation that flourishes in many of these wet climates makes competition for sunlight intense.

Heat and sunlight are obviously related. Heat is unevenly distributed around the planet. The amount of heat varies with latitude and altitude, as we discovered in Part 3. (See pages 32-33.) Every plant has an optimum temperature range where it thrives. Trees, for example, need at least one month with a mean temperature above 10°C.

The other plant requirements—nutrients, support, and space—are all related to the lithosphere. Soil provides essential nutrients, including nitrogen, potassium, phosphorous, and organic substances. Wet climates often leach the water-soluble minerals out of the soil. The infertile soils of many equatorial ecosystems severely limit agriculture if the forest cover is removed. On the other hand, local deposits in drained swamps and river deltas may provide rich soils that often support extremely productive farmland.

Soil supports the roots of the natural vegetation. Thin soils provide poor support for some types of trees, while deep, well drained soils are excellent for plants. Aquatic plants are supported by the water in which they grow.

Naturally plants need space in order to grow. This commodity is in short supply in many forests. In the rainforest, many plants have adapted by climbing up other plants. Only the strongest plants are able to survive as other saplings compete for room and light.

Each ecosystem provides a unique mix of plant essentials. The plants respond by evolving in such a way that they can take advantage of these ingredients while compensating for deficiencies in other areas. Consider the cactus. The desert provides abundant heat and light but not much water. As a consequence, moisture is stored in fleshy stems. Leaves are tiny to reduce transpiration, and often serve merely as protection from animals thirsty for the stored water. By contrast, in the tropical rainforest little sunlight breaks through the dense canopy overhead. Leaves are large in the

FIGURE 35.2 *The Tree Line*
The tree line is clearly visible on this BC mountain slope. The line indicates the 10°C warmest month isotherm.

gloomy understorey to catch any light that filters through. The large surface area of the leaves also allows moisture to transpire readily from the plants in this humid realm. Plants that are unable to adapt to this environment disappear from the region and in some cases become extinct.

Changes in Natural Vegetation

Change is happening all the time in the ecosphere. It may take place quickly as the result of natural and human disasters. But it also happens slowly as part of natural evolution. Take a walk in a forest and you can see changes taking place everywhere. A tree may have fallen during a wind storm. As the tree decomposes, the soil is enriched by its recycled nutrients. Mushrooms and toadstools sprout along the tree trunk; termites, millipedes, and other animals move in to feast. Thus the death of one plant provides many opportunities for life in other forms.

Plants must adjust to changes. New species may be introduced into the area. Competition for essential needs may result in individual plants or even whole species dying out in a certain area. If all the essentials are provided, then the species will survive. But if just one essential element is reduced to a point below the minimum requirement then the plant will die. If this happens on a large scale a whole species will become extinct. Other plants will take over if they can tolerate the change in environmental conditions.

You've heard of animals migrating, but plants migrate, too. Seeds are spread by winds, animals, and other natural processes. If conditions are right, the plant will flourish. If conditions are poor, it will not reach maturity or be able to reproduce. The coconut palm, for example, grows in virtually every tropical island in the Pacific Basin. Coconut, the seed of the tree, floats and is carried by ocean currents to colonize islands where conditions for success are possible.

Vegetation Succession

New land is often created either naturally or through human activity. When this happens, natural vegetation soon colonizes it. When a volcano erupts, for example, lava and ash create a new ground surface. Hardy plants are able to grow on these lava fields even while they are still warm to the touch.

Over time vegetation changes. At first only lichens and mosses may grow on a rocky sur-

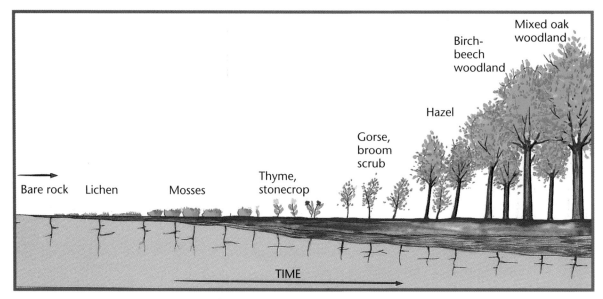

FIGURE 35.3 *How an Ecosystem Develops from Pioneer Species to Climax*

face devoid of soil. In time these **pioneer species** decompose the rock, forming soil minerals and providing **humus**. Eventually soils start to form and satisfactory growing conditions for other plants develop. Grasses and flowering annuals may begin to grow in the thin soil. Now there is no room for the pioneer species, and they die out. In time, small shrubs and even trees develop. As the soil becomes fully developed, even the microclimate changes. Trees reduce the amount of sunlight reaching the soil; humidity increases and soil moisture rises. Eventually the plant community reaches a stage of stability. No new species move into the area. The natural vegetation has reached its mature stage, called **climax vegetation**.

The concept of climax vegetation is a simplification of chaotic natural processes. It is unlikely that a natural system will ever reach its climax because of changes in the atmosphere, lithosphere, or hydrosphere. Still, it is a level to which all ecosystems move towards if left undisturbed. For example, in areas of uplift, like the Andes Mountains, tectonic changes may occur over a relatively short period. In other areas where the geology is stable, conditions may remain the same for millions of years.

Change Caused by Human Activity

People greatly affect the ecosphere. Modifications made by humans cause the ecosphere to change rapidly. Frequently these changes are more far-reaching than we had anticipated. Consider these examples.

A massive irrigation project near the Aral Sea was established by the government in what was then the Soviet Union. Water was used to irrigate the land so that the country could be self-sufficient in cotton. The project was a great success. The land was more productive and farmers prospered. But today the region is a desert wasteland. So much water was taken from rivers flowing into the Aral Sea that it has been reduced to a fraction of its original size. The once-thriving fishing and resort industries have died. The climate has changed, too. The shrinking sea resulted in reduced evaporation. This in turn led to a decrease in precipitation. Strong winds blew salt from the dried seabed over the land. Poisoned by this salt, the land could no longer support cotton crops. Moreover, the natural grassland vegetation did not return. The land has become a barren desert. This shortsighted project destroyed the local ecosystem for the immediate future, although it could recover if water levels in the Aral Sea are restored.

In southern Africa, the San, a race native to the Kalahari Desert, has lived in harmony with the environment for thousands of years. The San used to hunt for meat and gather nuts, berries, and roots from the plants that grew naturally in their semi-arid home. About thirty years ago, they came in contact with Bantu-speaking peoples who taught them to herd cattle, an age-old activity that was sustainable in the lush savanna of southeast Africa. But the fragile Kalahari ecosystem could not support large herds of animals. The grass was overgrazed. Roots were destroyed in the process, so the grass did not grow back. The climate became drier as a result and the region was unable to support domesticated animals. After twenty to thirty years of grazing cattle, the land was so dry no grass remained. The cattle all died. The San were unable to return to their tra-

> **GEO-Fact**
>
> The Mediterranean Sea was once much smaller than it is now. During the last ice age, sea level was so low that the Strait of Gibraltar formed a land bridge between Africa and Europe. The Atlantic Ocean and the Mediterranean Sea were not joined as they are today. Geologists believe this vast inland basin was a desert much like the Sahara. As the glaciers melted, about 14 000 years ago, sea level started to rise. An enormous waterfall over the Gibraltar land bridge allowed water to start filling the great depression and eventually created the Mediterranean Sea. Today the climate and natural vegetation of the region have been changed from their earlier form because of this great body of water.

Clear-cut logging

Urban expansion

Terrace farming

Hydroelectric power development

Industrial emissions

FIGURE 35.4 *How People Affect the Environment* Identify how people have changed the environment and evaluate whether the change has improved or harmed each of the ecosystems shown here.

ditional way of life. Today many of the San have been assimilated into mainstream society, working as miners or farm labourers. Thus the abuse of the environment led to the decline of one of the world's oldest cultures.

Ecological disruption doesn't occur only in distant places. Modern developments here in Canada have caused significant environmental damage. In suburbs across the country housing subdivisions have transformed natural environments. Before any houses are built, the land is stripped of every living thing. When the trees, grasses, and flowering plants disappear, so do animals like deer, muskrats, and squirrels. Next to go is the topsoil. Huge machines scrape off the top layer, leaving only subsoil. Once the houses are built and the roads are constructed, some topsoil is returned and lawns are planted. Does this improve the environment? It does for people, but in the process the natural ecosystems are destroyed forever.

Things To Do

1. Find and describe examples in your community of a) climax vegetation, b) pioneer species colonizing "new" land, and c) changes in natural vegetation.

2. Study Figure 35.5. Explain how competitive advantage and competitive disadvantage affect ecosystems.

3. Explain what climax vegetation is and why it is seldom found in natural systems.

4. Find newspaper articles about people modifying the ecosphere in your community. Discuss whether the change is positive or negative. Display your results on a bulletin board or in a scrapbook.

5. Many environmental groups wish to preserve natural environments. Debate the value of wilderness preservation in modern society.

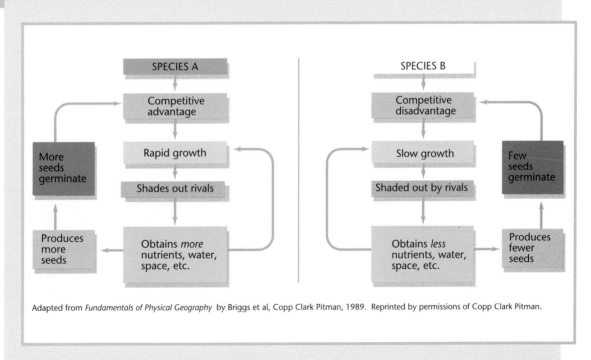

Adapted from *Fundamentals of Physical Geography* by Briggs et al, Copp Clark Pitman, 1989. Reprinted by permissions of Copp Clark Pitman.

FIGURE 35.5 *Effects of Competition Between Species of Different Competitive Abilities*

Communities of Natural Vegetation

The Forest Community: coniferous forest (for example, pine, cedar, fir), deciduous forest (for example, oak, maple, linden), mixed forest (both coniferous and deciduous), tropical and subtropical dry forest (for example, acacia, eucalyptus, baobab), tropical rainforest (for example, teak, mahogany, rosewood).

The Grassland Community: temperate grassland, tropical grassland, tropical steppe.
The Scrub Community: schlerophyll scrub (Mediterranean).
The Desert Community: tundra, alpine, temperate desert, hot desert (for example, cactus, spinifex)

Classifying Natural Vegetation

While there are millions of different plants on earth, it is possible to classify terrestrial ecosystems into four main groups: forest, grassland, scrub, and desert. Within each of these communities are several sub-groupings representing the possible climax vegetation. These are indicated in the feature Communities of Natural Vegetation.

The Forest Community

There is a strong relationship between climate and natural vegetation patterns. To be successful in a climate, trees generally need abundant precipitation and at least one month with a mean annual temperature above 10°C. Coniferous trees have adapted to long periods of drought. During winter in the middle and upper latitudes, water is frozen and so is unavailable to plant life. The needle-like leaves, thick bark, and conical shape allow coniferous trees to thrive in climates where other trees would perish. These types of trees are common in the boreal forest or **taiga**, an unbroken band of forest that covers much of northern Canada, Russia, and Scandinavia.

In the middle latitudes, deciduous trees have adapted to a cold winter season, although they are generally not as hardy as cone-bearing trees. Deciduous trees lose their leaves in autumn and remain dormant throughout the winter. These mixed forests form a transition zone between the boreal forest to the north and the deciduous forests closer to the Equator.

FIGURE 35.6 *The Boreal Forest of Northern Canada*

FIGURE 35.7 *Mixed Forests of Western Canada*

Moving towards the Equator, seasonal variations are less pronounced. Patterns of natural vegetation are more often related to patterns of rainfall rather than variations in temperature. Where there is irregular rainfall close to savanna regions, tree growth is limited. In equatorial regions where rainfall is heavy for much of the year, extensive forests flourish. In these tropical rainforests, broadleaf evergreen trees are joined by palms and other tropical plants.

Biomass is the amount of living material produced by a given hectare of land. The amount of biomass indicates that some regions are better at producing plants and animals than others. Forests generally have more biomass than other ecosystems. The height of the natural vegetation means that there is more organic matter concentrated in a given area.

FIGURE 35.8 *Grain Fields on the Prairie Grasslands*

The Grassland Community

Grasses are usually found where trees cannot grow. Dry, cold winds cause wind burn that kills trees. In places like the Canadian prairies, most trees cannot survive the bitterly cold blasts of winter and the low rainfall of summer, so grasses predominate.

When forests are cleared, grasses replace the trees until the trees are able to reestablish themselves. European settlers on the North American prairies in the late 1800s believed the land was unsuitable for farming. Coming from heavily forested western Europe, they thought the only good farmland was land that had once been covered with forest. They were wrong! The great grasslands of Canada and the United States have become the most productive grain-growing regions in the world. Although there may not be enough rainfall to support trees, there is plenty of rain for grass and grain crops.

Grasslands also exist in the tropics. The savanna climate has a dry winter season. The drought is often so severe that trees are unable to survive from one season to the next. In Africa, vast tropical grasslands provide the natural habitat for such animals as rhinoceroses and lions. In the rainy season there is so much rain and heat that grass can grow as high as 4 m. This "elephant grass" is so tall it can literally hide an elephant!

FIGURE 35.9 *The Tropical Grasslands of Australia During the Dry Season*

In the savanna, trees are often found scattered throughout the region. Local groundwater supplies and variations in seasonal rainfall make this tree growth possible. People have occupied tropical grasslands for so long that some authorities believe the forest has disappeared because of human activities such as gathering fuel wood and the grazing of domestic animals. Others contend that deforestation is the result of long-term climate change.

The semi-desert or tropical steppe that marks the transition between the savanna and the desert is also a grassland. The grass is shorter and thorn scrub is common in drier areas. In the rainy season the region can be as rich in plant growth as the mid-latitude prairies. But when the drought hits, the high evaporation rate makes this region more like a desert than a grassland. Because of this grasslands often have considerably less biomass than the forest community.

The Scrub Community

Schlerophyll scrub has to survive a dry summer and a wet winter. Most plants need moisture during the summer growing season, not in the winter when growth is limited by low temperatures. Scrub plants, however, have adapted to the dry growing season. They appear to be dried out. They have small leaves and seem dwarfed in size. The Mediterranean climate zone is noted for this type of vegetation, but scrub also grows in other areas where conditions are inadequate for grasses or forests. Many parts of the Mediterranean region have been populated with people for so long that little is left of the natural vegetation in many

FIGURE 35.11 *Tundra in the Northwest Territories*

places. Often irrigation in the summer has enabled much of the region to look greener than it would normally and has allowed for the introduction of **exotic species**. Undisturbed natural vegetation can usually be found only in areas where human access is difficult because of topography.

The Desert Community

As conditions become even drier, scrub turns to semi-desert and eventually to desert vegetation. Plants in these regions have adapted to the arid conditions. The broad band of desert that girdles the globe at about 25° latitude is dominated by **xerophytes**, plants that can survive drought. In areas like the Namib and Atacama deserts where rain is almost nonexistent, there is almost no plant life at all.

Tundra and alpine vegetation are well adapted to the cold, dry conditions found in the polar regions and high in the mountains. Mosses and lichens can survive where no other plants can. These arid lands have less biomass than other regions because the lack of water limits the amount of organic matter produced.

Local Variations

Figure 35.12 shows the general locations of ecosystems. While a particular ecosystem may dominate a region, other forms of natural vegetation may exist. For example, within a mixed forest zone, there may be places where

FIGURE 35.10 *Scrub Vegetation in the Australian Outback*

conifers prevail and other locations where there are only deciduous trees and virtually no conifers. Still other places may be dominated by grassy fields or swampy wetlands. These local variations are not shown on the map so that the general patterns can be seen and order can be perceived in the apparent chaos of the real world.

The reasons for such variety within regions are many. Differences in microclimates restrict the range of some species while allowing others to flourish. For example, southern slopes with their more abundant heat and sunlight are preferred by some plants over northern slopes. Soil and drainage conditions also affect the success of plant species. Poorly drained soils are unsuitable for most trees but cedars, for example, like to have wet roots. In places where there is more water than soil, aquatic plants such as water lilies, bulrushes, and swamp grasses predominate. Sandy soils are preferred by some plants over poorly drained clay soils. Variety to some extent exists in all natural vegetation regions.

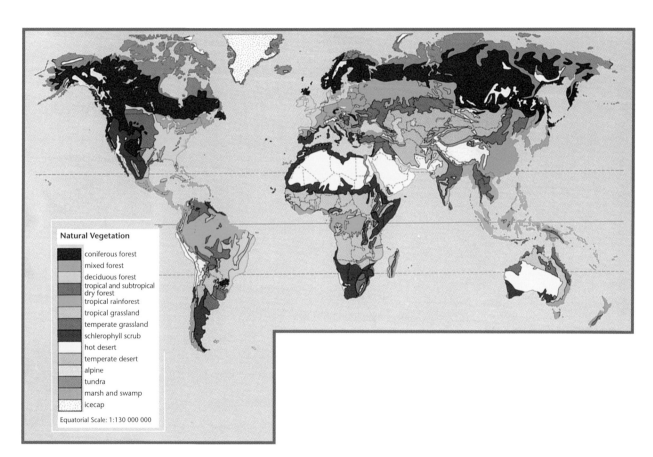

FIGURE 35.12 *Natural Vegetation Regions of the World*

Things To Do

1. Refer back to Figure 8.8 on page 64, which shows the net amount of organic matter, or biomass, produced by various ecosystems in one year.
   a) Why is the measurement of biomass a good indicator of how well an ecosystem meets plant needs?
   b) What other measurement of natural vegetation could show the effectiveness of ecosystems in meeting plant needs?
   c) Which column in the table best shows the effectiveness of the environment in meeting plant needs?
   d) Prepare an appropriate graph to show the results of the column selected in c).
   e) Does the graph confirm your speculation in c)? What inconsistencies are there?

2. a) Prepare a comparison organizer like the one shown below. Show how well plant needs are met in each of the natural vegetation zones by indicating G for good and P for poor in each category. The Tropical Rainforest has been completed for you.
   b) Based on the information in the organizer, rank the ecosystems in order from "Best Provider of Plant Needs" to "Worst Provider of Plant Needs."

3. Using Figure 35.12 and old geographic magazines, make a collage illustrating vegetation patterns for the continent of your choice. On a large cut-out map, identify the different natural vegetation zones. Select representative pictures for each zone and glue them at the appropriate spot on your map.

4. Study Figure 35.12 and Figure 16.2 (page 116).
   a) Prepare an organizer showing the types of natural vegetation found in each climate region.
   b) Draw a cross section at 30°E longitude, marking the latitude in 15° intervals. Estimate the profile of the land along this meridian using a relief map. Use symbols to illustrate the different types of natural vegetation.
   c) Determine the limits in latitude for each vegetation zone.
   d) Relate the patterns you found for Africa and Europe to the Americas. Account for differences between these continents.

Vegetation Zone	Air	Water	Light	Heat	Nutrients	Support	Space
Tropical Rainforest	G	G	P	G	P	G	P
Tropical and Subtropical Dry Forest							
Mixed Forest							
Coniferous Forest							
Tropical Grassland							
Temperate Grassland							
Schlerophyll Scrub							
Hot Desert							
Temperate Desert							
Alpine							
Tundra							

Legend   G=Good   P=Poor

Chapter 36

Soils

INTRODUCTION

Soil is vital to the development and preservation of healthy ecosystems. Plants and animals are dependent on soil for most of the nutrients needed to sustain life. In fact, the entire food chain is dependent upon healthy soils. Yet the study of soils is often undervalued. Perhaps it is because soil is covered by natural vegetation that we often forget about its importance.

The many different types of soil are created by complex chemical and biological processes. Variations in climate and bedrock result in the enormous variety of soil types. In order for soils to develop several conditions must exist.

Soil creation takes time. It is estimated that many of the deep soils in our most fertile agricultural regions have been thousands of years in the making. There must also be a rock surface as rocks and minerals are the raw material from which soil is created. Another requirement of soil creation is water. Moisture must be present to break down the rock into smaller particles. Finally, there must be soil organisms. Plants and animals mix and break up the rudimentary soil and add organic compounds that give the soil its life-sustaining properties.

A_o — Loose leaves and other litter

A — Dark colour—rich in humus; zone of leaching

B — Minerals from A horizon deposited

C — Weathered parent material

Bedrock

FIGURE 36.1 *Soil Profile*

The Properties of Soils

Soil Horizons

Soils develop three **horizons**, or layers. At the surface, where soils interface directly with the biosphere, there is a concentration of decomposing organic material. This layer of humus-rich topsoil is called the **A horizon**. Where the soil interfaces with the bedrock, there is little humus but much broken up rock of different sizes and shapes. This **C horizon** contains the **parent material** from which the soil is made. Between the A and C horizons is the transitional **B horizon**. It is here that nutrients are often stored in humid climates where leaching occurs.

Soil Composition

Soil is a complex mixture of four elements: humus (decomposed plant and animal matter), parent material (broken-up pieces of rock), moisture, and air. Individually, each element does not make up part of the soil. Only when these are mixed together are they transformed into a new substance—soil. These four elements are mixed together by several forces. Water from a rainstorm penetrates the soil and allows chemical processes to transform it. Plant roots and burrowing animals like worms and ants mix soil materials together. Shifting caused by freezing and thawing and even slipping down slopes all contribute to the development of soil.

Humus

Plants and animals depend on soil for their basic needs. In turn, soil is a product of the natural vegetation and animals that live in and on it. Even the most primitive soils have microscopic plants and animals living within them. Most also contain worms, insects, and other more complex animals. On the surface, there exists a wide range of plants and animals. When these life forms die, they decompose. Tree trunks, leaves, roots, animal droppings, dead animals, and dead microorganisms make up the litter that forms in the top layers of most soils. Termites, fungi, worms, and other decomposers reduce this organic matter to a dark, absorbent layer of humus. This part

Mollisol

Spodosol

FIGURE 36.2 *Soil Profiles*
Compare the soil horizons of a grassland (left) with a boreal forest (right). How are they similar? How are they different?

of the soil is rich in nutrients such as nitrogen, potassium, and phosphorous. Natural vegetation influences the type of humus that is created. When there is a lot of dead organic material, rich, deep, absorbent soils develop. The temperate grasslands of Russia, Argentina, and North America produce so much humus that a deep, black soil is formed. This fertile soil is one of the reasons why these grasslands are prime grain-growing regions.

Other vegetation types, such as the boreal forests of northern Canada and Russia, create highly acidic humus. This is because of the high concentration of acid in the coniferous needles that litter the forest floor. Plants and animals in this ecosystem have adapted to the acidic soils. The effects of acid rain in this region are often magnified because of the high acidity that already exists.

The lack of plants in desert ecosystems limits the development of decomposing litter. As a result, most desert soils lack humus and so do not produce the nitrates needed for plant development. As you can see, there is a delicate interrelationship between soil and the vegetation that depends upon it.

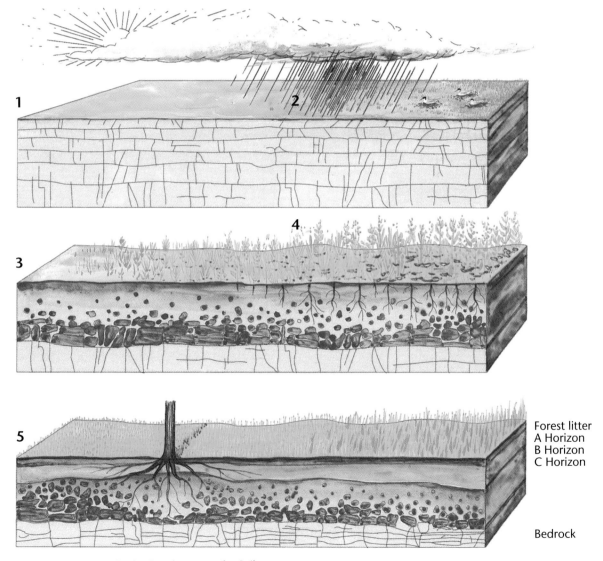

FIGURE 36.3 *Stages in the Development of a Soil*
Study these illustrations and describe how the soil changes in each of the five stages.

Parent Material

Soil is also affected by the rock over which it develops. At the base of the soil lies the broken rock debris formed from weathering. This parent material is what the soil is made from.

Mechanical and chemical processes work on the bedrock to change particle size and chemical composition. Running water, freezing and thawing, wind, wave action—the forces of weathering—shatter bedrock into increasingly smaller particles. Chemical action combines minerals and humus. It is only as rock fragments decompose that pioneer plant species are able to colonize the surface of the soil. The humus enables the soil to hold more water and increases fertility for more advanced plant species that follow in the natural succession. After many hundreds of years, a climax vegetation evolves in a fully developed soil.

The characteristics of the parent material determine what the soil is like. Soils that develop from rocks that easily erode, like certain limestones, are often deep and fertile. Because the rock is easily broken down, soil formation is facilitated. It is no accident that many of the best farming regions of the world develop where the bedrock is relatively soft. Rocks like the granites and gneisses found in the Canadian Shield and the Western Cordillera are so hard that weathering and soil formation is slow. Thin soils with shallow horizons result. Often the parent material is imported from another region. Rivers, glaciers, and other gradational processes carry parent materials far from the bedrock where they originated. The soils that form in river valleys develop from materials that often come from weathered rock high in the interior of the drainage basin.

Moisture and Air

In addition to organic and inorganic materials, a large proportion of most soils is made up of air and water. As much as 50 per cent of a soil can be made of these two substances, depending on how much space is available. As the amount of soil moisture increases, the amount of air decreases. This is why these two materials are studied together.

The amount of air and water a soil contains is determined by its **porosity** and **permeability**. Permeable soils, such as sand, have large particles that do not bond together. There is plenty of room for air and water. However, the spaces between grains are connected, so water flows readily through the soil, leaving it dry. These types of soils are preferred by plants that like a well drained, dry soil.

On the other hand, soils with small particles that bond together, such as clay soils, are porous but not permeable. Moisture enters the soil, but once there the particles bond together to make an impermeable structure that water cannot penetrate. Many crops flourish in these soils because they hold water received in the spring well into the summer when evapotranspiration is at its greatest. Clay soils can become water-logged. They contain little air for plants, which is essential for nutrients such as nitrogen.

Nitrogen-fixing bacteria on the

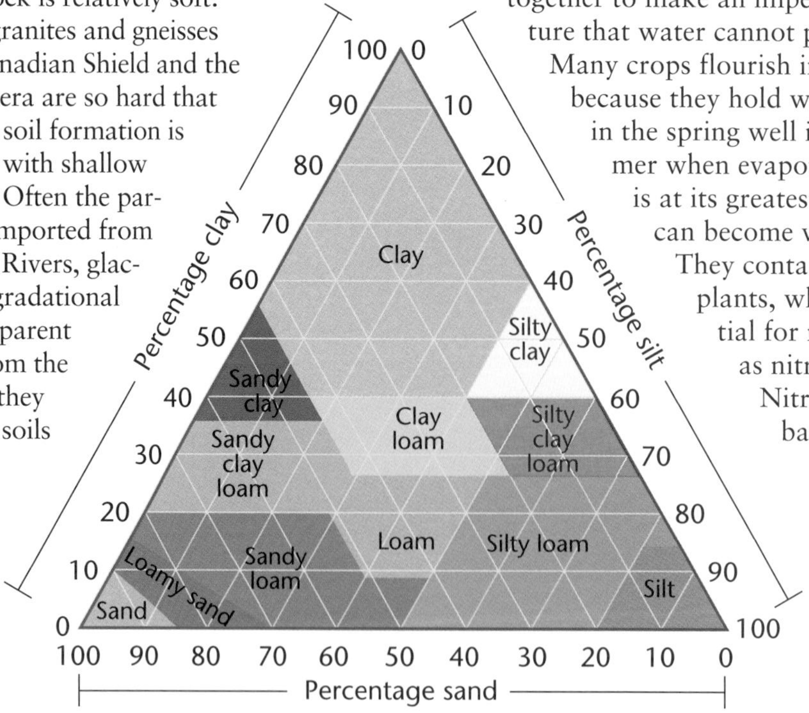

FIGURE 36.4 *Soil Texture*

roots convert the nitrogen directly to a form that plants can metabolize. If the clay is broken up or cultivated, the soil is aerated and can be extremely fertile. The best soils for vegetation have a combination of particle sizes. Loam, the most fertile of all soils, is made up of approximately one-third sand, one-third clay, and one-third silt (a particle size in between clay and sand). The clay particles hold the moisture essential for plant growth, while the sand and silt allow for better drainage and air to enter the soil.

The Effect of Climate on Soils

Both precipitation and temperature influence soil development. Soil moisture is obviously dependent on precipitation. If the amount of precipitation is greater than the rate of evaporation, soluble minerals are washed out of the upper layers of the soil. This process is called leaching. Alkaline elements such as silica, magnesium, calcium, sodium, and potassium are removed. The characteristic red soils of the tropics contain a large proportion of iron oxide. The other elements are gone, leaving the soil with a rusty appearance.

In areas where the rate of evaporation exceeds precipitation, **translocation** occurs. This is the opposite of leaching. Minerals are brought to the surface of the soil rather than being washed out of it. **Capillary action** draws moisture to the surface from deep within the soil. Water soluble minerals are dissolved in this water. When the moisture reaches the surface and evaporates, the minerals are left in the form of surface deposits. These remain on the surface of many deserts as non-porous salt pans or alkaline flats.

In many parts of Canada and other temperate lands, both leaching and translocation operate on a seasonal basis. The soil is leached in the spring as snow melts and soils thaw out. In the summer, high rates of evaporation cause translocation as the moisture is drawn back to the surface. In a way these temperate lands have the best of both worlds: adequate rainfall with limited leaching.

Temperature affects soil development in two ways. Chemical and bacterial activity are greatest where it is hot. These activities decrease as the temperature drops, until they cease altogether when the soil is frozen. In hot climates, chemical activity is greater than in

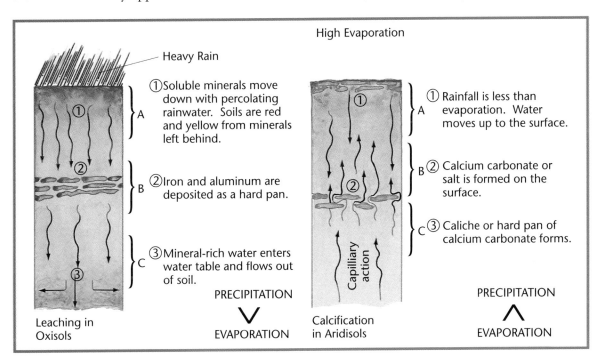

FIGURE 36.5 *Water Movement in Oxisols and Aridisols*

cooler regions. Parent material is chemically altered so rapidly there is practically no C horizon. Bacterial activity is also greater where it is hot. Forest litter is consumed immediately, so there is virtually no litter or humus in the tropics. As a result, tropical soils are composed of very deep red oxide clays with little horizontal development.

In cooler climates, chemical activity is much slower. Parent material is converted to soil so slowly that temperate soil horizons often have thick C horizons, with rocks of varying shapes and sizes. In addition, organic matter takes longer to decompose into humus and become a part of the soil because bacterial action is slower. As a result, temperate soils have three well defined horizons: a thick A horizon containing lots of humus, the transitional B horizon, and a C horizon containing parent material awaiting chemical decomposition by weathering agents.

Polar soils seem to be all C horizon. The lack of chemical action and the virtual absence of bacteria result in little horizontal development. The lack of vegetation means there is little humus and so there is a poorly developed A horizon. However, mechanical weathering, such as frost action, causes the bedrock to break into jagged rocks of varying shapes and sizes. Without warm temperatures, with little vegetation, and with moisture frozen for most of the year, polar soils are the least developed of all. Parent material is often exposed at the surface.

Soil Management: A Fine Balancing Act

Managing soils is a lot like balancing a bank account. In order for the account to stay healthy, the amount of money leaving it must not exceed the amount going into it.

> **GEO-Fact**
>
> Ethiopia is a grain-producing nation. Yet often it cannot produce enough food to feed its citizens. One of the problems lies with the new hybrid grains introduced after the Second World War. In many environments these products of the Green Revolution produce more grains of higher quality. Yet they are not well suited to north African soils. As a result, they require abundant fertilizer, which the poor farmers of Ethiopia cannot afford. Local scientists have found that grains indigenous to the grasslands of Africa grow better naturally than these scientifically engineered grains.

Otherwise, you will become bankrupt! The same principle applies to soils. If more nutrients are removed from the soil than are put back in, eventually the soil becomes bankrupt.

Left undisturbed by human activity, nature frequently maintains the balance between what is removed from the soil and what is returned. The **inputs**, or deposits, come from several sources: the natural vegetation in the form of humus, the parent material as the bedrock gradually decomposes, water from rainfall or surface streams, and air. All these inputs provide essential ingredients for a healthy ecosystem. There are often no outputs in a natural system—that is, nothing is removed from this closed system. Plants live, die, and decompose in the same place. Their nutrients are recycled through the soil from one generation to the next.

This balance is broken when either inputs exceed outputs or more nutrients leave the soil than can be replaced naturally. These imbalances occur frequently. For example, drought reduces the soil's ability to support natural vegetation. One of the inputs—moisture—has been reduced, so the soil's ability to support life is reduced. A forest fire, on the other hand, may increase soil fertility. When trees are burned, nutrients stored in dead wood are freed rapidly. The increased fertility in the soil ensures that a new forest can grow where the old one had been.

People are the greatest reason for changes in soil fertility. Whenever a crop is harvested, nutrients are removed from the soil permanently. Each spring farmers across Canada apply fertilizers to replace nutrients lost in the previous year's harvest. It does not matter whether the crop is wheat, dairy cattle, Christmas trees, or lumber, inputs must equal

FIGURE 36.6 *The Effects of Crop Harvesting*
When a crop is harvested (left), nutrients are lost from the soil. Fertilizers must be applied to restore the soil to a productive state (right).

the nutrients removed if the operation is to be sustainable. Consider forestry as an example. Trees are harvested and the wood is transported out of the region. If the trees had died naturally, they would have decomposed, thereby returning nutrients to the soil. But with the trees removed from the forest, artificial inputs of nutrients are needed if soil fertility is to remain stable. This is why foresters and farmers need to apply fertilizers to their fields.

The study of soils is fascinating, complicated, and important to the success of our species. Each year we rely on the harvest to provide us with food. If soils are allowed to deteriorate and food production drops, the results would be horrendous for the 6 billion human inhabitants of the planet. We must remember that modern farming practices have only been used for the past fifty to sixty years. Soils that have sustained natural vegetation for thousands and even millions of years are being taxed more today than ever before. Fortunately, modern farming and forestry techniques take into consideration the need for artificial fertilization to replace nutrient outputs.

Things To Do

1. Explain the importance of the following spheres in the development of a soil: a) the biosphere; b) the lithosphere; c) the hydrosphere; d) the atmosphere.

2. Explain the importance of soil texture to soil fertility.

3. Study Figure 36.4. Follow the horizontal and diagonal lines to determine the approximate breakdown of clay, silt, and sand for a) loam, b) sandy clay, and c) silty loam.

4. For each of the following natural and human examples decide whether the soil fertility has increased or decreased and explain your decision.
   a) irrigating a golf course
   b) harvesting a soybean crop
   c) wind erosion
   d) applying fertilizer
   e) grazing cattle
   f) raking leaves
   g) clear-cutting forests
   h) cultivating soil

5. Choose one example from activity 4 that you believe reduces soil fertility. Provide a strategy to increase the inputs so that balance is restored.

6. Find examples in your community where soils are used as a resource. Determine whether the soils are being utilized in a sustainable way. Report your findings to the class.

FIELD STUDY

Work with your group on a local field study. Continue with the site you have studied throughout the course or choose another site with a variety of ecosystems. (A small river valley is a good place.) Each person in the group should select a different ecosystem to study. If your site is a river valley, suggested sites might include: marshland that is frequently flooded, part of the flood plain, a forested northern slope, a forested southern slope, or a grassy meadow above the river valley.

Mapping the Site
1 Measure off an area about 10 m on each side.

2 Make a grid of 1 m squares using string and stakes driven into the ground. (This may be difficult to do in forested areas.)

3 Draw a map to a scale of 2 cm to 1 m [1:50]. Show the location of major features, plants, and animals. Number the plants and animals and use a field guide to identify plant and animal species. Use a legend to help.

SAFETY FIRST

Follow these simple guidelines to prevent injury.
1 Never work in a hole deeper than your own height.

2 Assess the area for dangerous situations before you start.

3 Never work alone.

4 Avoid conducting this field study in early spring when soil is saturated with water.

5 Conduct field studies under the supervision of your teacher.

6 Use protective clothing and boots where necessary.

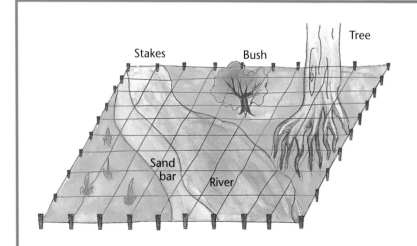
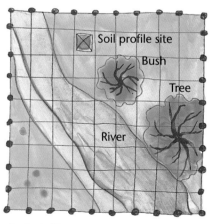

Scale 1:50

FIGURE 36.7 *Field Study Map*

Studying the Soil

1. Carefully clear away some of the natural vegetation and litter. Dig a hole or pit 50 to 100 cm deep.

2. Look for layers in the soil. Make a scale drawing to show where each layer occurs in the profile. Colour each layer appropriately, trying to copy the texture of the soil at each layer. Indicate the locations of pebbles and larger stones.

3. Alongside your soil profile, write a description of each soil layer. Include the following details:
   a) *Particle size or texture (sand, clay, or silt):* For larger particles, use a ruler or sieve to measure exact size.
   b) *Water content:* Weigh some of the material you excavated back in the classroom. Bake the soil to remove as much moisture content as possible. Weigh the soil again and estimate the percentage of the soil that was water.
   c) *Humus content and colour:* Describe the litter on the surface and the humus in the A horizon. Estimate the amount of the A horizon that you think is organic material.
   d) *Parent material:* Use a field guide to determine the type of rock the regolith is derived from.

4. Use clues such as natural vegetation to determine the fertility of the soil. Make recommendations of inputs that would improve soil fertility.

5. Fill in the hole and restore the natural vegetation as best you can to its original state.

6. Account for differences in natural vegetation and soil for each of the study sites in your group.

Using the Land

1. Describe the way the land is being used. For example, is it wilderness? Is it pasture?

2. In what positive or negative ways has the land been modified by people?

3. What suggestions would you make for the sustainable development of the field study site?

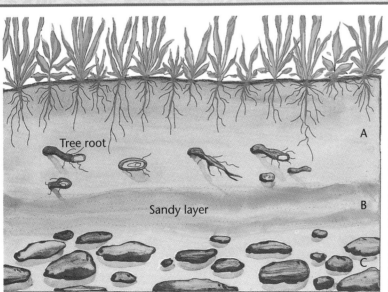

A Horizon
Sandy brown soil with high humus content. Moist from recent rain.

B Horizon
Clay with cobbles of different sizes.
A sandy layer probably where the river once flowed.
High water table.

C Horizon
Heavy clay and rocks. Too difficult to dig lower.

FIGURE 36.8 *Field Study Soil Profile*

Soil Classification

Devising a standard soil classification system is difficult because of the many variables that affect soil development. As a result, there are many systems in the world today. You saw in the field study that even in a relatively small area there is considerable variation in soil types. To map soil types on a global scale is a great generalization. Nevertheless, it is justified so that we can study patterns and relationships. It is important to remember that in any one general soil type there are many differences in water, porosity, permeability, fertility, texture, and so on.

The study of soils, or **pedology**, is a fairly recent discipline. Early attempts at soil classification were based on the descriptive work of the Russian soil scientist V.V. Dukuchaev in the late 1800s. In the past thirty years, the American Soil Conservation Service has developed a detailed classification system. Since its inception, the system has been refined and revised seven times until the present model was published in 1975. This system is based not on generalized descriptions but on actual soil analysis in the field. Figure 36.9 details soil characteristics under the Soil Conservation Service system.

Order	Derivation of Prefix	World Land Area [%]	Description
Oxisols	oxide (compound of oxygen)	9.2%	• red, yellow, and yellow-brown A horizon • heavily leached with iron/aluminum accumulations in B horizon • low mineral and humus content but soaked with water • thin A horizon but deep C horizon – ancient soils
Aridisols	arid (dry)	19.2%	• A horizon thin – little humus due to lack of vegetation • salt and calcium concentrations on surface and in B horizon • movement of moisture is towards surface • parent material coarse-grained sands and larger particles
Mollisols	mollis (soft)	9.0%	• black or dark brown A horizon • deep and fertile with much humus from accumulations of grass • well-structured horizons – soft and easily ploughed
Vertisols	inverted	2.1%	• clays expand >30% in rainy season • black, grey-brown A horizon – fertile but heavy clay • huge cracks develop in dry season and humus falls through to B horizon, so the soil seems inverted

FIGURE 36.9 *Soil Taxonomy*

Order	Derivation of Prefix	World Land Area [%]	Description
Ultisols	ultimate	8.5%	• red-yellow, yellow A horizon due to leaching • more fertile than oxisols due to seasonal rainfall • heavy clay in B horizon – tropical forest cover
Alfisols	alfalfa (hay crop)	14.7%	• grey or grey-brown slightly acid A horizon • moderately leached but fertile due to forest cover • high clay content
Spodosols	ash wood (Latin)	5.4%	• highly leached, acidic A horizon due to needle leaf litter • sandy parent material – thin soils • low in fertility – ash-grey colour
Inceptisols	beginning (Latin)	15.8%	• weakly developed, early soils • little development of horizons • infertile, little weathering • shallow – found on new land (lava flows)
Entisols	recent	12.5%	• sandy, porous soils • tidal mud flats, sandy deserts, marshland, alluvium • poor structure, infertile • undeveloped horizons
Histosols	tissue (Greek)	0.8%	• muck soils with heavy humus and clay • no horizons

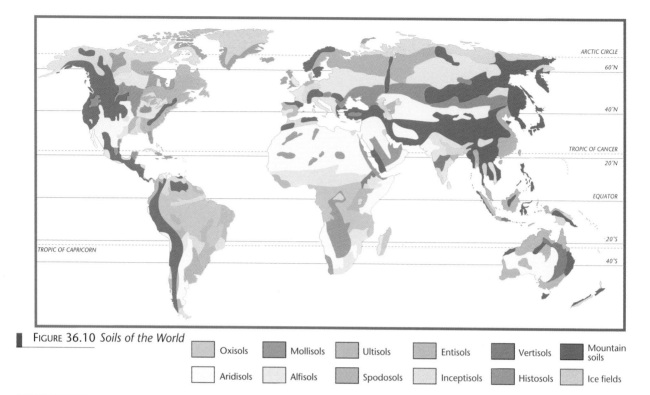

FIGURE 36.10 *Soils of the World*

Oxisols | Mollisols | Ultisols | Entisols | Vertisols | Mountain soils
Aridisols | Alfisols | Spodosols | Inceptisols | Histosols | Ice fields

Things To Do

1. Using Figure 36.9 determine the soil type in each of the sites in the field study.

2. a) Prepare an organizer to show the natural vegetation and climate zones for each of the ten soil orders.
   b) What other variables determine the location of the soil types identified in b)?

3. Suggest soil types that would support the following land uses and explain your choices: a) lumbering; b) grain farming; c) pastoralism; d) vegetable farming.

TECHNOLOGY UPDATE: THE UNITED NATIONS SOIL SURVEY

A United Nations team of over 200 analysts conducted a survey of world soil resources. Remote sensing and sophisticated computer analyses as well as extensive field studies have produced the following alarming statistics:

- 13 billion hectares—about 11 per cent of the earth's surface— is arable. The rest is unsuitable for agriculture.
- Over the last fifty years, more than 7 billion hectares of agricultural land (an area equal to the size of India and China combined) has been damaged by human activities.
- Each year the world's farmers have to feed 92 million more people with 24 billion tonnes less topsoil.
- Since the Second World War, 54 million hectares of the world's land has lost its ability to support vegetation.
- Over 5.7 billion hectares of land require costly reclamation programs. Most of the countries involved cannot afford the cost.
- In the United States, 25 per cent of cropland is eroding faster than it can be preserved.
- Livestock over-grazing accounts for 35 per cent of the soil degradation worldwide. Soil is exposed to erosive forces once the vegetative cover is removed. It is compacted and moisture content reduced. This is most common in Australia and Africa. However, corrective measures have lessened the problem in Australia.
- Deforestation accounts for 30 per cent of soil damage. This occurs most often in South America.
- Poor agricultural practices have caused 28 per cent of soil degradation worldwide. Much of this involves irrigation practices in desert regions like southern California. As the water evaporates, a residue of alkali builds up. This problem is most prevalent in North America.

Things To Do

1. Prepare graphs showing some of the statistics in the United Nations study. Suggestions include:
   - a circle graph showing the leading causes of soil degradation
   - line graphs projecting population growth and soil degradation for the next 50, 100, and 1000 years
   - a circle graph showing the area of farmland lost over the past fifty years compared to the total amount of farmland on earth
   - a circle graph showing land that can no longer support vegetation.

2. Study Figure 36.11 on page 286. Select one of these areas of significant soil erosion. Research to find out more about this problem, including its causes, the current situation, and what actions are being taken, or could be taken, to alleviate the problem.

3. Find examples in your community of poor soil conservation. Discuss what can be done to correct the situation.

286 PART 6 THE ECOSPHERE: ALL THINGS LIVING UNDER THE SUN

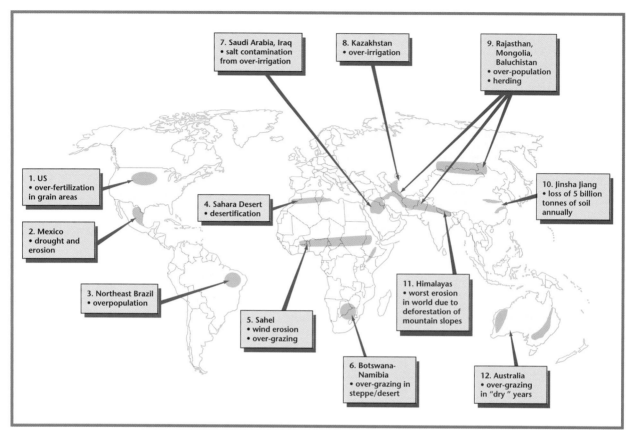

FIGURE 36.11 *Areas of Significant Soil Erosion*

FIGURE 36.12 *The Australian Outback*
Use Figure 36.9 to identify the soil shown in this photograph.

Part 7

People and Their Ecosystems:
Integrative Studies

Chapter 37

You, Your Lifestyle, and Nature

INTRODUCTION

People look to the ecosphere to provide their basic physical needs—food, shelter, water, air, fuel, and space. When there were few people on the planet, it was not hard for the ecosphere to provide these needs. The damage people inflicted was minimal. But times have changed! With the population of the world approaching 6 billion, more **natural resources** are being utilized than ever before. There are limits to the number of people the earth can support. Is the earth's **carrying capacity** 10 billion people? 20 billion people? Or have we already reached the maximum number of people the earth can support and remain a healthy place to live?

Population growth is not the only pressure being placed on our planet. People are demanding an increasingly higher standard of living. What were once human wants are now human needs. We hear people say things like "We need a second car" or "I need a vacation." Fulfilling these needs puts pressure on the earth. A new car requires many natural resources to manufacture and operate it. Travelling involves cars, trains, boats, and planes, all of which consume natural resources, not to mention the pressure tourists place on delicate ecosystems. It is therefore important to keep our consumption in check. There is a limit to how much the planet can provide.

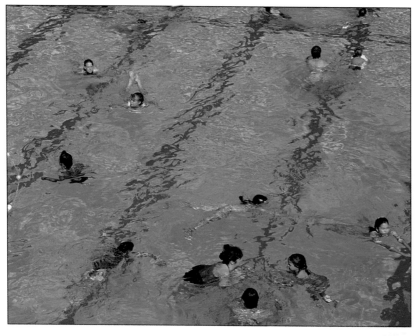

FIGURE 37.1 *Water Consumption*
North Americans use more water than any other people in the world. Is our rate of water consumption necessary? Explain your answer.

Sustainable Development

The key to preserving the planet is to use the earth's resources without permanently damaging the ecosphere. If we develop resources intelligently, we can continue to harvest the earth's renewable resources indefinitely. This process is called **sustainable development**.

Consider Canada's Atlantic fishery as a case in point. These ocean waters were once thick with cod. Fleets from many nations came to fish the Grand Banks. Today the cod fish stocks have been drastically reduced. Many people who live in the villages that dot the shores of Newfoundland have lost their livelihood. With sustainable management, it would have been possible to harvest some of the fish every year while still allowing the stocks to replenish themselves.

Many of the resources we use are **renewable**. Fish, trees, field crops, even milk and meat are all resources that regenerate themselves. Other resources are finite. Once we use them, they are gone forever. These **non-renewable resources** were formed millions of years ago and are in limited quantity. Natural gas, oil, gold, zinc, and many more minerals and metals are non-renewable resources. Only through recycling and careful management can these valuable commodities be sustained.

Some resources are so common they are found virtually everywhere. These permanent resources include air, space, and, in most ecosystems, water. While it is unlikely that we will run out of these essentials, we must be careful not to poison them. To do so would not only be harmful to the resources, it would have a devastating effect on the ecosphere and in turn on ourselves.

Of course, the most valuable of all resources is the human mind. If we use our ingenuity and technology, we have the power to sustain ourselves for generations to come.

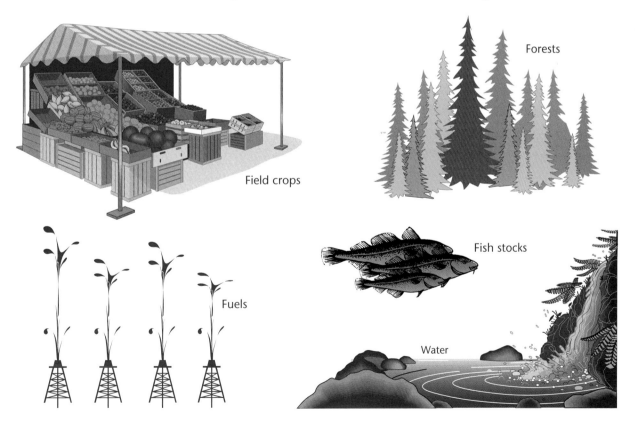

FIGURE 37.2 *Natural Resources*
Classify each natural resource as renewable, non-renewable, or permanent.

Things To Do

1. a) Write your own definition of a human *need* and a human *want*.
   b) Explain how a human want becomes a human need in today's society.

2. a) Prepare an organizer like the one shown here comparing the following needs of people in the past, present, and future: water, air, food, clothing, shelter, fuel, space. Include in your comparison a description, the source, and the cost for each need. The first one, water, has been done for you.
   b) Circle each negative item in the Future column.

   c) Develop a strategy whereby human needs in the future could all be positive.

3. a) Work with your group to prepare a decision-making organizer for one of the following lists of consumer habits. Include such criteria as cost, convenience, waste, type of resource, personal image, comfort, and reliability.
   - home heating using firewood, natural gas, oil, or solar energy
   - lunch in a disposable bag, a reusable lunch box, or from the cafeteria
   - using cloth diapers, disposable diapers, or a diaper service
   - travelling to school by bus, bicycle, car, car pool, or on foot
   - shaving with an electric shaver, a disposable razor, or a straight razor
   - eating out in a restaurant, cooking from scratch, ordering in, or microwaving a prepared meal.

   b) Report the findings of your group to the class. Decide whether or not each choice is good for the environment.
   c) In what ways could you change your lifestyle to make the world a better place?

Need	Past	Present	Future
Water			
Description:	fresh pure	treated chlorine taste	chemically produced
Source:	wells or rivers	taps and treatment plants	pure bottled
Cost:	free	cheap	expensive

Studying Different Ecosystems: An Independent Study

People live in a variety of different ecosystems. Each one provides for the needs of those who live in it in different ways. The ways in which people obtain the necessities of life determine how successful they will be and how healthy their ecosystem will remain. Some fragile environments have been so badly abused that people are no longer able to obtain life's necessities. Incredible suffering is the result.

Things To Do

In a group, select one of the following ecosystems: tropical rainforest (pages 292–301), boreal forest (pages 302–313), tropical grassland (pages 314–324), temperate grassland (pages 325–331), maritime (pages 332–337). Divide your investigation equally among the members of your group. Use the outline below to guide you through the assignment.

Present your information in an interesting way. You might consider:
- an oral presentation with slides and overheads
- a written hand-out with follow-up activities such as a board game, a word search, or a crossword puzzle
- a simulated debate among students role playing the inhabitants of the ecosystem
- a videotape documentary.

Outline

Climate
1. Show the location of your ecosystem on a world map.
2. Review the Köppen Climate Classification system on pages 108–113 to establish a statistical definition of your ecosystem.
3. Give a detailed description of the climate.
4. Select three or four locations in your ecosystem and prepare a climate graph for each one using climate statistics.
5. Conduct an inquiry into why the climate is the way it is.

Plants and Animals
1. Give a detailed description of the natural vegetation. Include the names of different plant species and how they have adapted to the climate.
2. Evaluate the climate to determine which plant needs are well met and which factors limit plant growth.
3. Give a detailed description of some of the animals native to the ecosystem. Include a food web and the animals' behavioural and physical adaptations.

Soils and Landforms
1. Give a detailed description of the different soil types. Explain how they are influenced by natural vegetation and climate.
2. Evaluate the soils to determine how they meet plant needs.
3. Describe the dominant landforms. Explain how they influence the development of the ecosystem.
4. Describe gradational processes and how they influence soil formation (see pages 274–277 for details.)

Indigenous Peoples
1. Describe how indigenous peoples adapted to their ecosystem.
2. Describe how indigenous peoples modified their ecosystem.
3. Evaluate the lifestyle of indigenous peoples to determine:
   a) the impact they had on the environment
   b) how they used the resource base
   c) the success of their way of life from an environmental perspective.

Resource Development
1. a) Describe the ways in which resource development has modified the environment.
   b) Determine the impact these modifications have had on the ecosystem and if these modifications are sustainable.
2. Recommend ways in which resources could be better managed to ensure they are sustained.
3. Create an action plan to correct environmental damage caused through mismanagement.

Chapter 38

Living in the Tropical Rainforest

INTRODUCTION

Tropical rainforests are the most complex ecosystems in the world. The multitude of plants and animals that inhabit these verdant regions are so numerous and diverse that new species are being found every year. These rich biological storehouses are found straddling the Equator in South America, Africa, and South East Asia. The only equatorial regions that do not have tropical rainforests are those areas of high elevation found in east Africa and along the west coast of South America. Where prevailing winds bring plentiful rain all year, the tropical rainforest can extend as far as the tropics of Cancer and Capricorn, usually along east coasts. No matter where tropical rainforests are found, they are a wonder of nature.

FIGURE 38.1 *The Tropical Rainforest*
Viewed from the air, the rainforest takes on a different appearance than it does at ground level.

Climate

As you have already discovered in Part 3, this tropical climate has warm temperatures every day of the year. Unlike the middle latitudes, one air mass dominates the weather all the time. Temperate air masses seldom move into equatorial lands. Tropical rainforests are very wet and experience convection thunderstorms almost daily. Nighttime temperatures are not much cooler than daytime ones. This is because the high humidity and extensive cloud cover prevent the heat that builds up during the day from escaping at night.

Prevailing winds are non-existent; this is the area of doldrums. The equatorial low is produced as air is heated and rises. The only place where local winds occur on a regular basis is along the coast. This is because the difference in temperature between the water and the land causes air to flow from land to sea at night and from sea to land during the day.

Plants and Animals

The plants and animals of the tropical rainforest are incredibly diverse. More species have been counted and catalogued here than in all other ecosystems combined. Moreover, each hectare of the rainforest is unique; no other place in the world is like it. Rainforest flora and fauna are often uniquely interdependent. It is possible to find animal species, such as certain types of tree frogs, that can survive in only one specific plant species!

There is tremendous vertical development of species in the tropical rainforest. Plants and animals exist only in certain levels of the ecosystem. Most land ecosystems do not vary with height. In this way the tropical rainforest ecosystem has been compared with marine ecosystems, since they, too, contain different plants and animals in each layer.

Plants have adapted to the unique conditions of the rainforest. The climate provides

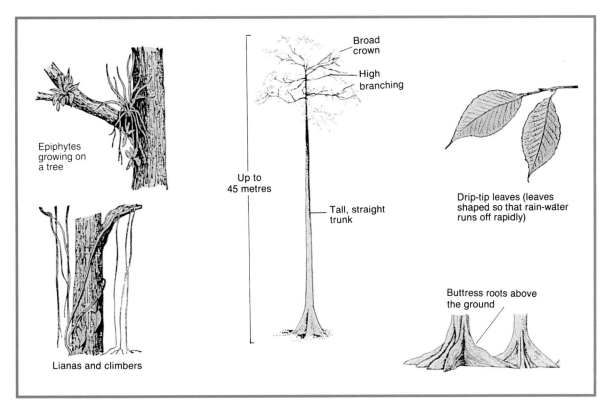

FIGURE 38.2 *Vegetation of the Equatorial Rainforest*
Much of the plant life in the rainforest has adapted by growing upwards towards the sunlight that seldom reaches the forest floor.

abundant heat and rain. Space and sunlight are limited, however, because of the tremendous competition among species. Many plants have adapted to these limiting factors by growing very tall. Forest giants are trees that grow above the canopy of the forest. Because rainforest soils are often marshy and water-logged, these giant trees have developed buttress roots to support them. These roots spread the trunk and root over a large area of the ground, providing support for the mass of the tree. This enables these trees to grow to great heights.

Some plants like lianas use other plants for support in their quest for sunlight. These vines wrap themselves around the host plant. They lack the trunk of larger plants but travel up the trunks of giant trees towards the light. **Epiphytes** such as orchids grow in the crooks of tree branches high above the forest floor where they are able to reach the sunlight. **Parasites** also live in the canopy. They sap the nutrients from the host plant while receiving life-giving light high above the forest floor.

Below the canopy, plants such as the elephant-ear split-leaf philodendron have adapted by growing huge leaves that catch whatever sunlight filters through the gloom.

The leaves also allow plants to transpire the large quantities of water that are needed to bring nutrients from the soil to the plant cells. Whatever their level in the rainforest, these tropical plants are uniquely suited to this diverse ecosystem.

Contrary to what many people believe, the floor of the rainforest is almost free of plant material. The lack of sunlight is such a limiting factor that dense underbrush exists only around a clearing or along a river where sunlight can penetrate the gloom.

There are several myths about rainforest animals. For one thing, the number of animals per

FIGURE 38.3 *Layers of the Rainforest*

hectare is lower than for more temperate forests, especially on or near the ground. Jungle rivers do not teem with fish and other animals. The lack of nutrients in the water and near the ground makes life difficult for most jungle dwellers. These creatures are also relatively small. The largest animal in the Amazon is the tapir, a relative of the pig family. It has a mass of 100 to 300 kg. In Africa, forest elephants and pygmy hippos, smaller than their grassland relatives, inhabit the fringes of the jungle. Otherwise small animals are the general rule. This adaptation facilitates feeding on the meagre plants available at ground level.

Animals are also adapted to the rainforest

in other ways. They tend to be **arboreal**—that is, they live in or among trees. Because of this they have many adaptations that help them get from place to place. Birds and insects fly. Mammals, reptiles, and amphibians climb. Camouflage enables many of the smaller animals to protect themselves by looking just like the surrounding plants. This is particularly true of insects and reptiles that cling to the tree branches or roam the forest floor.

Soils and Landforms

The lush vegetation of the tropical rainforest leads many people to think that the soils are very fertile. In fact, the opposite is true. The A horizon offers few nutrients; most are contained close to the bedrock. To compensate, plants have adapted. They have developed incredible root systems that travel as deep as 5 m into the soil. There they enter the active layer of bedrock and obtain the nutrients that are being released as the bedrock decomposes. The fruits of many rainforest plants are large and robust. This allows them to store enough food energy to send a root many metres into the ground. This is how plants flourish in the tropical rainforest despite the nutrient-poor soils.

Oxisols have developed in old landscapes where the parent material has been stable for hundreds of thousands of years. High temperatures and moist conditions have greatly weathered the bedrock, creating soils that may be many metres deep. These soils are heavily leached. Their red or yellow colour comes from the iron and aluminum oxides that are left behind when other nutrients are removed in solution. There is often a hardpan layer deep in the soil where iron and aluminum deposits are concentrated. There is usually little humus in the jungle. Once a plant or animal produces organic matter, it is quickly broken down by the many decomposers. The nutrients that are released are quickly reabsorbed by the surrounding plants.

The soils of the tropical rainforest are mainly clay. Once they dry out, they can be broken down easily. While the soil can be tilled, the lack of nutrients in the upper horizon is one of the main problems with

FIGURE 38.4 *Rice Terraces in the Tropical Rainforest The rich volcanic soils of Indonesia allow extensive farming despite leaching.*

agricultural development in the region. Local variations in soil fertility are caused by differences in the underlying bedrock. If the bedrock is composed primarily of silicates like quartzite or sandstone, the soil is infertile and plant growth is limited. On the other hand, limestones and basaltic bedrock provide the soil with many important nutrients.

Young fertile soils are found wherever there are volcanoes. Volcanic ash breaks down quickly, leaving an abundance of nutrients readily available for plants. In Java, an island in Indonesia, the soil was formed from volcanic ash. It is so rich in nutrients that leaching has not reduced the fertility to the same extent as it has with other soils of the tropical rainforest. The fertile soil enables this island to support the densest rural population of any place in the world.

Most tropical rainforests are found in great river basins. The best known are the watersheds of the Zaïre and the Amazon rivers. These ancient river systems have developed extensive but poorly drained flood plains. Much of the water lies stagnant in great marshes and bogs.

Indigenous Peoples

Indigenous peoples have lived in the rainforest for millennia. Throughout this time their economic systems were sustainable and unchanging. They were a part of the ecosystem. Whether it be the Yanomami of the Amazon, the Bambuti of central Africa, or the Gimi of New Guinea, all lived in harmony with the rainforest. It is only in recent years that the way in which they live has changed.

Traditionally, people of the rainforest took only what they needed for survival. The extreme climate, tropical diseases, and the shortage of natural resources kept aboriginal populations low. Because their populations were small they had little impact on the environment.

The people have adapted to the environment in many ways. The hot, humid conditions make clothing uncomfortable, but biting insects are a common problem. The usual response is to paint the body using organic dyes that act as insect repellents. This leaves the skin exposed to the air but prevents the bugs from biting. Housing is made of materials readily available in the rainforest. Walls are often open to allow ventilation. Steep-sloping roofs shed the heavy rainfall. Many dwellings are built off the ground on stilts, a good feature if you live in a wet environment where flooding can be a problem. Food is not abundant in the jungle. There are few animals to hunt and most plants are not edible. Life is difficult for those people who still rely on hunting and gathering.

Slash and burn agriculture is practised by most people of the tropical rainforest. Foods such as corn and manioc can be grown close to home. This is much easier than having to search through the forest for food staples. Although agriculture does not usually succeed in the rainforest, this type of shifting cultivation is sustainable. A small portion of the jungle is cleared, usually when rainfall is at its lowest. Trees are "girdled"—that is, the bark is cut around the whole trunk. The tree soon dies. Then it is set on fire. With the underbrush and foliage gone, the sun can reach the forest floor. Crops are planted in the ashes. Although the soil is infertile, the ashes provide enough nitrates for two to three years. Once the land becomes infertile, it is abandoned and a new site is selected. The plants surrounding the clearing soon take over the land again and the forest is restored. Many of the people are nomadic. They move to a new location once all of the good land has been exhausted.

Death by disease and injury is common in the tropical rainforest. The average life span is only thirty to forty years. In the past, people had little contact with the outside world, so literacy was very low. There are no modern conveniences, although incongruous pictures of aboriginal people wearing wristwatches or listening to portable radios indicate that industrialized society is encroaching everywhere.

Today the number of indigenous people living in tropical rainforests is declining. Many have died from diseases introduced when outsiders entered their land. Others have been assimilated into mainstream society where they work as labourers in the towns and villages that line the rivers and new highways crossing these vast areas.

FIGURE 38.5 *The Yanomami Indians of the Brazilian Amazon*

YEAR 1 — The jungle is cleared, resulting in ash and stumps.

YEARS 2–4 —"Crops" are grown in the clearing.

YEARS 5–7 —The plot is not used for planting any more because the soil has lost most of its fertility. Small trees start to grow.

YEAR 8 — The jungle grows back over the plot.

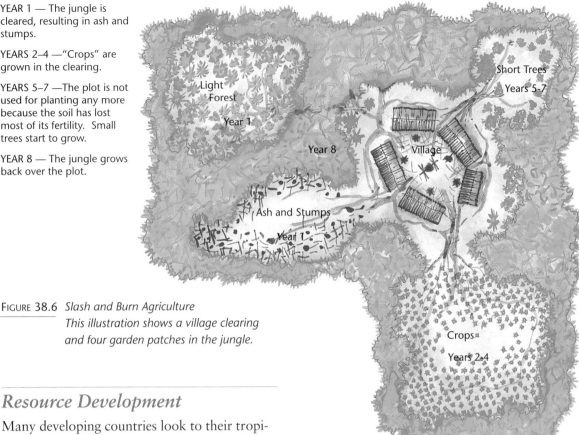

FIGURE 38.6 *Slash and Burn Agriculture*
This illustration shows a village clearing and four garden patches in the jungle.

Resource Development

Many developing countries look to their tropical rainforests to improve their standards of living. Brazil is one such country. Much of its huge and rapidly growing population is from the economically depressed northeast. These *nordestinos* lack formal education and live in relative poverty. With limited opportunities in their home state, they migrate to wherever they can earn a living. In addition, Brazil has a massive foreign debt. The interest payments alone are so staggering that repayment of the principal is impossible. To service its debt and to improve the quality of life for its citizens, Brazil needs to make the most of its natural resources, including those found in its rainforests.

In Brazil, there are five ways in which the government seeks to develop the Amazon: (1) subsistence farming; (2) lumbering; (3) industrial development; (4) cattle ranching; and (5) extractive reserves.

Subsistence Farming

Brazil is a country with many landless peasants. These are people who support themselves either by working in the plantations of wealthy landowners or by **sharecropping**. Sharecroppers rent land from the landowners and pay their rent with some of the proceeds from their harvest. This system has been practised in many countries, including the United States after the Civil War, East Africa, the Indian subcontinent, and some Latin American countries. Because the farmers are very poor and the land is not theirs, sustainable farming methods are seldom practised. The soil is overworked, fertility drops, and crop yields decline. Eventually the land is unable to support the peasant family. The farmers must then move on to new regions where they can rent more fertile land. As the land runs out, they often move to the cities of the south in search of a better life.

As one solution to unemployment, the government introduced a plan to give the peasants

FIGURE 38.7 *A Favela in Brazil*
To improve the quality of life, Brazil needs to exploit the wealth of its natural resources.

their own farmland. The plan involved opening up large tracts of the Amazon rainforest for agricultural settlement. Roads were built deep into the western Amazon. People moved into this wilderness with the goal of transforming it into productive farmland. Theoretically, it appeared to be a great idea. After all, it worked in North America when European settlers transformed the prairies into the world's most productive grain-growing region.

The problem with the Brazilian resettlement program was that the developers did not fully understand the complex nature of the rainforest. As we have discovered, the soils are not fertile. Once the trees are cut down and the underbrush is burned, the soil is exposed to leaching and erosion. For the first year or two of farming, adequate yields can be harvested.

By the fourth or fifth year, the land becomes so infertile that nothing will grow. The settlers move further up the highway, clearing new land and beginning the cycle all over again. Today few of these subsistence farm communities remain in the western Amazon. Most of the people have returned to the cities of the southeast.

Contrary to what many people believe, the clearing of the rainforest has not destroyed it forever. Eyewitnesses returning from the region report that the jungle is quickly reclaiming land cleared in the early 1980s. So while the resettlement program was not successful, it appears that the Amazon is more resilient than we had thought.

Lumbering

The rainforests have always been considered an inexhaustible source of wood. Many of the trees possess unique qualities that make them much in demand. Teak, mahogany, rosewood, and ebony are easy to work and yet are durable and attractive.

Logging is common in all the world's rainforests, especially in areas that are accessible either by water or road. Harvesting the trees of the Amazon is much more difficult than logging in Canada. Marketable species are usually spread throughout the jungle. It is not possible to find a whole stand of ebony in one spot, as white pine might be found in a Canadian forest. In fact, there may be several kilometres of jungle between each ebony tree!

On average only about 5 per cent of the trees in the rainforest have any logging value. In the past, **selective cutting** was practised. Only the commercially valuable trees were cut down. To get to the trees, however, much of the other vegetation was destroyed, leaving the soil exposed to the physical elements. Methods are less wasteful today, although much of the forest is still disturbed. Modern equipment has made it possible to utilize the non-commercial wood. Machines grind the wood into chips, which are then made into pulp for use in the manufacture of paper.

With new roads into the Amazon, forests that were once inaccessible are now being logged. These logging operations often operate in conjunction with other land uses, such as mining and farming.

Logging in the tropical rainforest was often an extractive operation in the past. Reforestation rarely happened. An attempt at reforestation was made in the eastern Amazon in the mid-1960s. Along the Jari River, a tributary of the Amazon, a single species was planted over an area the size of Prince Edward Island. But diseases were able to destroy the entire plantation.

Introducing a monoculture in this genetically diverse ecosystem was doomed to failure. Plants of the same species have a better chance of survival if they are spread out. This prevents pests that live off of one particular species from moving from plant to plant. Today, however, the Jari Project has been transformed into a great success. Scientific and technical knowledge have improved so that sustained yields of reforested pulpwood are being produced year after year. Other advances have reduced the susceptibility of monocultures to infestations. Rows of native forest at least 400 m wide are planted between plantations. In addition, new blight-resistant eucalyptus trees have been introduced. Selective breeding and cloning have produced trees that are hardy and can be harvested every six years. Spraying has become a thing of the past. In addition, eight genetic reserves totalling 20 000 ha shelter 530 different types of trees. Ecologically sensitive areas on steep slopes, around springs, and along rivers are also protected.

The Jari Project has been an economic success as well. A pulp mill using ecologically friendly processes manufactures over 1000 t of pulp each day. The region now has a population of 60 000 people, most of whom are directly or indirectly employed in the pulp industry. There is a modern hospital, four schools, and recreational facilities. Jari is a model for lumbering operations in Brazil and around the world.

FIGURE 38.8 *A Section of the Jari Project*
Trace this oblique air photo. Colour the modified area one colour and the natural area another colour.

Industrial Development

Soil and trees are not the only resources to be found in the Amazon. The region contains vast mineral deposits. Precious metals like gold, gems such as diamonds, and industrial metals like iron, manganese, and bauxite are all found within the Amazon watershed. In addition, the region contains enormous amounts of fresh water. If harvested, the rivers could provide hydroelectric energy for the growing cities of southern Brazil, Paraguay, and Argentina.

One massive development scheme in Brazil is the Grande Carajás Project. This open-pit mine produces over 50 million tonnes of iron ore each year from the world's largest iron ore deposit, estimated in excess of 17 billion tonnes. Financed by the World Bank and US Steel, the development has created controversy. On the one side, there is condemnation of the environmental damage. Natural vegetation had to be cleared to get to the mine area. A train connecting the mine to port facilities 1000 km away disrupted the jungle ecosystem and disturbed its 4500 inhabitants, many of whom had never been in contact with the outside world. A hydroelectric plant was built on the Tucuruí River to provide power. The dam created a reservoir over 2000 km in area. As the vegetation rotted in the water, the oxygen was removed and a huge dead lake replaced the rainforest. (Today the lake has recovered and it is used for commercial shrimp fishing.) Trees were cut down and burned to produce charcoal needed for smelting the ore. Those who oppose the project argue that the damage to the ecosystem is not worth the profits made by the developers.

On the other hand, some people argue that the claims about environmental destruction have been exaggerated. They claim charcoal smelting is carried out on a small scale by local people and does not have a major impact on the ecosystem. The lake formed behind Tucuruí Dam has created a new aquatic ecosystem that benefits local people. Perhaps more importantly, they argue, the development has created jobs and opportunities in an economically depressed region, just as the Jari Project did.

Cattle Ranching

After a region has given up its trees to loggers and its minerals to miners, often all that the land is good for is grazing. As with other agricultural activities, the amount of land required to sustain an animal increases each year as the soil becomes increasingly less fertile. At first 1 ha of land per animal is needed. After five years, this figure may increase to as much as 5 ha. At this stage, the land is completely exhausted. The forest cannot regenerate itself because the cattle eat anything that grows before it has a chance to reach maturity.

In spite of the environmental consequences, cattle ranching continues in the Amazon. Most of the beef these cattle produce is destined for the export market. Much of it finds its way to North America where it is used to make hamburgers in some fast-food chains. Patrons concerned about protecting the rainforest should investigate where the meat used by their favourite hamburger chain comes from.

Extractive Reserves

The extractive reserve is the only modern development strategy that seeks to preserve the rainforest in its entirety while still utilizing its valuable resources. The idea is not a new one. A hundred years ago, rubber tappers made small fortunes collecting latex from the rubber trees that are indigenous to the Amazon. They collected a little of the precious sap each day, much like Canadians collect maple sap. Because the tree is not killed, it is possible to collect the latex indefinitely.

Rubber trees are now grown in many tropical areas. In addition, synthetic rubber made from oil has replaced Brazilian rubber in the

FIGURE 38.9 *A Rubber Tapper in the Brazilian Rainforest*

manufacture of tires. Although there is still a market for latex, Brazil no longer monopolizes the industry. To compensate for the reduced demand for latex, Amazonian rubber tappers expanded their activities. They now collect other valuable resources such as fruits and nuts. Originally these were for local consumption only. Today Brazil nuts, mangos, and other rainforest products are sold in the villages to trades people who export them to the rest of the world. Along the forest trails, bananas, cassava, taro, and other tropical vegetation have been planted. The fruit can be harvested along with wild forest products. Because the forest cover is not disturbed, this type of extensive horticulture does not deplete soil the way intensive agriculture does.

The extractive reserve is the legacy of the rubber tapper Chico Mendes. He became a martyr of the cause after he was murdered by cattle ranchers in 1988. After his death, the Brazilian government gave its support to extractive reserves. Up until then, it had only supported large-scale projects like Jari and Grande Carajás. By the mid-1990s, four such extractive reserves had been established in the Amazon. The largest is 6 million hectares. No mining, forestry, or other development is permitted. These areas are reserved strictly for gathering forest products and small-scale horticulture. Supported by the World Bank, deforested parts of the rainforest are recovering with the hope that in time they can also be developed into extractive reserves.

There are some disadvantages to extractive reserves, however. The financial rewards are limited; the rate of return is much lower than it is for other development projects. Farming, ranching, and in particular mining all offer opportunities to make much more money much faster. (Of course, in the long run extractive reserves will generate revenue long after the mines are closed and the farms are abandoned.) Another disadvantage is that because the forest products are so spread out, there is a limit to the number of forest collectors that can be supported. For these reasons, this development strategy will not help to solve the problem of resettling the millions of Brazil's landless peasants.

Things To Do

1. Prepare an organizer summarizing the characteristics of the tropical rainforest under the following headings: location, climate, plants and animals, soils and landforms, indigenous peoples.

2. Identify some of the myths or inaccuracies that people hold about the tropical rainforest and determine why these myths developed.

3. Many impassioned articles have been written about the destruction of the rainforest. Find four to six such articles. Highlight the factual statements that can be substantiated and circle unsubstantiated opinions. Determine which article is most accurate by comparing the number of facts to the number of opinions.

4. Prepare a decision-making organizer to determine the land use most suitable for Brazil's rainforest. Make sure criteria reflect a balance between environmental concerns and economic considerations.

5. Research another rainforest area to determine the success of development strategies there.

Chapter 39

Living in the Boreal Forest

INTRODUCTION

In the northern latitudes of North America and Eurasia lies a continuous belt of coniferous forest. Like the tropical rainforest, this is a vast, sparsely settled region with an abundance of resources. Unlike equatorial regions, however, these northern forests are relatively young and are still evolving. Until 10 000 to 15 000 years ago, most of the land was covered with glaciers. Therefore these ecosystems have had less time to develop than those of the tropical rainforest. Fewer species are found here and these forests are less fragile than rainforests. Nevertheless, careful resource management is needed if these ecosystems are to maintain their present healthy state.

FIGURE 39.1 *The Boreal Forest*
Canada's boreal forest is a vast, unbroken belt of forest and lakes across the northern regions.

Climate

As you have already discovered in Part 3, the boreal forest has a temperate climate, with warm summers but cold, often bitter, winters. Precipitation is adequate and may fall year round or seasonally, depending on location. The only large expanse of boreal forest that experiences winter drought is in eastern Siberia. Even where there is no dry season, winter precipitation is often limited because the cold weather reduces evapotranspiration and air masses are more stable when it is cold. The region has adequate precipitation and warmth for coniferous trees.

Great coniferous forests also grow on the west coasts of Canada, the United States, Chile, and New Zealand. While strictly speaking these are not boreal forests, the natural vegetation is similar. These west coast forests grow in a climate moderated by oceans. They have cooler summers, milder winters, and greater annual precipitation than the boreal forest.

The growing season in much of the boreal forest is relatively short, lasting only two to four months. During the summer, days are long and temperatures are remarkably warm for such high latitudes. When winter comes, precipitation is stored as snow and ice. Accumulations are often so great that the ground is insulated from the cold and does not freeze to the great depths it does where there is less snow. In spring, the concentration of water in the soil increases when the snow and ice melt. This surge of moisture helps seeds to germinate and gives spring growth a revitalizing boost. Regular precipitation is important during the growing season because the soils are too thin to hold much groundwater. Precipitation comes frequently during mid-latitude cyclonic storms that move into the region regularly. These climatic conditions are excellent for growing coniferous trees.

Plants and Animals

Unlike the tropical rainforest, there are few plant species in the boreal forest. While some deciduous bushes and trees like birch and aspen grow in sheltered areas, most species are coniferous, with pine, larch, spruce, cedar, and fir trees dominating. Almost all trees share similar characteristics, such as needle-shaped leaves and seed cones. With the exception of larch, they all remain green in winter.

Coniferous trees are well adapted to this climate. Adequate sunlight, heat, and precipitation are provided for part of the year. In winter, heat and sunlight are lacking and precipitation is trapped in its frozen form. To compensate, moisture is stored in the tree trunks and roots for use in the winter. The thin, needle-like leaves limit the amount of moisture that is lost through evapotranspiration. These adaptations allow conifers to grow slowly during the winter months.

The roots of black spruce are often shallow and extend over a large area to support the tree, especially in the shallow, poorly developed soils that are often found here. Conifers have a distinctive conical shape. This allows them to shed heavy loads of snow. Otherwise the limbs could snap under the snow's mass. The coniferous tree is well adapted to this zone.

Animals have also adapted to the severe winters. Some animals, such as bears and groundhogs, hibernate in winter. In this sleep-like state, metabolism and heart rates slow to incredibly low levels. This physical adaptation enables these animals to live off the body fat they stored during the summer. The hoards of annoying insects that abound in summer lie dormant in winter. Some animals, such as wolves, foxes, and lynxes, have heavy fur coats that protect them from the cold and allow them to remain active all year. The many birds that inhabit the wetlands of the boreal forest in the summer migrate south when winter comes. Some hoofed herbivores, such as elk, deer, moose, and caribou, also migrate with the seasons. While the number of animal species is less than in tropical rainforests, wildlife is abundant and varied.

Soils and Landforms

During the last ice age, the original soil was greatly disturbed. As the glaciers flowed south, bare rock was left in many places while hollows were created in others. As the glaciers

304 PART 7 PEOPLE AND THEIR ECOSYSTEMS: INTEGRATIVE STUDIES

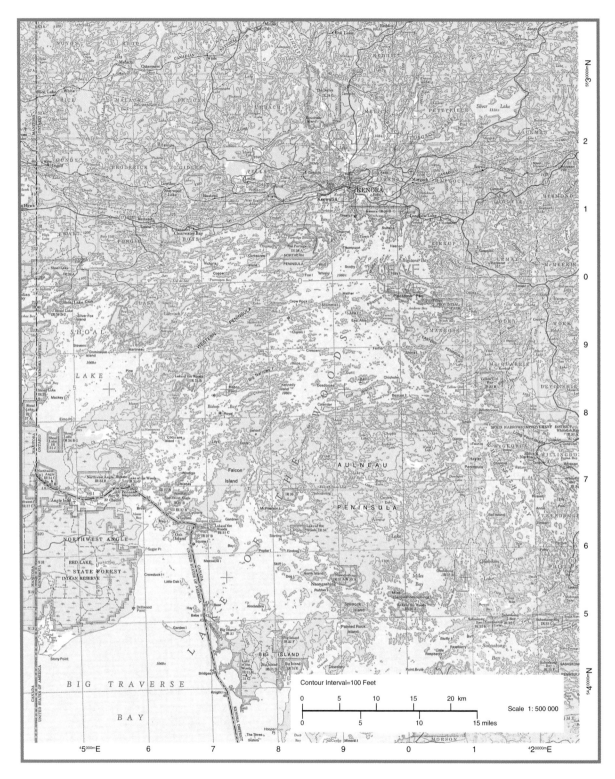

FIGURE 39.2 *Lake of the Woods in the Canadian Boreal Forest*
 a) *In what ways is this a good example of the Canadian Shield?*
 b) *How is the topography especially significant for forestry, hunting and fishing, and tourism?*

retreated, the hollows were filled with water. Over the years a succession of tundra, marsh plants, and eventually trees grew in the thin soils. Gradually, many of the hollows are being transformed into swamps and peat bogs as aquatic plants die and other organic matter finds its way into the water. The histosols that develop here are generally more fertile than the spodosols that dominate the region. As a result, hardwoods and other vegetation anomalies grow in these low-lying features.

Hollows that are still filled with water form the many lakes of the boreal forest. On the rock that lies between the lakes, soils are gradually developing. Many of these are so thin it is a wonder they can sustain vegetation. Yet trees of remarkable height grab onto the rock with their roots, drawing whatever nutrients they can from the thin soil. The low temperatures reduce the amount of chemical weathering. Instead, bedrock is broken up by mechanical weathering such as frost shattering.

Spodosols are infertile soils. They are highly acidic, mainly because of the needles that make up most of the forest litter. The acid removes many essential nutrients and clays from the A horizon, leaving an ash-grey sandy soil. If they are treated with limestone to reduce acidity and potash to increase fertility, these soils can support agriculture. However, the climate and the lack of soil restrict agricultural possibilities.

Drainage systems in the boreal forest are poorly developed. The land has not had time for these systems to develop since the last ice age. Great rivers such as the Lena, Ob', and Volga in Russia and the Mackenzie and Ottawa in Canada drain these immense wilderness areas.

Indigenous Peoples

Like the inhabitants of the tropical rainforest, the indigenous peoples of the boreal forest felt a spiritual oneness with nature. This formed the basis of their relationship with the environment and led to the development of a sustainable lifestyle. Traditionally they were hunters and gatherers. Animals provided meat as well as warm furs and supple skins to make clothing or to offer in trade. Their diet was supplemented with crops grown in forest clearings. Tools and weapons were created out of bone, wood, and rock. The main means of travel was by canoe. These were often made of birch bark, which was lightweight and enabled the canoes to be carried easily from one lake to another. Buildings were simple lodges made from the forest trees. Traditionally the people of the boreal forest used the resources available to them well and their activities caused little disruption to the environment.

Resource Development

The boreal forest is a vast storehouse of natural resources. It provides many of the raw materials needed in the industrial world. But like the rainforest, the northern woods need to be protected from exploitation. The fragile soils, the abundant wildlife, and the valuable forests must be managed in a sustainable way. Canada contains vast expanses of boreal forest. We have long been developing the natural resources this region offers. The most important of these are forestry, mining, and hydro-electric development.

Forestry

The forest industry in Canada is more than just cutting down trees. It includes related activities such as forest administration, planning, harvesting, protection, and regeneration, as well as the manufacture of lumber and wood products.

Coniferous trees have long been cut down for commercial use. The wood is used in construction and the pulp and paper industry. Pulp is made from smaller trees and the wood chips that are by-products of lumber. Trees are chopped up and made into a slurry or pulp, the raw material for paper. Lumber is made from larger trees so that planks, boards, timbers, posts, and beams can be cut from them. The trees of British Columbia are the largest in Canada, so they are often cut into stock suitable for construction. In Newfoundland and Labrador, the trees are much smaller, so pulp and paper is the main activity, as it is in Nova Scotia and New Brunswick. In central Canada,

Hand planting ensures that a new crop will be available in thirty to eighty years.

Transporting trees out of remote areas is a difficult task.

New harvesting machines have simplified lumbering.

Lumber is processed at a planing mill.

Pulp mills process forest products into paper products.

FIGURE 39.3 *The Many Aspects of Forestry*

there are both lumbering and pulp and paper industries.

Proper forest management implies that the resource is utilized in a sustainable way. This has not always been the case. Before Confederation, lumber from Nova Scotia and New Brunswick was used to build the ships of the British navy. The huge trees of the untouched maritime forests provided wide planks and long poles for masts. The forests seemed endless, so forestry practices were wasteful. Little thought was given to reforestation. Later these unsustainable forestry practices expanded into Quebec and Ontario. The really big trees were cut down to build the cities of an emerging nation before the twentieth century. Today there are few ancient stands of trees left in Canada.

Like all living things, trees have a lifespan. As they grow old, they are more susceptible to disease, insect infestation, and fire. Thus harvesting old growth is not necessarily bad. The important thing is that new forests are able to replace the old trees, either through natural regeneration or reforestation.

Frequently, there is tension between forest workers and preservationists in parts of northern Ontario and British Columbia. Foresters in both regions want to utilize ancient forest stands. They argue that they should be allowed to cut down the trees to boost local economies. Furthermore, they claim to have the legal right to harvest the trees because they have leased the land from provincial governments. Finally, they argue that old growth stands need to be replaced with new trees. The preservationists disagree. They argue that these old growth areas are living national treasures and as such should remain untouched. This kind of conflict is common in resource management industries.

A great deal of Canada's northern forests are, in fact, secondary growth. Trees grew back in areas that were harvested. In time, landscapes look just as they did when the trees were first removed. After years of exploiting the forests, forestry companies now use sustainable methods. Forestry management decisions are made on an individual basis depending on the forest site. No one method is used on all stands.

There are two forest management systems in Canada today: **clear cutting** and **selective cutting**. With the clear-cutting system, all the trees are harvested and the area is then regenerated. This technique can be used successfully in areas where stands of timber contain trees of the same age and size. This is frequently the case in the boreal forest, so clear cutting is often used here.

Clear cutting is much like farming. The only difference is that the growing season is much longer. In farming, it may take three to four months from the time the seeds are sown until the crop is harvested. Trees take much longer to reach maturity—from thirty to eighty years.

To regenerate land after clear cutting, the brush and logging debris are burned off. Then tree seedlings are planted or seeds are sown. In some cases there are enough seeds or seedlings left on the site after harvest to create a new stand of trees naturally. Unwanted competing vegetation must be removed, either manually or with herbicides. This allows the seedlings a greater chance of survival.

The clear-cutting system is often misunderstood. Once an area has been clear cut, the landscape is unsightly compared with the unspoiled wilderness. Many people believe that the ecosystem has been destroyed forever. While it is true that poor forest management can have a devastating impact, especially on wildlife habitats, natural ecosystems usually recover in a relatively short time. Of course, there are exceptions. Forest lots that have steep slopes and heavy precipitation are especially vulnerable to erosion. If the soil is stripped away, the ecosystem may suffer irreparable damage.

Canadian foresters make every effort to restore the forests. To ensure that this is the case, the provinces award woodlot leases on crown land on the condition that sustainable forest management is practised. It makes good business sense, too. If forest companies didn't act responsibly, they would eventually be out of business.

Wildlife is often well adapted to periodic disruptions to their environment. When forests are cleared, they migrate to other areas. Sometimes they even benefit from the process. In Ontario's Algonquin Park, for example, the deer population has risen since the area was first harvested. The new forest growth is preferred food for these browsers. Under proper forest management the boreal forest is able to regenerate and produce wood resources indefinitely. From a resource development perspective, forest management improves the value of future harvests by ensuring that healthy forests continue to be available.

Strip cutting is a form of clear cutting. Strips or blocks 20 to 100 m wide are clear cut through the forest. Equivalent areas of uncut forest are left between the cut areas. This reduces the disruption to wildlife. Wind and water erosion are minimized because of the undisturbed forest. Even replanting may be unnecessary. Seeds from the trees in the remaining forest find their way into the cleared area, so the forest regenerates itself naturally. Of course, much more area is required to obtain the same amount of wood, which means more roads must be cut through the forest.

There are other sustainable ways of harvesting trees. It is critical that the method of harvesting is compatible with the types of trees and the site conditions. Selective cutting means only a few trees in a given area are harvested at any one time. While this sounds like a good idea, it has its problems. Like clear cutting, selective cutting can be carried out only in certain areas. It works best in deciduous or mixed wood forests where the differences in the age and size of the trees makes clear cutting impractical. The large old trees are selected for harvest, leaving the younger small trees to grow to replace them. Selective cutting is not widely practised in the boreal forests of Canada.

The forest industry is a major contributor to the Canadian economy. Sustainable forest management practices that ensure the health and value of Canada's forests will result in the forest industry continuing to play an important role in Canadian society.

CAREER PROFILE: FORESTER

Do you like hiking, canoeing, or other outdoor activities? Do you consider yourself a conservationist? Do you want a respected career that pays well? If you answered "yes" to these questions then you may want to consider a career as a forester.

The Nature of the Work
Robert Tomchick is chief forester with QUNO Corporation, the fifth largest producer of newsprint in Canada. An avid conservationist, Tomchick entered the industry to "make sure the company treated the forest right." He found that the forest industry was not as bad as he had thought. QUNO Corp., for example, has an excellent record. The company reforests more area than it cuts each year.

Today Tomchick works in an office planning forestry policy and managing the company. But he spent many years in the field, collecting data from air photos, satellite images, topographic maps, and ground observations. From this information, decisions about managing the forest were made, such as which method of cutting to use, how to regenerate the forest, and how to incur the least disruption to wildlife habitats.

An important part of Tomchick's job was preparing detailed plans and reports of the company's activities for the Ontario Ministry of Natural Resources. Often he was called upon to talk with community groups and conservationists about the company's plans.

Tomchick enjoys being a professional forester. The analytical work is challenging and stimulating. "But what is best," Tomchick states enthusiastically, "is you personally get to do something positive to protect the environment."

Qualifications
Tomchick has a science degree in forestry from the University of Toronto. While U of T no longer has a forestry program, Lakehead University, the University of New Brunswick, Laval University [French], the University of Moncton [French], the University of Alberta, and the University of British Columbia all offer forestry programs. Students interested in forestry should take senior high school courses in geography, biology, chemistry, and mathematics. A knowledge of computers is useful and good communication skills are essential. Canada is a world leader in forestry management, so there are opportunities for success in this field.

If you are a "hands-on" type of person, you might consider a job as a forestry technician. These professionals gather and interpret data, participate in wildlife management, and work in harvesting and regeneration. Programs in forest land management, timber management, fish and wildlife management, and environmental studies are offered at community colleges.

Mining

Much of Canada's richest mineral deposits are found in the boreal forests. In Ontario, mining involves only a small part of the total area of the province—less than 0.47 per cent. By comparison, the proportion of Ontario that is covered by highways and roads is ten times as large—4.7 per cent. Nevertheless, mines can be unsightly. When the shafts or tunnels are dug, the debris litters the landscape, as do **tailings**, the waste product of mining. Today mining companies are required to return the area to its natural state once mining operations are over.

Several methods are used to extract minerals. These include **dredging, open-pit mining,** and **shaft mining**. The choice depends on the nature of the deposit.

Placer deposits are deposits found in rivers. These are best removed by dredging. Sediments are dug from the river bed by giant conveyers and mechanical shovels. The ore is processed, often using the water from the river in a **sluice**. Nuggets are heavier than the sediment so they settle to the bottom. The waste material is returned to the river. Gold, silver, and diamonds are mined in this way.

Dredging sometimes means that aquatic ecosystems can be disrupted downstream because of the increased sediment in the water. In the past mercury was used as an agent to separate the mineral from the ore. Inevitably some of this heavy metal got into the river. When fish ingested the mercury, it built up in their bodies and eventually poisoned them. People eating the fish can also be poisoned. Today the mining industry in developed countries no longer uses mercury.

Open-pit mining is often used when deposits lie close to the surface. The vegetation, soil, and **overburden** are removed. The mine is excavated in **benches**, or giant steps that spiral to the bottom of the pit. Explosives break up the ore body so that bulldozers can load the ore onto conveyers and trucks, which travel up the benches to the surface. The ore is transported for processing to the mill, which is usually close to the mine. Often crushers and other mechanical equipment are found on the mine site to facilitate transport. The Grande Carajás mine discussed on pages 299–300 is an example of an open-pit mine in Brazil, but many are also found in Canada.

After several years, the hole made by the mining operation can be very deep. The deeper it becomes, the wider the hole extends across the countryside. Eventually the hole is so deep that it is too hard to reach the deposit. The mine either closes or a different method of extraction is used. Minerals such as limestone, iron, and coal are often mined in this way. Deposits are usually large and can cover many hectares. For this reason, ecosystems can be temporarily disrupted. Once the mine closes, the overburden and tailings are returned to the hole. The soil is replaced and the natural vegetation restored. While this procedure is costly, mine operators and governments are committed to preserving ecosystems.

Shaft mining is used to extract ore bodies lying in veins deep underground. A shaft connects the surface with the ore body. Tunnels follow the veins of ore through the rock. Holes drilled into the ore body are packed with explosives that loosen the ore so that **scooptram** operators can load it onto waiting conveyers and dump trucks. The ore is then transported to the surface where it is milled. Many shaft mines are unassuming structures on the surface but underground they consist of many kilometres of tunnels and shafts.

As with open-pit mines, shaft mines produce a lot of waste. In Sudbury, tailings and overburden are found everywhere. Mine operators are working hard to improve their image in this northern community. A tree nursery actually produces seedlings underground in one of Inco's mines. Soil is scattered on the mine waste and the trees are planted.

Perhaps the most serious problem caused by mining in the boreal forest is **acid precipitation**. When ore is smelted, impurities are removed. These enter the atmosphere as sulphur dioxide. Water vapour reacts with the sulphur dioxide to produce a weak solution of sulphuric and sulphurous acid. When it rains or snows, this acid enters the already acidic soil, causing damage to plants and animals.

Smelting is not the only cause of acid precipitation, however. Nitric oxide is formed when any fossil fuel is burned. Every time you drive a car or light the barbecue, you are producing nitric oxide. Like sulphur dioxide, nitric oxide combines with water vapour to produce nitric acid, which returns to the earth as precipitation. Plants and animals have adapted to the natural acidity of the spodosols but they cannot cope with the increased acidity caused by acid deposition. The metabolism of some plants and animals is disturbed and they die.

Aquatic animals are particularly affected in the spring. When the acid snow melts, the acidity levels in lakes and wetlands rises significantly. At the same time, amphibians and fish hatch. The newborn animals are particularly susceptible to the acid and die from **acid shock**. Today there are many dead lakes in northern Ontario and Quebec where there is virtually nothing living. It has been estimated

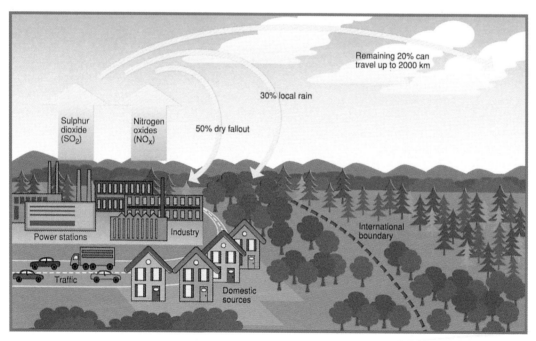

FIGURE 39.4 *The Causes of Acid Deposition*
Using this diagram, summarize the causes of acid deposition.

that in Canada 15 000 lakes are dead and another 40 000 are threatened because of acid precipitation.

Acid precipitation is most common in Canada's boreal forests. Other regions have soil that is less acidic and so the risk of acidification is lower. The regions most prone to the problem are downwind from major mining centres such as Sudbury, Ontario, and urban centres. Acid rain is also a problem in Europe. Parts of Germany, Norway, Sweden, and Switzerland experience acidification. Downwind from industrial centres like the Ruhr in heavily populated western Europe, forest regions share many of the same problems caused by acid precipitation.

There are solutions to acid precipitation but they are costly and require a concerted effort by citizens, governments, and industries alike. The problem will be reduced if the consumption of fossil fuels is reduced. People can accomplish this by using public transit, car pooling, and burning fewer fossil fuels. Governments have increased emission control standards on automobiles. Auto makers have responded by developing catalytic converters to reduce harmful gases. The development of alternate fuel vehicles could also help solve the problem. Cars that run on hydroelectricity stored in batteries, hydrogen-powered cars, and even solar-powered cars could help to reduce acid precipitation.

In the mining industry, there are a number of innovations that could reduce acid precipitation. Treating the mineral before it is smelted can remove much of the sulphur so that it never gets into the atmosphere. When the ore is smelted, scrubbers remove the sulphur from the smoke before it is sent up the smokestack. Much of the sulphur is reclaimed and used to make sulphuric acid. Adding limestone to increase the pH levels of acid lakes is one solution, but it is extremely costly. The limestone is alkaline, so when it dissolves into the lake water the acidity is reduced. Millions of tonnes of crushed limestone would be needed to buffer even one moderate-size lake. The best solution is to reduce the harmful emissions of nitric oxide and sulphur dioxide.

Hydroelectric Development

The third valuable resource in the Canadian boreal forest is hydroelectricity. Wherever there is abundant water and a slope, this sustainable source of power can be harnessed. It is sustainable because no fuel is used. As long as the water is flowing downhill, power can be generated. Hydroelectricity is an environmentally friendly alternative to burning fossil fuels for energy. So why have so many people protested the building of the James Bay II hydroelectric project in Quebec?

The first James Bay Project was approved in 1975. La Grande Rivière was dammed to create a reservoir so a hydroelectric plant could be built. The dam and reservoir provided a regular flow of water that did not vary seasonally. In this way water flows constantly through turbines to create a steady flow of electricity. The river was a good choice for development. It had an average discharge rate of 17 000 m^3/s and dropped 548 m over its 800 km course. To make the flow even stronger, two adjacent rivers were diverted into La Grande. Native groups and environmentalists protested but the project went ahead as planned. Native peoples were given cash settlements and guarantees of fishing, hunting, and trapping rights.

> **GEO-Fact**
>
> Egyptian hydrologists hope to develop a hydroelectric project in reverse. Instead of having the water run from the land downhill into the ocean, they plan to have sea water flow into the land! The Al Qattarah Depression is over 120 m below sea level and covers an area of several hundred square kilometres. Tunnels will be drilled from the Mediterranean Sea 15 km away through the desert to the depression. As the water fills up this huge hole in the ground, turbines will generate hydroelectricity. The sheer size of the hollow and the high evaporation rates could mean that power will be generated for a hundred years or more.

When rivers are dammed trees are drowned and animal habitats are destroyed. The drowned vegetation behind dams creates an unhealthy environment for aquatic animals. When the trees decompose in the water, they remove much of the oxygen. Without oxygen, most life forms die.

In 1989, the Quebec government announced that an even larger project called Great Whale was being planned for the rivers that flow into James Bay. Three new dams and a massive network of reservoirs would have changed the forests and tundra of northern Quebec forever. The people who live in the region expressed concern about the adverse effects of the development on the ecosystem. Late in 1994, a new government in Quebec announced its intentions to cancel the Great Whale project. Native and environmental groups claimed a victory in the ongoing debate over economic prosperity versus environmental preservation. (See "Quebec Shelves Great Whale Project" on page 313.)

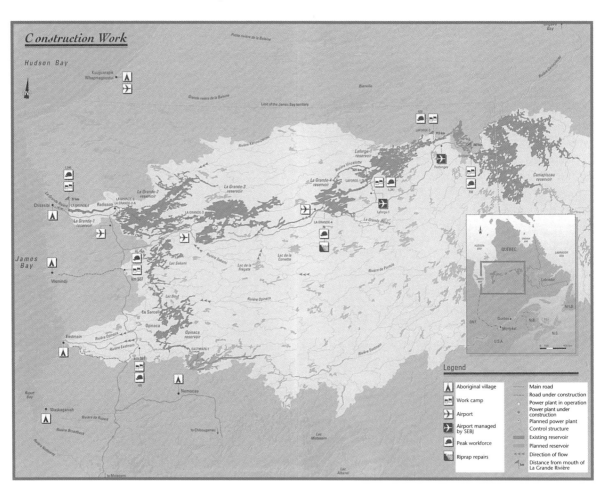

FIGURE 39.5 *The Extent of the James Bay II Power Project*

Quebec Shelves Great Whale Project

By Rhéal Séguin, Ann Gibbon, and Graham Fraser

The Quebec government is pulling the plug on the $13-billion Great Whale hydroelectric project in Northern Quebec.

Premier Jacques Parizeau said yesterday that while the massive project has not been totally abandoned, it will not be constructed in the foreseeable future and is not a priority for the Parti Québécois government.

The Premier was responding to remarks made yesterday in Washington by the Grand Chief of the Grand Council of the Crees in Northern Quebec, Matthew Coon Come, who accused the Quebec government of destroying native land with the construction of the Great Whale.

"We know what is happening in our lands," Mr. Coon Come told an academic conference. "We know that rivers have been made to flow backward, that camping grounds have been destroyed. We know the spawning grounds of fish have been destroyed and contaminated with mercury. We are the ones that know that. And we want to come here and tell our story of the impacts we felt before they go ahead on Great Whale."...

Later, Mr. Coon Come lauded the Premier's decision.

"I would like to extend my congratulations to Premier Parizeau and his government for a courageous decision," he said.

"I think Great Whale was never viable, it was never economically sound, it was never environmentally sound.

"It's a victory for the environment. It's a victory for Quebeckers."...

With growing international opposition to Great Whale generated in large part by the Crees' successful public-relations campaign in Canada, the United states and Europe, coupled with energy surpluses in North America, interest in further hydroelectric development has waned considerably.

"There was a time when public interest was considered as linked to Great Whale. As far as we're concerned, it isn't." Mr. Parizeau said. ...

"It's good news," said François Tanguay of Greenpeace Quebec. "It's a very sound economic and social decision...and it's more money to do more intelligent things in the economy than pour cement all over the place."...

Not everyone was jubilant. Clément Godbout, president of the Quebec Federation of Labour, said he was disappointed about the decision to put on ice thousands of jobs and economic spinoffs the project could have generated at a time when Quebec's unemployment rate is "scandalously high."...

Reprinted by permission of The Globe and Mail, Toronto.

FIGURE 39.6 *Cancellation of James Bay II*

Things To Do

1. Prepare an organizer summarizing the characteristics of the boreal forest under the following headings: location, climate, plants and animals, soils and landforms, indigenous peoples.

2. Explain the difference between a *preservationist* and an *environmentalist*. How is it possible for a resource developer to be an environmentalist?

3. a) Assume one of the following roles: aboriginal Canadian, resource developer, labour union official, government official, tourist agent, local shop owner, conservation club president, or president of an association of hunters and anglers.
   b) Prepare an argument for or against each of the following development strategies: forestry, mining, hydroelectricity.
   c) Have a formal debate with other members of your class or group to determine which form of resource development is most appropriate for the boreal forest.

4. a) In small groups watch a variety of nature documentaries. Can you detect any biases in these programs? Report your findings to the class.
   b) What political purposes do you think these biases serve? Explain your answer.

5. What is your stand on wilderness preservation? Is it important to preserve all the world's wilderness areas or do you think some resource development is essential? Support your arguments with facts.

6. Research another region of boreal forest in Europe or northern Russia to determine the success of development strategies there.

Chapter 40

Living in the Tropical Grassland

INTRODUCTION

The tropical grasslands of Africa, South America, and Australia encircle the earth between the tropics and the Equator. This is a transition zone between the rainforest and the tropical desert. This interesting ecosystem has many of the characteristics of both the rainforest and the desert.

When population densities are low and the environment is carefully managed, the region has great potential for agriculture. The Australian outback is noted for the beef and sheep that can be successfully raised. Similarly the giant ranches of southern Brazil and northern Argentina have prospered since early in the nineteenth century. The same situation does not exist in the tropical grasslands of Africa. Overpopulation and unsustainable practices are creating an ecological disaster.

Anthropological evidence indicates that the savanna of East Africa is the birthplace of the human species. Since that time, people have continuously modified this fragile environment. The Sahel south of the Sahara Desert is probably the most threatened. The region is plagued with problems, including deforestation, overgrazing, drought, and civil war. It is not enough for people in developed countries to offer food aid. Sustainable development strategies must be adopted if the people are to survive and their environment is to be preserved.

FIGURE 40.1 *The East African Savanna, Mozambique*
The grasslands of Africa are typical of those found in the southern continents.

Climate

Two Köppen climate zones make up the tropical grasslands. The savanna region encircles the tropical rainforest. As it approaches the tropical desert the savanna gradually gives way to the steppe climate. This short grassland region is drier and has many of the characteristics of the desert, especially in the dry winter season.

Temperatures in the savanna are hot, with monthly averages exceeding 18°C. In summer the region experiences heavy convectional rainfall similar to that in the tropical rainforest; the savanna becomes a wet, green world, alive with plants and animals. In winter, however, the savanna is dominated by a high pressure cell, the same one that makes tropical deserts dry. Descending dry air is rapidly heated by the hot tropical sun. The evaporation rate increases and the landscape is transformed from lush greenness into a dry, brownish-yellow hue.

The steppe climate has total precipitation below 500 mm per year but above the 250 mm of the desert. As with the savanna, there is a dry season in the winter. Summers have enough rainfall to produce abundant grass for the many grazing animals. The main problem with rainfall in both regions, but especially in the steppe, is that it is unreliable. There can be many years of good rains. Plants and animals flourish and the inhabitants thrive. But there can also be periods when the rains fail. Then there is starvation, misery, and death.

The tropical grasslands of Africa were once much larger than they are today. Archaeological evidence indicates that the Sahara Desert has taken over some of the savanna. Did the desert expand to its present size because of the way people used the land? Or has the climate changed because of natural fluctuations related to glacial periods? (See page 140.)

Plants and Animals

Savanna grasses are well suited to this climate. They grow rapidly when there is rain but lie dormant for months during the dry season. The grasses have developed extensive root systems that draw moisture from deep in the soil. In humid regions the grass can be extremely long. "Elephant grass" is so called because it is so tall that an elephant can hide in it! In drier steppe regions, the grass is short. It often grows in clumps, leaving bare patches of ground between plants. There is not enough moisture to produce an unbroken blanket of grass. The roots extend beyond the leaves above the ground, drawing moisture from a larger area than the plant seems to occupy. The only limiting factor to plant growth is water. Other plant needs are well provided for.

While grass is the most common plant in tropical grasslands, trees are also a part of the ecosystem. Moving from the tropical rainforest to the savanna, the number of trees gradually decreases. In some areas, local features such as springs and rivers often provide enough water in the dry season to maintain dense forests.

FIGURE 40.2 *Adapting to the Environment*
How have these acacia trees adapted to seasonal drought?

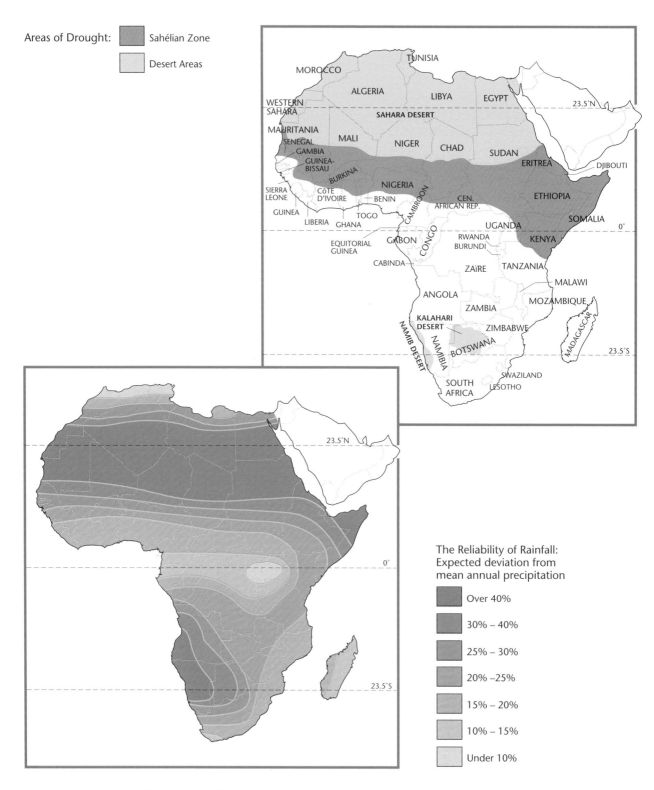

FIGURE 40.3 *The Reliability of Rainfall and Areas of Drought in Africa*
What connection can you make between the desert and Sahélian zones and expected deviations from mean annual precipitation?

Trees have also adapted to this region. The acacia tree of east Africa is a good example. It has small, shiny leaves, covered with a waxy coating, which are shed in the dry season to reduce water loss. It also has extensive root systems to draw moisture from the soil. They grow far apart to reduce the competition for water. The bark is thick and does not burn readily. This provides protection from the fires that frequently plague the region during the dry season. Some savanna trees, such as the baobab and bottle trees, store water during the rainy season and draw on these reserves during the dry season. Many other unique trees and thorn bushes break up the monotony of the grassy plains.

The animals of the tropical grasslands are among the most interesting and diverse in the world. Large herbivores such as elephants, wildebeests, antelopes, and giraffes populate the region. Smaller animals are grazers. Taller herbivores eat leaves from short bushes. Giraffes feast on leaves high up in the acacia trees. Each animal has a **niche** where it can survive without competition from other animals.

Many herbivores have long legs and strong hoofs that suit their migratory habits. During the dry season, they roam the grasslands in search of food and water. Migration is becoming difficult, however, as fences are being erected across vast stretches of the countryside. These barriers sometimes trap the animals in parched lands where they cannot survive. Wildlife is also threatened by loss of habitat. As increasing amounts of land are being developed, less and less wilderness survives. A further threat is caused by poachers eager to profit from the illegal slaughter of wildlife. For their protection, many animals are now contained in game reserves and national parks. While these provide a safe haven from poachers, they also prevent the animals from following the rains as they always have. In addition, because park animals are isolated from other members of their species, they become **inbred**. The **gene pool** is limited, so certain traits become intensified. Eventually the animals in each ecological island become different from their counterparts in other isolated parks. With an uncertain future, it is unlikely that many of these animals will survive in the wild into the twenty-first century.

Long legs and strong hoofs also help herbivores escape from their natural enemies. Carnivores such as lions, cheetahs, and hyenas are well adapted to hunting. Strong legs, sharp claws, and powerful jaws help these animals catch their prey. Healthy adults can easily escape, so predators concentrate their efforts on weak and young animals. Only the fittest survive. The predators keep the number of herbivores down, which ensures that the vegetation is not over-grazed. Like the herbivores, carnivores are often camouflaged so that they can creep up on unsuspecting victims! Lions and cheetahs, for example, blend in so well they are almost invisible in the tall yellow grass.

Scavengers also roam the grasslands. These animals are the garbage collectors of the savanna. They eat whatever the large carnivores leave behind. Flies, bacteria, fungi, and other decomposers clean up the remains and add essential nutrients to the soil. The natural vegetation, the herbivores, their predators, and the decomposers work together to make the tropical grasslands a stable ecosystem.

Soils and Landforms

Savanna soils are remarkably fertile. They contain much more humus than the aridisols of the desert and are less leached than the oxisols of the tropical rainforest. Nevertheless, they are moderately leached and quite acidic. During the dry season vegetation dies, producing a mat of litter that decomposes to form humus.

There are several different soil types in the region. The predominant soils south of the Sahara Desert are alfisols. With adequate moisture this soil provides the essentials for the natural grass cover. This deep, moderately rich soil has well developed horizons. The A horizon is reddish-brown and easy to work. In the B horizon there is a concentration of clay that absorbs moisture and retains it during the rainy season. With applications of calcium to reduce acidity and fertilizer to increase nutri-

FIGURE 40.4 *Soil Profiles*
(left to right) alfisols; vertisols; aridisols

ents, these soils can be excellent for agriculture. When vegetation is removed, however, alfisols are easily eroded. Wind blowing out from the desert can strip topsoil and produce blinding dust storms. When the rains come, the unprotected soil erodes, forming deep gullies and outwash fans where the silt settles. These precious soils need to be carefully tended if they are to be sustained.

Vertisols are found in the grasslands of Australia, central Ethiopia, and India. Like alfisols, these black clay soils are fertile when there is adequate moisture. Vertisols develop huge cracks in the dry season. Humus falls down these cracks. This results in a soil that seems to be upside down because the humus is low in the A horizon. When it rains the soil expands and the cracks disappear.

Other soils found in this region include entisols and inceptisols. These undeveloped soils result from poor drainage and waterlogged swamps and riverbanks. While they are relatively infertile and underdeveloped, they have the one thing other tropical grassland soils lack—water. The Okavango Swamp in Botswana, the Pantanal of Brazil, and the wetlands of northern Australia are alive all year round. They do not have to depend on the rainy season for water. In dry years, wild animals by the thousands find sanctuary in these areas.

On the border of the desert regions, aridisols are common. Sand blowing from the desert buries productive grassland soils. This is a serious problem since the natural vegetation cannot survive in the infertile desert sand. Because the natural cover is removed, the climate is changed. Evapotranspiration is reduced. In addition, the albedo increases. The lower humidity and cooler temperatures close to the ground result in less rainfall. Eventually the region turns to desert. This problem is common in the Sahel, the Middle East, and other regions that border deserts.

Landforms in the grasslands vary. In Africa, much of the land is either high and flat or high and rugged. The plateaus of equatorial East Africa are cooler than the steamy equatorial forests of the lowlands. In northeast Africa, the highlands of Ethiopia and Kenya are wetter than the surrounding lowlands. Orographic precipitation occurs as air is forced to rise and water vapour condenses in these highlands. In South America and Australia, high interior plateaus and poorly drained wetlands make up most of the tropical grasslands.

Indigenous Peoples

People have inhabited the grasslands of Africa longer than anyplace else on earth. They developed a lifestyle that enabled them to live successfully in a fragile, unpredictable ecosystem.

Long ago the people were hunters living off wildlife native to the region. Hunting became a pastime, however, once animals were domesticated. Depending on the people, cattle, goats, sheep, and camels were herded. These animals were rarely killed for their meat. Instead they were maintained for their milk. In some cultures, such as the Masai, the animals were also bled; the combination of milk and blood was a traditional staple. As long as the grass could support the animals that lived off of it, this was a sustainable economy.

The people of the grasslands were nomadic, moving from place to place following the rains. This lifestyle provided their animal herds with fresh pastures and protected the land from over-grazing. The usual migratory route was circular. In the dry season the people travelled south into the wetter regions near the Equator. In the wet season they moved north closer to the desert. The pattern closely followed the migratory patterns of the wildlife of the savanna.

All aspects of life were affected by the nomadic lifestyle. People had few personal possessions because they were difficult to transport. The idea of ownership was foreign. The only exception was livestock, which represented wealth and status. Large herds were also essential for survival. In bad years when the rains failed and animals died, the family with the most animals had the best chance of survival.

Housing was portable or temporary. Tents made from animal hides were often used in dry areas where the people were forced to move every day or two. Further into the tall grasslands, huts of grass, mud, and animal dung were built. These were abandoned when it was time to move on. The lifestyle was hard, but it functioned successfully for thousands of years.

Resource Development

In the grasslands of Africa, natural resources are being depleted so rapidly that human and ecological disasters are the result. The area experiencing the greatest challenges in resource management is the Sahel. There are three types of agricultural development: pastoral farming, subsistence farming, and commercial farming.

Pastoral Farming

When Africa was colonized by European nations, the lifestyle developed by many aboriginal peoples over thousands of years changed in a single lifetime. Much of the choice land was settled and owned by Europeans. The aboriginal peoples were no longer allowed to migrate freely. Instead they were forced to settle in one place. Because the concept of ownership was foreign to many African cultures, they did not understand the Europeans' claims of land ownership. Conflicts and rebellions broke out between the colonial governments and the aboriginal peoples. Eventually many African cultures were forcibly resettled in **sedentary** villages, usually located in marginal areas that were unfit for anything other than **pastoralism**. When individual African nations gained independence, the new governments continued many of these colonial policies.

The second major factor that affected the environment was demographic. Traditionally African people had high birth and high death

FIGURE 40.5 *Traditional Masai Herders of Kenya*

Technology Update: Using Satellite Images to Measure Desertification

For years there has been concern that the Sahara Desert was expanding south into the pasture lands of the Sahel at the rate of 5 km per year. Now satellite images show that this is not quite the case. The desert is only expanding at certain points; in other places it is stationary or even receding. The desert moves south some years and recedes in other years. The line between the desert and the steppe moves as rainfall patterns change. In wet years the Sahara gets smaller, while the opposite occurs in times of drought.

US meteorological satellites have enabled analysts to view the entire southern Sahara over a ten-year period. The line between the desert and the grassland roughly corresponds to the 200 mm **isohyet**. Places that receive less than 200 mm of rainfall per year have little vegetative cover, while those that receive more are grasslands. **Infrared satellite images** show where the grass is and where the ground is exposed. Chlorophyll in the plants absorbs the red light; bare ground is shown as dark red and grassy plains are a lighter shade. By comparing the satellite images from season to season and year to year, it is possible to plot how the border has moved.

Satellite Observations
- From 1980 to 1984 the border moved south a total of 240 km in the worst drought of the century.
- From 1984 to 1985 the trend reversed and the border moved north, 110 km the first year and 30 km the second year.
- In 1987 the border moved south 55 km.
- In 1988 the border moved north 100 km.
- In 1989 to 1990 the border moved south 77 km.

In order to plot a long-term trend, measurements need to be taken over several decades. Only then can the full extent of desertification in North Africa be determined.

Country	Population Projections (millions)			Growth Rate (% annually)
	1988	2000	2020	
Burkina	8.5	11.8	20.6	6.4
Ethiopia	48.3	71.1	127.6	5.1
Kenya	23.3	38.3	79.2	7.5
Mali	8.7	12.3	21.1	6.5
Niger	7.2	10.6	18.9	5.1
Nigeria	111.9	160.9	273.6	6.5
Somalia	8.0	10.4	18.6	6.5
Sudan	25.0	33.3	48.5	2.9
Uganda	16.4	26.7	46.4	5.7
Canada	26.1	28.4	30.2	0.5

Source: Population Reference Bureau

FIGURE 40.6 *Population Growth in the Sahel, 1988–2020*

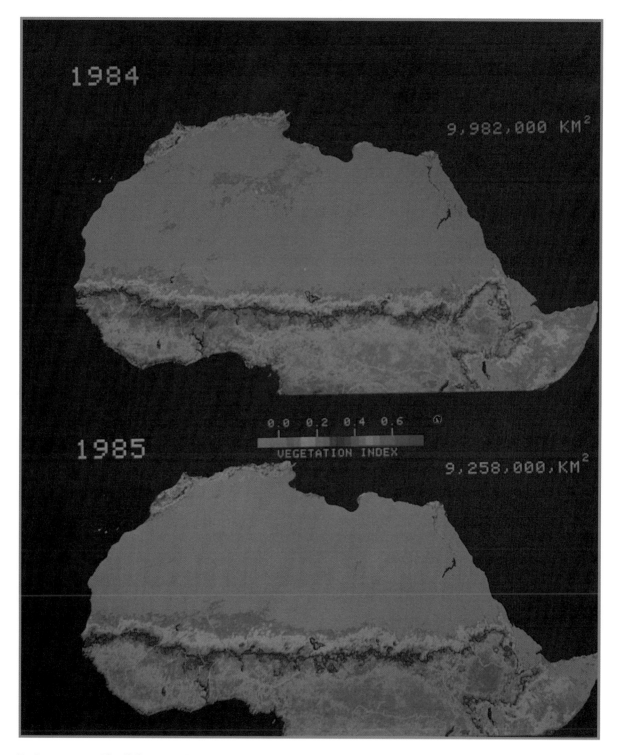

FIGURE 40.7 *The Shifting Sands of the Sahara*
The Sahara Desert expanded to 9 982 000 km² in 1984 but shrunk to 9 258 000 km² in 1985.

rates. This caused the population to remain fairly constant. After the Second World War, medical advances resulted in longer life expectancy and a lower mortality rate among children. These factors caused the population to increase rapidly.

The growing population and the concentration of domesticated animals were more than the land could sustain. The herds stripped the natural grasses, exposing the soil to the erosive forces of wind and water. Dust storms and gullies carried away millions of tonnes of precious topsoil. The constant pounding of millions of hoofs over the same ground compacted the soil, reducing air and moisture content. This made it impossible for small plants to grow. When packed down, the clay soils became **non-porous**. Rainwater was not absorbed by the soil. The lack of vegetation meant less evapotranspiration. The surface of the ground was lighter in colour and so it had a higher albedo. The lower humidity and increased level of reflection caused the climate to become increasingly dry, making the land even less productive. Today much of the grassland on the edge of the Sahara has become desert.

The ecological destruction was not limited to grasses, however. As the population continued to increase, more and more trees were cut down for firewood. Today much of Ethiopia is without trees. All of these factors have led to the ecological collapse of the sub-Saharan ecosystem.

Subsistence Farming

Subsistence farming has been practised for many years in East Africa. These farming regions are generally far removed from the deserts of the north. Rainfall of 300 to 500 mm per year provides adequate moisture for crops. Since the region is near the Equator, it is warm for most of the year; however, frosts can occur in highland regions.

Field crops like millet, sorghum, and other hardy grains are grown as cash crops as well as for food. In the past, the people used only a small part of the savanna at any one time.

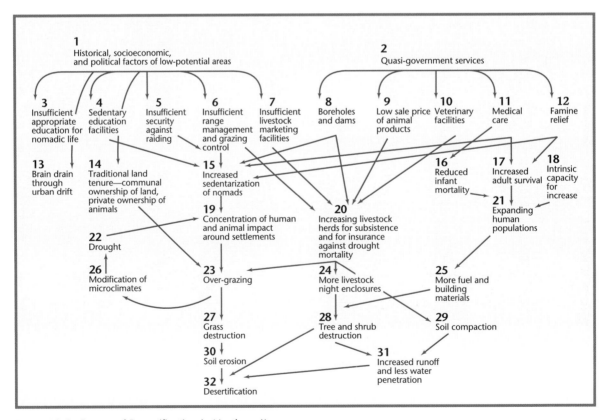

FIGURE 40.8 *Causes of Desertification in Northern Kenya*

Fields were abandoned when the soil became infertile. The savanna was then allowed to reclaim the land and the soil's fertility was gradually restored. During this period, the sap of acacia trees was tapped and **gum arabic** was gathered for use in the manufacture of adhesives and inks. Eventually the grasses and trees were burned to clear the land for agriculture once again. Each cycle lasted about one generation. This form of agriculture is similar to the sustainable slash and burn **horticulture** practised in the rainforest.

This type of subsistence farming can no longer be practised today. There are simply too many people. Kenya, for example, has the highest population growth rate in the world. Leaving much of the land unproductive for long periods of time is no longer feasible. Today farming practices are intensive. Intercropping— growing several crops in one field—is common. The soil is in constant use. A humid microclimate is often formed close to the ground when shade from one plant shelters another from the sun's harmful rays. The soil is never cultivated all at the same time, so wind erosion is reduced. Often rows of perennial bushes are used as windbreaks. Depending on the local climate, coffee, tea, and palm trees are grown. These simple farming practices are good but they drain the soil of its nutrients. Regular doses of expensive fertilizer are needed. Often the cost of the fertilizer is more than the farmer can afford. If the soil loses its fertility, crop yields decline and people go hungry. The system is no longer sustainable.

Commercial Farming

The best agricultural lands in north Africa are reserved for commercial farming. Governments often support these huge farms because they are profitable and are owned by influential citizens. Impoverished countries in the sub-Saharan region need the foreign exchange income provided by such cash crops as groundnuts, cotton, and coffee.

In the Jazirah region of central Sudan, huge plantations have been producing cotton since before the Second World War. The region is ideal for growing the long staple cotton for which the Nile Valley is famous. The climate is hot and sunny most of the year. The low-lying flat land, located on the interfluve between the Blue and the White Nile, receives abundant water through irrigation. Soils are deep and fertile from the thousands of years of accumulated **alluvium** that was deposited when the river flooded each year. All of these factors contribute to the region's ability to produce top quality cotton.

Farms in the Jazirah region are really agribusinesses. Tractors, harvesters, and irrigation pumps make the operation smooth and efficient. The owners are often wealthy business people who live in big houses in the capital of Khartoum rather than farmers who actually work the land.

Unfortunately the world cotton market has been in decline for years. Synthetic fibres and a glut of cotton have reduced the profitability of the crop. Some would argue that it is inappropriate to continue this type of farming in a country where many people face malnutrition and even starvation. Some planners believe this fertile farmland should be divided among the landless peasants instead of forcing them to eke out an existence on marginal lands. Not only would these farmers become self-sufficient, but the exhausted soils of the steppes would have a chance to recover from overuse.

This issue is not one that can be resolved easily. The reasons are primarily political. In north Africa, violence has erupted as different interest groups struggle to gain control of a region and take whatever wealth they can.

FIGURE 40.9 *Digging an Irrigation System on an Egyptian Farm*

Caught between warring factions are the poor and landless peasants who have no political power. In some north African countries, such as Uganda, Sudan, Ethiopia, and Somalia, civil wars have killed thousands and destroyed what was left of the already abused ecosystems. United Nations troops have been sent in to some of these countries to restore and maintain order. Some stability has returned over the short term, but permanent solutions are yet to be found.

Things To Do

1. Prepare an organizer summarizing the characteristics of the tropical grassland region under the following headings: location, climate, plants and animals, soils and landforms, indigenous peoples.

2. a) Assume one of the following roles: peasant farmer, wealthy landowner, government agent, foreign aid worker, conservationist.

   b) Prepare an argument for or against each of the following land uses: pastoral farming, subsistence farming, commercial farming.

   c) Role-play a discussion among these people over how the land should be used.

3. In what ways do population patterns contribute to north Africa's environmental problems?

4. Watch documentaries or news clips on famines in north Africa. What biases do you think exist in these programs? Explain your answer.

5. Investigate the role of the Canadian International Development Agency (CIDA) in north Africa. Do you think Canada's aid policies are effective? Give reasons for your answer.

Chapter 41

Living in the Temperate Grassland

INTRODUCTION

The temperate grasslands go by many different names. In Canada they are called the prairies. In the United States they are the great plains. South Africans know them as the veld. In Argentina they are the pampas. And in Ukraine and Russia they are the steppes. No matter where they are or what they are called, these fertile grasslands are the breadbaskets of the world.

Precipitation determines the location of the temperate grasslands. Most are found in the mid-latitudes in the dry interiors of large continents. The only exceptions are the pampas and veld, which are on east coasts where rainshadows occur. The region is too dry for trees; it is a transition zone between mixed forest and desert. The temperate grasslands are important agricultural regions. It is therefore important to understand how human modifications have changed them.

FIGURE 41.1 *The Badlands of Alberta*
These "badlands" were created partly as a result of resource mismanagement.

Climate

The continental effect creates extremes in climate. Temperatures range from hot in summer to bitterly cold in winter. The further from the Equator, of course, the greater the temperature range. Winters in Pueblo, Colorado, for example, are a lot less extreme than in Regina, Saskatchewan. The American city has a mean January temperature of about -1°C, while Regina is a frigid -18°C.

The growing season is short. Frosts occur in early to late fall and last well into the spring. This greatly affects the ecosystem because plants and animals must cope with frozen soil for much of the year. In winter, most of the precipitation falls as snow. When the spring thaw occurs, the water is released into the ground, providing moisture for germinating seeds.

Like the tropical grasslands, the mid-latitude prairies are made up of two vegetation zones—short grasslands and tall grasslands. The short-grass region, called **steppe**, receives between 250 mm and 500 mm of precipitation. This is the semi-arid zone that borders the temperate deserts. Moving east away from the interior of the continent, precipitation increases and taller grasses flourish. This tall grassland region ranges from the low latitude subtropics to continental and even subarctic climates. As with all ecosystems there is a great deal of variation from one region to the next.

Plants and Animals

As the name states, grasses are the main form of natural vegetation. In its natural state, the tall grass of the American Midwest reaches heights of over 2 m. Today, however, agriculture dominates the entire ecosystem, and the original tall grass prairie is difficult to find.

The natural grasses were well adapted to the climate. Buffalo grass, a short steppe variety, and various types of blue-stem grasses predominated. They had extensive root systems that could reach deep into the rich soil. These plants grew rapidly when it was warm and there was rainfall. They seeded quickly before frosts came in the fall. Most were perennial. They were dormant in the winter but came back to life when growing conditions returned.

Trees are seldom found in the region. The only exceptions are sheltered places where there is protection from dry winter winds and abundant groundwater close to the surface. River valleys and springs often have small deciduous forests.

The large herbivores that used to live on these grasslands have also vanished, except where protected in parks and game preserves. Bison, pronghorn antelope, and deer were all indigenous to the region. Long legs and hard hoofs allowed these animals to migrate in search of fresh pastures. The bison's long matted fur protected it from the severe winters. Its heavy bearded head was well suited to pushing through snow to get to grass in winter. Smaller animals, such as prairie dogs and gophers, as well as numerous insects live off the grasses. The coyote, the only major predator, preys on these smaller animals.

Soils and Landforms

There are three main soil groups in this ecosystem: aridisols, mollisols, and alfisols. In the steppe around temperate deserts, aridisols are common. Low water content, a thin layer of humus, and concentrations of salt in the A horizon make aridisols in their natural state unsuitable for most field crops. The region makes up the range country of the western prairies. If the soil is irrigated and well drained, the salts and other mineral deposits can be washed away, making the soil arable. Many American cattle ranches grow alfalfa—an excellent cattle feed—on irrigated sections of their farms. Water either comes from rivers or by tapping aquifers deep underground.

Mollisols are found in the core of the grasslands. These are some of the most fertile soils in the world. Extremely deep, with well developed horizons, these black soils have a high humus content. They absorb water when it rains yet do not pack down tightly like clay soils. Litter accumulates in deep layers because decomposition of the annual build-up of grass is slow in this temperate climate. When it rains, the mat of dead grasses and roots retains moisture and allows it to slowly percolate through the soil. It is this soil that makes the region a good agricultural area.

To the east lie the alfisols. These share many of the characteristics of the rich mollisols, but they have a higher water content because of the increased precipitation. Like mollisols, they are deep and easily ploughed once the natural vegetation is removed. Alfisols are moderately leached and slightly acidic. With the addition of lime and fertilizers they can be excellent agricultural soils.

While there are some exceptions, these temperate grasslands have developed on flat or gently undulating plains. The absence of trees and the level ground make it possible to see great distances in this vast, monotonous landscape. In North America, the region was a shallow inland sea in the Cretaceous and Tertiary periods. Sediment carried by rivers from the Canadian Shield and the newly forming Rocky Mountains was deposited in the shallow water. Over millions of years, layer upon layer of sediment accumulated until the sea was virtually filled in. When sea level dropped, the region found itself above sea level. Today's soils formed from the limestone regolith that remained. Today rivers erode through the plains creating some variety to the flat prairie.

Indigenous Peoples

In North America, many cultures lived sustainable lifestyles on these grasslands. Their populations were relatively small, so they had little impact on the land. Hunting and gathering were the main economic activities. The people relied on the abundant herds of bison and deer and used these animals to supply most of their needs. Meat was eaten, skins were used for clothing and shelter, and bones were substituted for wood to make implements in this treeless ecosystem. Berries, grains, and other wild foods supplemented the diet. Generally the people lived in harmony with the environment and were well adapted to it.

For centuries the herds were so large that hunting had no negative impact on animal populations. This changed, however, once the European settlers introduced horses and guns. Horses first came to the North American plains in the mid-1500s. These animals provided greater mobility and enabled both the settlers and the aboriginal people to hunt more efficiently. In the eighteenth and nineteenth centuries, guns were introduced to the process. These made hunting even easier and led to a dramatic increase in the number of animals, particularly bison, being killed. As a result, what had once seemed like an inexhaustible resource was almost completely wiped out in the nineteenth century. Today the North American bison can be found only in protected reserves. With the demise of the bison, the traditional lifestyle of many native peoples could no longer be sustained.

Resource Development

To appreciate how the grasslands of North America became the great agricultural regions they are today, it is important to understand the historical geography. To European settlers, the North American prairies appeared to be a wasteland. In Europe and eastern North America, the best farmland had to be cleared of trees before it could be farmed. Because the plains lacked trees, settlers believed the region was unsuitable for farming. In addition, the sod was so thick that ploughs could not break it. Even cattle ranching was difficult. The lack of trees meant there was no wood for building fences. Cattle roamed the grasslands freely, identified by the ranch's brand burned into the animal's hide. Because it seemed so inhospitable, the region came to be known as the Great American Desert.

Revolutionary agricultural inventions provided the technology needed to open the land to development. In 1825, John Deere invented the self-scouring steel plough. Now that the deep prairie sod could be turned, the ploughing of the American West began. The invention of barbed wire allowed the range to be fenced with a minimum of wood. Animals could now be contained and crops protected. Drilling equipment allowed water wells to be dug deep into aquifers that lay beneath the farmland. This provided settlers with an abundant supply of water.

In Canada, the prairies were developed after the American West was settled. The Canadian Pacific Railway opened the prairies to settlers from eastern Canada, Europe, and the United States. These pioneers were able to use the naturally fertile soils effectively because of the new technology. Today the grasslands support two main economic activities, farming and livestock ranching.

Farming

The temperate grasslands are the most modified rural landscapes in the world. Virtually every aspect of the region has been changed. The natural vegetation has been replaced with farm crops. The native animals have disappeared, while domestic herds have claimed their territory. The soils have been ploughed, treated, and irrigated. Drainage patterns have been controlled. Even the aquifers are being changed as wells remove water from them. Only the climate remains the same, and even this could be changing as a result of global warming.

Much of the world's population is being supported by the breadbasket regions of North America, Argentina, Russia, and Ukraine. But can we continue to use these grasslands without destroying them?

Whenever crops are harvested, nutrients are removed from the soil. This is true whether it is a grain crop or cattle. To bring the soil fertility back to its previous levels, nutrients must be added. For crop yields to remain high, soil fertility must be kept high.

By the 1930s, the soils of the Canadian prairies and the American plains were much less fertile than they were before intensive farming began. The amount of nutrients being removed far exceeded those being returned. In addition, the region was plagued by droughts. Marginal areas that were only suitable for farming when rainfall was good became **dust bowls**. The natural vegetation had been removed by the farmers. The soils became deeply eroded in places as dust storms ripped away millions of tonnes of topsoil. Crop yields dropped and the region's economy suffered.

These problems continued into the 1940s. After the Second World War, chemical fertilizers were introduced to increase soil fertility. The soils recovered and so did crop yields. As population increased so did the markets for grain and meat. To meet the demand, prairie farmers relied increasingly on fertilizers to revitalize soils. Over the next thirty years, farmers experimented with this chemical technology. Sometimes too much was applied to the land, especially in the spring when runoff is greatest. Phosphorous and nitrogen from the fertilizer entered groundwater and found their way into streams and rivers. Too much of these chemicals stimulated the growth of algae that used up much of the oxygen in the water. Thus natural ecosystems were damaged. Today farmers have a better understanding of soil dynamics. Fertilizers are used in moderation to ensure an abundant crop with little harmful effects on natural ecosystems.

Another problem is pest control. Grain farming is an unnatural process. To plant one or two species over thousands of hectares enables insects or plant fungi to travel from plant to plant until the entire crop is destroyed. In the natural ecosystem, a variety of plants is interspersed throughout the region. It is difficult for pests to spread across vast areas because some plants are naturally

> **GEO-Fact**
>
> We hear everyday that there are too many people to feed in the world for the amount of grain we produce. This is false! Every year more grain is produced than can possibly be eaten by all the people on earth. The problem is not that there is not enough food. The problem lies in the politics of farming. Mismanagement, inadequate transportation systems, and political unrest are the main causes of hunger, not a lack of grain. The issue is very complex. It is not enough to send food aid to developing countries. They need tools, seeds, fertilizers, and training to improve their crop fields. But who will provide this aid and what is the best way to do it?

resistant to them. To compensate for this lack of diversity, artificial pesticides must be used.

Unfortunately, pesticides often kill other things, too. Sometimes beneficial insects are killed off in the process of destroying harmful pests. For example, the preying mantis eats grasshoppers. Grasshoppers destroy grain crops but they are held in check because they are the favoured prey of preying mantises. If these predator insects are inadvertently killed by pesticides then more grasshoppers will survive to eat the crop. Pesticides may also destroy organisms in the soil. These are responsible for breaking down litter and weathering parent material so that nutrients can re-enter the soil. Without these organisms, soils suffer.

In the 1950s, toxic sprays like DDT were used in farms worldwide. This insecticide not only built up in the soil but in animals and groundwater as well. DDT takes over two years to break down naturally. Each year more and more of the poison was added to the soil, thereby increasing the residual amount in the ecosystem. Animals high in the food chain, like hawks, fish, and even people, had DDT in their bodies. As with fertilizers, more ecologically responsible practices are being followed today. New pesticides break down in several weeks. They are much more toxic, so smaller amounts are needed and there are no residual chemicals.

Another common farming practice, especially in the southern plains, is irrigation. Higher evapotranspiration rates make rainfall less effective here than in the cooler north. Irrigating the fields produces higher yields and allows farmers to grow crops of greater value. Water for irrigation often comes from **aquifers**. These are layers of porous rock deep within the bedrock that contain water within their pores. They are underlain by an impermeable layer of rock that prevents the moisture from seeping lower in the rock strata. Water enters and leaves the aquifer naturally where the porous layer comes to the surface. When evaporation is high, water is taken out of the aquifer. In the spring when meltwater swells streams, the aquifer is replenished. This stored water is an abundant resource just waiting to be tapped.

The Ogallala aquifer stretches from Wyoming to Texas. It holds about the same volume of water as Lake Huron. This seemingly endless supply of groundwater has been tapped for forty years. Today, however, advanced pumping systems are removing the water faster than it can be replenished. There is a danger that so much of the water will be removed that the aquifer will no longer be able to provide water for irrigation. This water supply can be sustained only if people do not overuse it.

Sustainable farming methods are being rediscovered by many farmers. Crop rotation improves soil fertility. Instead of growing one crop year after year on the same field, the crop is changed. Growing a legume like soy beans or alfalfa actually increases soil fertility. Bacteria on the roots of the legume fix nitrogen directly from air in the soil. Leaving a field fallow for a year allows the soil to recover lost nutrients. Allowing crop stubble from harvests to remain

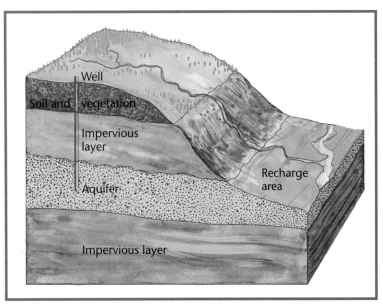

FIGURE 41.2 *How Aquifers Work*
A porous layer is trapped between impervious layers. If a well is drilled, water can be drained out of the aquifer.

in fields until new seeds are planted maintains soil moisture and reduces wind erosion. **Zero tillage** takes this idea one step further as seeds are planted on the fields without any ploughing. Whichever methods are used, sustainable agricultural practices are essential if the breadbaskets of the world are to remain productive.

Cattle Ranching

Cattle ranching in the temperate grasslands today relies on high technology. Ranches are often immense, covering thousands of hectares. Although still used, horses are no longer as feasible as they once were because of the sheer size of the farms. Helicopters are often necessary to monitor operations. All-terrain vehicles carry feed out to animals stranded in winter blizzards. They assist in rounding up cattle and are used to maintain fences, pumps, and other farm equipment that may be in remote locations.

Marginal lands unfit for field crops are usually reserved for cattle and sheep ranching. As long as there is enough land for the pasture to regenerate itself, this type of farming is sustainable. Problems result when too many animals are kept on too little land. If the land is over-grazed, grass cover is reduced. Wind erosion increases. The microclimate is altered and the process of desertification is launched. This has not been a serious problem in North America, but over-grazing has caused ecological problems in Australia where rainfall is less reliable.

Problems occur when precipitation fails. In periods of drought, the amount of land needed to support an animal increases. When this happens, there are three options available to ranchers. They may do nothing and hope that the animals survive, although this option is not usually chosen because many animals would die and the pasture would be destroyed. Another option is to import feed to keep the animals alive. This is expensive since feed prices rise as the demand for feed increases. The third option is for ranchers to send their cattle to market. The pasture is saved but there is often a glut of beef on the market as many ranchers try to reduce the size of their herds at the same time. Beef prices drop and great financial losses can be incurred.

Cattle are often shipped to feedlots in the eastern prairies. Here they feed on locally grown grain so they will earn top dollar when

FIGURE 41.3 *Preparing Cattle for Market*
What are the environmental problems associated with feedlots (left) and grassland ranges?

they are sold for meat. These feedlots concentrate many animals in a small area. This has a great impact on the land. Feedlots are often covered with an impermeable layer of clay. This prevents the manure from running into the water table. When it rains the polluted water is collected in reservoirs and used as a natural fertilizer. Without this precaution, runoff from feedlots could be unhealthy because of bacteria.

Things To Do

1. Prepare an organizer summarizing the characteristics of the temperate grasslands under the following headings: location, climate, plants and animals, soils and landforms, indigenous peoples.

2. Outline the geographic factors that led to the economic success of the North American temperate grasslands.

3. a) List the technical innovations that led to the agricultural development of the temperate grasslands.
   b) Explain how each innovation allowed people to develop the resource base.
   c) Determine the negative environmental effects of each innovation and suggest strategies that would minimize these effects.

4. The extraction of minerals such as potash, natural gas, and oil are other human activities found on the temperate grasslands of North America. Research to find out how mineral extraction affects grassland ecosystems.

Chapter 42
Living in the Maritime Ecosystem

INTRODUCTION

The previous four ecosystems have all been on land. This final study is of a marine ecosystem, so comparisons cannot be made as easily. In Canada we refer to New Brunswick, Nova Scotia, and Prince Edward Island as the Maritime provinces. But in physical geography, *maritime* has a wider application. Maritime regions are those places that depend on resources from the sea. Thus maritime regions include not only the Maritime provinces, but the coastal areas of Newfoundland, British Columbia, Alaska, New England, Europe, and Japan, to name but a few. While our discussion focuses on eastern Canada, many of the problems faced here are global ones.

FIGURE 42.1 *Peggy's Cove, Nova Scotia*

Climate

The climate of Canada's Atlantic provinces is greatly influenced by the ocean. Generally speaking, winters are milder along the coast than further inland, while summers tend to be cooler as ocean breezes moderate temperatures. The cold Labrador Current brings frigid arctic waters south along the coast of Labrador and Newfoundland. It is not uncommon to see icebergs in the waters off eastern Newfoundland even in the hottest of summers. There can be a difference of 5° to 10°C from the coast to inland regions of the big island.

Precipitation is also influenced by the ocean. In the fall tropical storms originating in the Caribbean sometimes reach as far north as Canada's eastern shores, bringing drenching rains and strong winds. In winter, storms originating over Labrador frequently flow out over the region. Pleasant weather is common in PEI, Nova Scotia, and New Brunswick in July and August.

Plants and Animals

Mixed forest in the south gradually gives way to boreal forest and finally to tundra further north along the east coast of Canada. The reason for these gradual changes is the decreasing temperature from south to north.

Southern forests are well adapted to the climate. Deciduous trees are found in sheltered valleys where temperatures are moderated. In winter, these trees shed their leaves to reduce water lost through transpiration in this dormant period. When spring comes, energy stored in the trunks and roots is used to produce the leaves needed for continued growth. Coniferous trees grow on the slopes of the Appalachian Mountains. Unlike deciduous trees, these trees grow all year. In winter, growth is slower because of the low temperatures and the lack of moisture. The thin needles of these trees reduce transpiration, especially in winter when so much of the moisture is frozen. The boreal forest is discussed in detail on pages 302–313.

The real bounty, of course, lies in the ocean. The condition of the water and the plants that live there determine where fish stocks lie. The fishing grounds off Canada's east coast are located in two main areas. Along the shore, the inland fisheries lie over the continental shelf. Further out to sea are the shallow fishing banks. The most famous of these are the Grand Banks, Sable Island Bank, and the St. Pierre Bank.

The coastal waters have always supported an abundance of fish. At the bottom of the food chain are the microscopic plants and animals called **plankton**. The water acts like soil; it provides the nutrients phytoplankton need. Many of the minerals that are dissolved from organic matter and parent material in the huge Great Lakes-St. Lawrence drainage system eventually end up in the Gulf of St. Lawrence. There they enrich the waters in the river estuary and far out to sea.

Another important factor is the cold Labrador Current. The solubility of most minerals is greater in cold sea water than in warm water. In addition, sea water near the freezing point holds approximately twice as much oxygen as sea water at 20°C. This steady supply of nutrients and oxygen allows plankton to flourish, so there is abundant food for fish.

Soils and Landforms

Spodosols are the predominate soil type. Formed under coniferous forests, these soils tend to be acidic and highly leached. Nevertheless they can be fertile depending upon the parent material from which they originate. Prince Edward Island and adjacent shorelines of New Brunswick and Nova Scotia have deep red soils unlike those found in the mountainous parts of the region. These unusual soils formed from red sandstones and shales in the Carboniferous period. Called placosols, these soils support a rich agricultural industry.

Landforms are also varied. In the south, broad coastal plains and wide valleys facilitate settlement and agriculture. Inland the old fold mountains of the Appalachians dominate the landscape. These ancient mountains have been eroded for so many years they look more like rolling hills than mountains. Soils are often thin on the slopes but accumulations in the valleys are common. In Newfoundland and Labrador, the landscape becomes increasingly rugged as the Appalachians give way to the Canadian Shield. The thin soils and cold climate make agriculture a marginal activity here.

Indigenous Peoples

Eastern woodland peoples had a varied diet. They gathered seafood and waterfowl from the lagoons and estuaries along the shore, but they did not venture far out into the North Atlantic. Most of the fish stocks remained untouched until Europeans discovered their abundance. Subsistence agriculture accounted for about 50 per cent of the food supply in southern regions where the climate was suitable for farming. This was supplemented by the gathering of wild berries, fruits, and nuts. Moose and deer were hunted for meat and skins, as were seals and other sea mammals in the northern regions. As with other indigenous cultures, the people maintained a sustainable relationship with the environment.

Resource Development

Although maritime regions provide a variety of natural resources, the one most commonly associated with the ecosystem is the fishing industry.

Fishing

Europeans began fishing the waters around eastern Canada over 500 years ago. They sailed from Portugal, England, and other European countries during the summer to take advantage of the bountiful fish stocks. Because there was no refrigeration, the fish were dried and salted to preserve them. No permanent settlements existed on Newfoundland for many years until Irish and British settlers populated the island. The other provinces were settled by American and British settlers in the eighteenth and nineteenth centuries. Even from earliest times the region depended on the sea, not just for fish but for trade as well.

To say that the fish resources off the coast of Atlantic Canada are managed today would be inaccurate. The resource has been exploited so much that the abundance of the past is gone. It could be restored if the fish were allowed to reproduce undisturbed for several years. The reasons for the destruction of fish stocks are many and include politics, biology, and technology.

The federal government has established the border between Canadian and international waters as 370 km from the shoreline. All fish within this limit are under Canadian jurisdiction. This region includes the inland fishery as well as most of the outer banks. The government establishes annual quotas on the number of each fish species that can be caught by Canadian fishers. Foreign fleets are allowed to fish within the 370 km limit if these quotas are not met by Canadians. Beyond the 370 km limit, there are no quotas. Here over-fishing prevails. Fish swim from Canadian waters into international waters, which are less crowded so there is less competition for food.

Over the years, Canadian fishers have protested over-fishing by foreigners. The inshore fishery accounts for about 75 per cent of fishing by Canadians. Northern cod, the most valuable fish stock, migrates from the

> **GEO-Fact**
>
> The first Europeans to settle in Newfoundland and Labrador were the Vikings. Archaeological evidence suggests that the first Viking settlement occurred around 1003 at L'Anse aux Meadows. Excavations at the site, which began in 1960, have revealed the remains of eight houses, a sauna, a blacksmith, and an ore deposit. In time, the harsh conditions of winter drove the Vikings to abandon the settlement. It would be 500 years before Europeans would settle in Newfoundland again.

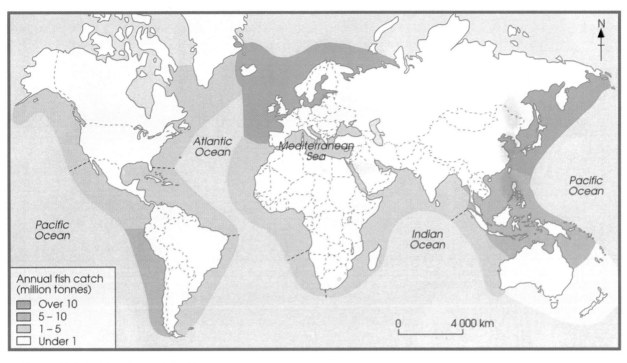

FIGURE 42.2 *Fishing Areas of the World*

Labrador Shelf to Newfoundland shores in the summer. If foreign trawlers over-fish in winter, there is little stock left for small Newfoundland fishing operations in summer. This problem would not exist if the quotas were low enough to enable fish stocks to grow back. Unfortunately, for several years the quotas were set too high. Fish stocks were unable to regenerate as quickly as they were being depleted.

Sampling techniques used to estimate fish populations are crude and inaccurate. One method involves towing a net behind a boat in several locations at the same time every year. Estimates of population for each species are made based on the samples these nets provide. This method fails to take into account the fact that fish do not stay in one place but move about in schools. If the net finds a large school then the figures for that species will be inflated; if it misses the school then the numbers will be low. The second method calculates the "catch per unit of effort." The number of fish a ship catches in one hour is compared to the hourly catch the year before. If the number remains about the same, it is determined that the fish are reproducing fast enough to replace those lost the year before. But this method is also inaccurate. As equipment becomes increasingly sophisticated few fish escape, therefore the size of the catch may remain constant even though there are fewer fish in a given region. The numbers of fish seem more than they actually are because of the improved technology.

By the late 1980s, fishers, biologists, and government officials realized that their estimates of existing fish stocks were too high. Quotas were substantially reduced from 266 000 t in 1988 to 197 000 t in 1990. By 1994, an indefinite **moratorium** banned any fishing of northern cod to allow fish stocks to recover. The effect of the moratorium on the small inshore fishing operations has been devastating.

Over-fishing may not be the only reason for the depletion of fish stocks. Biological factors may also be at work. The population of harp seals has increased rapidly since the seal hunt was banned in the mid-1980s. The seals eat cod, so if their population increases more cod will be eaten. Another factor is the temperature of the sea water. The seabed water temperature off the coast of Newfoundland has dropped

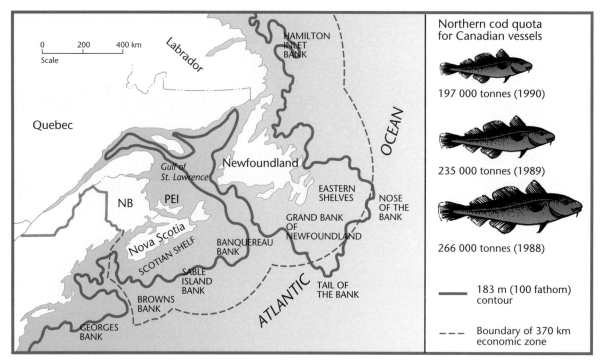

FIGURE 42.3 *Canada's East Coast Fisheries*

1-2°C from 1988 to 1993. The lower temperature results in limited spawning, poor hatching, and slow growth. Fish are cold blooded and activity rates are closely linked to water temperature. The lower temperatures result in lower recovery rates of fish stocks.

It is perhaps technology that has affected fish harvests more than any other factor. Few fish escape the dragnets of the multimillion-dollar offshore draggers. Modern ships with locational sensors can find fish stocks, determine the species, and steer the nets. These huge nets scoop up everything, from mature fish to small stock. If the immature fish are removed, they cannot replenish stocks. This practice has been compared with the clear-cutting of forests. But there is one significant difference: forests have been regenerated. The oceans have not.

So what is the solution? The moratorium on fishing is a first step. Unfortunately it causes severe economic hardships for communities that have relied on fishing for hundreds of years. Seal hunting has been restored as an economic necessity for unemployed fishers. The reduction of harp seals may help cod recover. Further research into fish stocks is essential so that biological changes in the ecosystem can be monitored. Advances in

FIGURE 42.4 *Newfoundland Fishers Protesting the Cod Moratorium, 1993*

FIGURE 42.5 *A Cod Fish Farm at Sea Forest Plantation, Bay Bulls, Newfoundland*

electronic tracking using satellites could help scientists follow tagged fish.

Perhaps the days of harvesting fish in the oceans are numbered. Fishing is the only large-scale hunting operation still in existence. No other form of wildlife is hunted in the same proportions. In Asia and northern Europe, **aquaculture** is becoming a common practice. Fish and shellfish are raised in lagoons where fish farmers can monitor growing conditions and harvest the stock, much like cattle ranchers keep steers in a feedlot. The world leader in salmon farming is Norway. In 1989, this seafaring nation raised 140 000 t of salmon, almost twice as much as British Columbia produces through traditional fishing in a year. In Canada, aquaculture has been introduced in Prince Edward Island to produce lobster and in New Brunswick to produce salmon. In addition, 95 per cent of Atlantic mussels and 25 per cent of oysters consumed by Canadians are the products of aquaculture. There is great potential for this new sustainable industry in the future.

Things To Do

1. Prepare an organizer summarizing the characteristics of the maritime region under the following headings: location, climate, plants and animals, soils and landforms, indigenous peoples.

2. a) Assume one of the following roles: fisher, biologist, fish plant worker, government official.
   b) Prepare an argument for or against the lowering of fish quotas.
   c) Role-play a debate these different people might have over the fishing moratorium.

3. Discuss the reasons for the current decline in fish stocks off the east coast of Canada.

4. Create a plan for sustained economic prosperity in Atlantic Canada.

5. Research another maritime region and compare it with Canada's east coast.

GLOSSARY

abrasion: erosion that occurs when sand, rocks, or some other material rub away the exposed layer of the earth's surface by the action of wind, water, or ice

absolute age: the age of a geologic phenomenon measured in present earth years rather than its age relative to other geologic phenomena

absolute humidity: the amount of water vapour present in the air measured by volume

absolute location: the place where a feature exists as measured on the earth's surface

absolute time: time that is measured in units such as hours, days, millennia, epochs, and eras

acid precipitation: precipitation that contains acid from either natural or artificial processes

acid shock: the extra infusion of acid into ecosystems when winter snow melts

adaptation: the ability of life forms to change in order to survive

aeolian landscape: a region that relates to or is formed by wind

air mass: a large air mass of more or less the same temperature and humidity

air pressure: the measurement of an air mass in a vertical column above the earth's surface

albedo: the percentage of incoming solar radiation that is reflected by surfaces on the earth; the higher the per cent the higher the albedo and the lower the solar energy that is converted to heat

algebraic grid: a number and letter grid commonly used on road maps

alkaline flat: a level area in an arid region overlain by salt deposits

alloy: the mixing together of two or more metals to create a new one

alluvial fan: a fan-shaped landform composed of fine-grained deposits dropped by a river after it loses momentum as it enters a broad valley from a narrow upland course

alluvium: the deposit of graded particles produced by rivers and streams

alpine glacier: glaciation that occurs in mountainous regions

altitude: distance measured in metres above sea level

altostratus clouds: a cloud type characterized by a continuous sheet or veil that often obscures the sun or moon; often accompanied by precipitation

aneroid barometer: an instrument used for measuring air pressure

angle of repose: the greatest angle or slope on which rocks or soil will remain stable

anomaly: in statistics an example that does not have the same characteristics as the rest of the data

anthracite coal: the metamorphic form of coal

anticline: the crest or upfolded part in rock strata

anticyclone: a region of high pressure usually thousands of kilometres in area; also known as a high

anvil heads: broad, massive clouds with flattened crowns that form over cumulonimbus clouds after a heavy thunderstorm

aquaculture: the raising of fish and shellfish on aquatic farms

aquifer: a porous rock under the earth's surface that holds large quantities of water

arboreal: of, living in, or connected with trees

Archaean crust: the oldest rock layers found on earth created when the planet was formed

arête: a steep knife-edge ridge between cirques in a mountain region

aspect: the direction in which a valley slope faces

asthenosphere: the part of the upper mantle that lies beneath the crust; composed of hot, semi-molten rock

atmosphere: the layer of air surrounding the earth

atoll: a circular island with a central lagoon formed when coral deposits fringe a volcanic crater

aurora borealis: coloured and white flashing lights in the atmosphere north of the Arctic Circle created by charged electrical particles in the magnetosphere; also called the Northern Lights

barchan: a crescent-shaped sand dune with the horns of the dune projecting down wind

barometer: a device used for measuring air pressure

barometric pressure: see air pressure

basal slippage: the advance of a glacier by slow movement along the valley floor; occurs in warm glaciers where the ice is at its melting point

base level: the lowest level to which a stream can erode its bed

batholith: a dome-shaped mass of igneous rock formed from an igneous intrusion

bayou: a marshy creek in flood plains of rivers in old age

bearing: the direction measured in degrees from one place to another, north being 0°

Beaufort Wind Scale: a system designed to estimate wind velocity through observation

behavioural adaptation: the ability of animals to change the way they do things in order to survive

bench: a step carved in an open pit mine to enable trucks to move in and out of the pit

big bang theory: the theory that the universe was created as a result of a massive explosion

biodegradable: the ability of matter to decompose and disappear naturally

biogenic sedimentary rock: sedimentary rock formed from once living plants or animals

biomass: the total mass of living organisms in a region

biosphere: the zone of the planet where life is found

bituminous coal: a hydrocarbon created from the fossilization of plants

blizzard: a severe winter storm where the temperature is below -12°C, the wind speed is greater than 40 km/h, and visibility is less than 1 km for a duration of no less than 1 h

butterfly effect: a theory that states that environments are so interdependent that an event as trivial as the beat of a butterfly's wings could affect the weather half a world away

camouflage: the adaptation that many animals have to blend in with their surroundings

canyon: a narrow, steep-sided gorge usually associated with aeolian landscapes

cap rock: an impermeable layer of rock that prevents fluids from escaping a porous layer

capillary action: the ability of water to move upwards through tiny pores because of surface tension

cardinal points: the major points of the compass—north, south, west, and east

carnivore: a meat-eating consumer

carrying capacity: the theoretical number of people the earth is able to support

cartographer: a person who makes maps

ceilometer: a device used to measure the height of cloud cover

cementation: the process in which a binding material precipitates around grains or minerals to form sedimentary rocks

chaos theory: the scientific theory that chaotic systems such as weather have patterns that are too complex for people to comprehend because of the huge number of variables that affect them

chemical energy: energy that is released when a chemical reaction occurs

chemical sedimentary rock: sedimentary rock created from chemical processes

chemical weathering: weathering that occurs as a result of chemical processes

chinook: a relatively warm, dry wind that occurs on the eastern slopes of the Western Cordillera

chlorofluorocarbon: a mainly synthetic chemical containing chlorine, fluorine, and carbon and used in propellants, refrigerants, and cleansers

circle of illumination: the line seen from space that separates day from night

cirque: a circular hollow with steep sides cut into bedrock during alpine glaciation

cirrostratus clouds: a milky, thin veil of high clouds that does not obscure the sun or moon

cirrus clouds: high, slender clouds with a feathery appearance

clastic sedimentary rock: sedimentary rock made from pieces of other rock

clear cutting: a harvesting technique used in lumbering where all trees are removed from a region

climate: the average weather a place has over a long period of time

climax vegetation: the final stage of development in an ecosystem in which no further changes occur

closed system: a system in which there is no transfer of matter into or out of a structure

coke: bituminous coal that has had impurities removed through heating

col: an erosional feature formed when two cirques join together to form a saddle-like pass through mountains

column: in karst topography the feature that is formed where a stalactite and a stalagmite join together

compaction: the pressing together of sediments to form rock

compass rose: an arrow indicating the cardinal points

condensation: the change of state from a gas to a liquid; the water that forms on cool surfaces

conical projection: the projection that is created when a map is shaped like a cone over a globe

consumer: an animal that eats producers and other consumers

contact metamorphism: metamorphic rocks that are formed in contact with igneous intrusions

continental drift: see plate tectonics

continental glacier: an ice sheet that covers a large part of a continent

contour interval: the vertical change between the consecutive lines that join places of equal elevation on a map

contour mapping: the technique of using lines to join points of equal elevation

convection: the process in which thermal (heat) energy is transferred from one place to another; hot air/water rises, cool air/water rushes in to fill the void, hot air/water cools and drops

convection current: the movement of a fluid as a result of heating

convection precipitation: precipitation that occurs as a result of condensation in convection currents

convectional flow: see convection current

co-ordinates: a pair of numbers or letters that designates the location of a place on a map

coral: mineral formations formed from tiny sea polyps in the ocean

core: the centre of the earth beginning at a depth of about 2900 km

coriolis effect: the apparent force caused by the earth's rotation which directs a moving body on the earth's surface to the right in the Northern Hemisphere and to the left in the Southern Hemisphere

crest: the top of a wave or dune

cross-sectional area: in stream dynamics the measurement of the average stream depth multiplied by the stream width

crust: the outer shell of the earth, including the continents and the ocean floor

cumulonimbus clouds: heavy dark clouds of great height associated with stormy weather

cumulus clouds: billowing white clouds with flat bases and rounded white crowns

cyclone: a region of low pressure in which winds move in a spiral around a central low; in the Northern Hemisphere winds move in a counterclockwise direction and in the Southern Hemisphere in a clockwise direction

cylindrical projection: the map projection created when a map is shaped like a cylinder around a globe

daily maximum temperature: the highest temperature recorded during a 24 h period

daily minimum temperature: the lowest temperature recorded during a 24 h period

decomposer: a plant or animal that breaks down dead organic material

delta: a low-lying area found at the mouth of a river and formed of deposits of alluvium

dendritic drainage pattern: a river system with a branching network similar to that of a tree

deposit: sediment that is laid down as part of a gradational process

deranged drainage pattern: a medley of islands, lakes, marshes, and streams that are poorly drained

dew point: the temperature at which water vapour in the atmosphere condenses

diastrophic processes: the faulting, folding, and plate movements that produce mountain ranges, rift valleys, continents, and ocean floors

dike: a vertical sheet of igneous rock formed from an intrusion of magma

discharge rate: the volume of water that flows past a given point on a river in a given measurement of time

diurnal temperature range: the difference between the daily maximum and daily minimum temperatures

doldrums: an area of low pressure associated with the thermal equator

doline: a sink hole found in karst topography

drainage pattern: a pattern that results as water flows over the earth's surface

dredging: a mining technique where underwater sediments are excavated

drumlin: a cigar-shaped hill associated with continental glaciation

dry adiabatic lapse rate: the speed at which temperature drops as a result of a change in air pressure; temperature drops about 1°C for every 100 m of increased altitude

duration: the length of time the sun is above the horizon in a 24 h period

dust bowl: a region of accelerated erosion that creates extreme dust conditions

dynamic metamorphism: metamorphism caused mainly by diastrophic processes

earthquake: sudden and unpredictable tectonic movements in the earth's crust

easting: a reference line drawn across a map from north to south

economic deposit: a mineral deposit that costs less to extract than its market value

ecosphere: see biosphere

ecosystem: a unique community within the ecosphere

eddy: a roughly circular movement within a current of air or water

El Niño: unusually warm ocean currents that occur off the coast of Peru

electron: the elementary particle of a mass

element: a basic substance that cannot be decomposed into anything lower

elevation: the height above sea level

emergent coastline: a coastline that rises out of the sea because of reduced sea level or tectonic activity

energy: the ability of a force to do work

environmental lapse rate: the rate at which air temperature drops as altitude increases

epicentre: the point of the earth's surface that is directly above the centre of an earthquake

epiphyte: a plant species that grows on another without drawing any nutrients from it

equatorial low: a low pressure cell found over the thermal equator

equinox: having a day and night of equal length

erg: a sandy desert

erratic: a large boulder that has been transported by a glacier from another geological region

erosion: the weathering and transport of weathered material

esker: a depositional glacial feature associated with flowing water in, on, or under a glacier

estuary: the area of a river mouth affected by sea tides

exfoliation: the weathering process in which layers of rock are peeled away

exotic species: plants or animals that have been introduced from another region

exposure: a valley slope that is unprotected from the elements

extrusion: see extrusive volcanism

extrusive igneous rock: a formation of rock made of magma that has erupted on the earth's surface as lava and has solidified

extrusive volcanism: volcanic formations formed from lava

fault plane: the surface where two faults meet

faulting: movement that produces displacement of one rock mass relative to another along a fracture

ferrous metal: a metal that contains iron

fetch: the continuous area of ocean over which the wind blows to create waves

finger lake: a lake formed from glacial gouging

firn: glacial snow

fjord: a u-shaped valley that is inundated by sea water

flood plain: that part of a river system that floods when a river overflows its banks

flow velocity: the speed at which wind, water, and ice move

fold mountain: a mountain formed by folding

folding: the distortion of the earth's crust by lateral pressure

foliation: the banding of rock minerals commonly found in metamorphic rocks

fossil: the remains or traces of plants or animals that have been preserved in the geological record

fossil correlation: the use of fossils to determine the relative age of sediments

fossil fuel: a fuel that has been stored in rock strata

four-figure grid reference: the use of eastings and northings to show location on topographic maps

front: the line at which two different air masses meet

frontal precipitation: precipitation that occurs as a result of cyclonic storms

frost shattering: weathering that occurs as a result of water freezing and thawing in tiny cracks within a rock surface

funnel cloud: see tornado

gangue: the waste material left after the ore in a mine has been extracted

gazetteer: an index listing all the places found in an atlas

gene pool: the population available for breeding that determines the genetic diversity available to a species

geological record: the information about prehistoric times that is preserved in rock layers

geological time scale: the scale that divides time in abstract terms and subdivides rocks in geological terms since the earth began

geology: the study of the formation, development, and nature of rocks and their landforms

geomorphic cycle: a theoretical model explaining how landforms develop

geosphere: the earth including the atmosphere, lithosphere, hydrosphere, and biosphere

geothermal energy: energy that originates deep within the earth

glacial retreat: the melting of snow in a glacier faster than it accumulates, causing the glacier to shrink

glacial surge: a relatively rapid advance of an alpine glacier

global warming: see greenhouse effect

gräben: a rift valley formed between two faults

gradational processes: the weathering, transportation, erosion, and deposition of material through the actions of water, wind, and ice

graded particles: particle sizes sorted by flowing water that gradually change from coarse to small grains as flow velocity decreases

gradient: the slope of some feature of the environment

gravity fault: a fault along which one block of strata has dropped in relation to another block

greenhouse effect: the absorption of heat energy by greenhouse gases and reradiation into the atmosphere

greenhouse gas: gases that allow light energy to pass through but that reradiate heat energy back to the earth

grid: a system of vertical and horizontal lines used to pinpoint a location on a map

gum arabic: the sap of the acacia tree used in the manufacture of adhesives and inks

half-life: the rate of decay of a radioactive isotope; the half life is the time required for half of the original parent isotope to decay

hanging valley: a glaciated tributary valley that is higher than the lower main valley

harmattan: a hot, dry wind originating in the Sahara that blows into southwest Africa bringing relief from equatorial humidity

Hawaiian volcano: a volcano characterized by extremely fluid basaltic lava

headland: a shoreline that projects into the ocean

herbivore: a consumer that eats vegetation only

high pressure: see anticyclone

hibernate: a dormant state experienced by some animals living in climates with cold winters

horizon: a layer in soil development

horn: an alpine glacial feature formed as a result of three or more cirques eroding the sides of a mountain

horst: a rock that is thrust up between two faults

horticulture: the intensive farming of fruits, vegetables, and plants

human time scale: the time scale measured in hours and years

humidity: the amount of moisture held in air

humus: decomposed plant and animal matter contained in most soils

hurricane: a disturbance several hundred kilometres across spinning around a central area of low pressure, with winds exceeding 140 km/h

hybrid: a plant or animal that is genetically engineered to produce a better crop

hydrocarbon: a fossil fuel containing only hydrogen and carbon compounds

hydrolysis: chemical weathering that occurs when carbonic acid dissolved in water acts upon silicates in rocks

hydrosphere: that part of the geosphere that is made up of water

ice age: the period in which temperatures plunged and large parts of the earth's surface were covered in glacial ice

igneous intrusion: see intrusive igneous rock

igneous rock: rock formed from molten magma beneath the earth's surface

impermeable: rock that does not allow fluids to flow through it

inbred: a genetic condition that accentuates inherited characteristics as a result of a reduced gene pool

infrared satellite image: a picture taken from space that shows the reflectability of surfaces

indicator minerals: less valuable minerals that are commonly associated or found with other more valuable minerals

inputs: the nutrients that are returned to soil either naturally or artificially

inselberg: a resistant rock fragment left after an extended period of aeolian erosion

intensity: the amount of solar radiation that reaches a place due to the inclination of the sun

intercardinal points: the parts of the compass that lie between the cardinal points; for example, northeast

intercropping: a horticultural technique in which two or more crops are grown interspersed among each other

interfluve: an area of land between two rivers

interglacial period: a distinct period of warmer conditions between glacial periods when the earth's glaciers have shrunk

interlobate moraine: a moraine that forms between two lobes of a glacier

International Date Line: the line following 180° longitude

intrusive igneous rock: a mass of igneous rock formed from magma that forced its way through existing rock and solidified beneath the ground

ion: an atom that has acquired an electric charge by the loss or gain of one or more electrons

isohyet: a line on a map connecting places of equal precipitation

isostasy: the theory that accounts for the upward thrusting of mountains

isotope: an element having the same chemical characteristics as another element but distinguished in other ways such as radioactivity

jet stream: a meandering band several hundred kilometres across of strong westerly winds blowing at 100 km/h or more in the upper air between 9000 and 15 000 m

kame: a small hill deposited by glaciation

karst topography: a landscape characterized by limestone rocks honeycombed with caves, caverns, and dolines and with little surface water

katabatic wind: a cold wind that flows down slopes in a mountainous area because of variations in mass

kerogen: a complex hydrocarbon from which oil evolves

kettle: a glacial feature formed when ice is buried in till creating a hollow often filled with water (kettle lake)

kimberlite pipe: an igneous intrusion that often contains diamonds

kinetic energy: the energy of motion

laccolith: an igneous intrusion that spreads and forces overlying strata into a dome

lacustrine: a deposit of fine sediment created under a now-dry lake

lahar: a mud flow on the side of a volcano caused by an eruption

landscape system: the way in which energy patterns and flows are organized in a particular environment

lateral moraine: a moraine created along the edge of an alpine glacier

latitude: circles drawn around maps of the earth parallel to the Equator and measured in degrees, minutes, and seconds

lava: molten rock found on the earth's surface

leaching: the movement of water down the soil profile that carries water soluble minerals to lower horizons

linear scale: a line indicating distance on a map in relation to actual distance

lithosphere: the solid part of the planet, including the core, crust, and mantle

load: the material transported by running water

longitude: the imaginary lines measured from east to west from the Prime Meridian in degrees, minutes, and seconds

longitudinal dune: a dune that runs parallel to the prevailing winds

longshore drift: wave action that transports sand along a beach

low pressure: see cyclone

magma: molten rock found beneath the earth's surface

magma chamber: a large chamber in the earth's crust occupied by molten rock

magma conduit: a duct through which molten rock flows in a volcano

magnetometer: a device used to find magnetic anomalies such as kimberlite pipes

magnetosphere: the full extent of the earth's magnetic field

mantle: the part of the lithosphere between the crust and the core

map projection: a method of creating a map

mass wasting: the gradational process that results from the force of gravity; includes soil creep, landslides, lahars, mud flows, avalanches, and solifluction

mean annual temperature: the sum of the average monthly temperatures divided by twelve

meander: one of a series of loops that form in an old river flowing at base level as a result of lateral erosion

meander scar: an oxbow lake that has dried up

mechanical weathering: the splitting of rock caused by the physical actions of wind, water, and ice

medial moraine: a moraine created between two coalescing alpine glaciers

Mercator Projection: the most common form of cylindrical map projection; it distorts high latitudes but keeps accurate direction

mesa: a table-shaped, flat-topped hill common in aeolian landscapes

mesopause: the dividing line between the mesosphere and the thermosphere

mesoscale eddy: a chaotic ocean current that behaves similarly to a storm in the atmosphere

mesosphere: the atmospheric layer between the stratosphere and the troposphere

metabolic processes: chemical processes in life forms in which nutrients form living tissue

metamorphic rock: rock that has been transformed by heat or pressure beneath the earth's surface

methane: a hydrocarbon formed from ancient organic deposits; also called natural gas

microclimatology: the study of climate variations on a small scale

military grid: a system used on topographic maps to show location

mineral: a naturally occurring substance made up of specific elements possessing definite physical and chemical properties

minute: one-sixtieth of a degree of latitude and longitude

molten rock: rock that is liquid because of heat and pressure

moraine: a glacial hill formed along the edge of a glacier

moratorium: the temporary suspension of an activity

muskeg: a swampy, poorly drained region often underlain by permanently frozen ground

natural resource: a naturally occurring substance used by people

natural selection: the natural process in which animals either adapt to changing environments or fail to survive

natural vegetation: those plants that grow naturally in a region without human interference

niche: a set of ecological conditions that provides a species with the energy and habitat that enables it to reproduce and survive

nimbostratus clouds: low-lying grey clouds of uniform base often associated with precipitation

non-ferrous metal: a mineral that does not contain iron

non-porous: not able to be permeated by air or water

non-renewable resource: a natural resource that is finite and cannot be restored

northing: on a map, a reference line running east-west and numbering up from south to north

nuée ardente: a blast of extremely hot, gas-filled air and volcanic debris that occurs when a blocked magma conduit explodes

occluded front: a stage in a mid-latitude depression where a cold front overtakes a leading warm front

omnivore: a consumer that eats meat and vegetation

open system: a system that is not separate from its environment but transfers matter into and out of it

open-pit mine: the mining of ore by excavating a giant hole in the ground

ore: a mineral deposit that is economically worth mining

organic matter: material that either is or was living

orographic precipitation: precipitation that occurs as relatively warm humid air is forced to rise over an elevated landform such as a mountain; also called relief precipitation

outputs: the nutrients that are removed from soils either naturally or artificially

outwash fan: see alluvial fan

overburden: the waste material removed from an open-pit mine to get to the ore body

oxbow lake: a lake formed in a meander that has been cut off from its river

oxidation: the absorption by a mineral of one or more oxygen ions

ozone layer: a form of oxygen 15 to 50 km above the earth's surface that absorbs harmful ultraviolet radiation from the sun

parabolic dune: a sand dune created when vegetation prevents erosion at each end, creating a crescent shape

parallel drainage pattern: a drainage system in which tributaries run parallel to the main river because of underlying topography

parasite: a plant species that grows on another by drawing nutrients from the host plant

parent material: the rocks, minerals, and other matter from which soils form

pastoralism: an economic activity based on the raising of cattle and other animals

pediment: the coalescing of several alluvial fans to form a gently sloping plain

pedology: the study of soils

permafrost: an area of rock and soil where temperatures have been below freezing for at least two years

permanent base level: see base level

permanent resource: a natural resource that is abundant in all ecosystems

permeability: the ability of soils to be drained of water

photosynthesis: the process whereby producers (plants) create carbohydrates from air, water, and sunlight

physical adaptation: the ability of animals to change their body structure in order to survive

phytoplankton: microscopic plants found in sea water

pioneer species: the first plants to colonize new land

pirated river: a river that has been captured by another more powerful river

placer deposit: a mineral deposit found in flowing water

planar projection: a map projection in which only one standard point touches the map surface; useful for mapping the poles

plankton: microscopic plants and animals found in the ocean

plastic: the ability of a solid mass to act like a fluid because of the great pressure acting upon it

plasticity: the property of a material that enables it to change shape without changing its volume or rebounding or cracking

plate: a portion of the earth's crust that is bounded by faults

plate tectonics: the theory that continents float on a less dense layer of the mantle

playa: see salt pan

polar easterlies: prevailing winds found between the poles and 66° that blow from the east

ponding: a feature of continental glaciation where meltwater accumulates between lobes of a glacier or between a glacier and a moraine

porosity: the ability of soils to hold water or air

potential energy: energy that lies latent waiting to be released

precious metal: a mineral with a high market value

precipitate: the process in which minerals separate out of magma as it cools; also the process in which water loses the load it transports

prehistoric: before people used writing to record events

prevailing winds: the usual direction winds flow at a given latitude

Prime Meridian: the 0° meridian located at Greenwich, England, from which all other lines of longitude are measured

producer: a plant that is able to create living matter from sunlight, air, and water

protoplanet: the precursor to a planet

psychrometer: a barometer used to measure air pressure

radial drainage pattern: a drainage system in which a river radiates out from a cone-shaped mountain to resemble spokes on a wheel

radioactive decay: the change of one element into another through the emission of charged particles

radiocarbon dating: the process by which the age of an organic substance can be determined by measuring the proportion of the isotope C14 in the carbon it contains

radiometric dating: a process by which the age of a substance can be determined by measuring the proportion of an isotope it contains in relation to the total amount of the given element

rainshadow: an area of relatively low rainfall on the leeward side of an elevated landform

reef: an offshore coral ridge that forms under tropical waters

reg: a rocky desert

regolith: the layer of loose rock that covers the surface of the land

rejuvenation: the process by which old streams and rivers regain energy due to the uplift of land through isostasy or a drop in the sea level

relative humidity: the ratio between the actual amount of water vapour in a given volume of air and the total holding capacity the air has at a given temperature

relative location: the place where a feature exists in relation to other features

relative time: time measured in terms of other events; for example, lunch is eaten before supper

relief: the relative difference between the highest and lowest point in a given area

remote sensing: a technologically advanced mapping technique that utilizes satellites, sensors, and digital technology

renewable resource: a natural resource that regenerates naturally

reservoir rock: a layer of rock strata that holds either methane or oil

resolution: the details of a digitized image; the higher the resolution the clearer the image

reverse fault: a fracture along which one block of rock has been raised relative to another

ria coastline: a submergent coastline formed when river valleys are flooded

Richter Scale: a logarithmic scale used to measure the magnitude of earthquakes

rift valley: see gräben

rise: the difference in elevation between two points on a slope

river capture: see pirated river

rock: a naturally occurring substance made up of minerals

run: the distance between two points on a slope

saltation: in river dynamics that part of the load that rolls along the stream bed

saltpan: see alkaline flat

scale ratio: the relationship between a distance on a map and the corresponding distance on the ground expressed as a ratio; for example, 1:12 000

scale statement: the relationship between a distance on a map and the corresponding distance on the ground expressed in a statement; for example, 1 cm represents 1 km

scattergraph: a graphic means of showing the correlation between two sets of data; two scales, one for each type of data, are drawn at right angles to each other and each set of data is plotted; if the line extends from the lower left to the upper right it is a positive correlation; if it extends from the lower right to the upper left it is a negative correlation

scooptram: a device used to remove ore in shaft mines

seamount: an extinct submarine volcano that provides a habitat for coral polyps

seconds: a unit of measure of latitude and longitude equal to one-sixtieth of a minute

sedentary: a system of land tenure in which people remain in one place

sediment: solid material that has been moved in suspension by water

sedimentary rock: rock formed from sediment

seismic wave: a wave travelling through the lithosphere caused by an earthquake

seismograph: a graph showing the earth's vibrations

seismologist: a scientist who studies earthquakes and other earth movements

seismometer: a devise used to graph seismic waves

selective cutting: the harvesting in intervals of mature trees in a forest of mixed age

shaft mining: a system of mining in which tunnels and shafts are used to access the ore body underground

sharecropping: a land tenure system in which a tenant farmer pays rent to a landowner in produce rather than in cash

sheeting: see exfoliation

sial: the layer of the crust under the continents made up primarily of silica and aluminum

sill: an igneous intrusion running parallel to the earth's surface

sima: the lower part of the continental and oceanic crusts made up largely of silica and magnesium

six-figure grid reference: an organized system of locating a place on a topographic map with greater accuracy than the four-figure grid reference system

skerry: a moraine that forms at the mouth of a fjord and appears as a string of small islands

slash and burn agriculture: the clearing of land, usually in the tropical rainforest, where trees are cut down, the land is cleared of the trunks, and the remaining vegetation is set afire

slipface: in aeolian erosion the side of a sand dune where sand slides down the slope

sluice: a mining device that uses water to separate minerals from sediment

smelting: removing impurities from metal ores using heat

snout: the foremost extent of a glacier

snow line: the level of altitude above which snow is found on a permanent basis

soil creep: the slow and gradual movement of soil down a slope

soil profile: the succession of horizons beginning at the earth's surface and extending down to the bedrock

solar wind: the flow of atomic particles from the sun

solifluction: mass wasting that occurs in areas where the soil becomes saturated because of permanently frozen subsoil; occurs in Arctic regions

solstice: the day when the sun is furthest away from the Equator and appears to stop before returning to the Equator

solution: in river dynamics that part of the load that is dissolved in the water

sorted sediments: layers of different-sized particles often found where ice age deltas once formed into ancient lakes or seas

sound: a body of fresh water found between mainland and barrier islands

source material: the material from which soil is created; similar to the regolith

source region: the area in which an air mass originates

spillway: a river valley carved by run-off from a continental glacier

stalactite: in karst topography an icilcle-like rock formation, usually of limestone, hanging from the roof of a cave

stalagmite: in karst topography, a rock formation, usually of limestone, formed on the floor of a cave from water dripping from the roof

standard line: a line of latitude used as a basis for a map projection or grid reference system

standard point: a point that is used as a basis for a map projection or grid reference system

star dune: a sand dune that is shaped like a star because of winds blowing from different directions

stratosphere: the part of the atmosphere between the troposphere and the mesosphere

steppe: a short grassland region that receives between 250 mm and 500 mm of precipitation a year

stratopause: the dividing line between the stratosphere and the mesosphere

strike-slip fault: see transform fault

strip cutting: a harvesting technique where strips of trees 20 to 100 m wide are removed from a forest

stromatolite: ancient sedimentary rock made from fossilized algae believed to be the oldest remains of life on earth

subduct: the movement of one plate under another

subduction zone: a region where one plate moves under another

submergent coastline: a coastline that is sinking into the sea either from increased sea level or from tectonic activity

subsistence farming: a form of agriculture in which almost all of the produce goes to feed and support the household and none is available for sale

superposition: the geological principle that states the age of sediments increases with depth

suspension: in river dynamics that part of the load that is carried by water without touching the stream bed

sustainable development: the modification of the environment in such a way that it may be used to produce raw materials indefinitely

syncline: a trough or downfolded arch in a folded landscape

taiga: an unbroken band of boreal forest covering much of northern Canada, Russia, and Scandinavia

tailings: fine waste material removed from a shaft mine or other mineral-processing plant that is too poor for further treatment

talus slope: a sloping structure made up of cobbles, boulders, and various other particles; a feature of mass wasting

tectonic process: the process whereby the land is changed due to forces within the earth

tectonic stability: the degree to which the crust stays still in a given region; a region lacking earth movement has high tectonic stability

temperature inversion: a condition where the temperature close to the ground is colder than the temperature at a higher elevation

temperature range: a calculation made by subtracting the colder temperature from the warmer temperature

temporary base level: a base level that only lasts for a short period of time because of the damming of run-off

terminal moraine: a moraine found at the furthest extension of a glacier

terrace: a step-like bench bounded on one side by an ascending slope and on the other side by a descending slope

thermal energy: energy produced by heat

thermal equator: the part of the earth that is directly under the sun—that is, where the angle of the noonday sun forms a 90° angle; it moves north and south between the Tropic of Cancer and the Tropic of Capricorn each year

thermal expansion: a force in weathering that causes rock layers to break off in sheets as a result of large daily temperature changes

thermosphere: the layer of the upper atmosphere above about 80 km from the earth's surface where temperatures increase with height

thrust fault: see reverse fault

topography: the shape of the land

tornado: a destructive, rotating storm under a funnel-shaped cloud that advances across land along a narrow path

trade winds: winds that blow from the northeast (Northern Hemisphere) and the southeast (Southern Hemisphere) located between the Tropic of Cancer and the Tropic of Capricorn

transform fault: a fault in which plates slide horizontally past each other without any change in elevation

transitional climate: a climate that takes on characteristics of the two climates found bordering it

translocation: the transfer of soil substances in solution or suspension from one horizon to another

translucent: allowing light to pass through without being transparent

transpiration: the process by which water escapes from a living plant and enters the atmosphere

transport: the process in which weathered material moves from one place to another

transverse dune: a giant ripple-like sand dune formed at a right angle to the direction of the prevailing wind

trellis drainage pattern: a drainage system in which tributaries join the main river at right angles, creating a structure resembling a garden trellis

triangulation: a method of surveying using a base line and two sight lines to form a triangle; an exact point can be determined given the distance to the point from three different locations

tributary: any stream that flows into another stream

tropopause: the area that divides the troposphere from the stratosphere

troposphere: the layer between the stratosphere and the lithosphere

trough: a shallow depression between the crest of waves

tsunami: a wave formed from tectonic movement under the ocean

u-shaped valley: a valley where the lower slopes have been scoured away by a glacier, creating a broad valley floor and steep slopes

unconsolidated material: rock particles that are not cemented together

uniformitarianism: a geological principle that states that the laws of nature operate today the way they have in the past

uranium-238: a radioactive isotope

vegetation succession: the evolution of natural vegetation from simple plant communities to complex ones

vein: an igneous intrusion of an economic mineral

ventifact: streamlined rock formations that were smoothed by the force of wind erosion

Vesuvian volcano: a volcano that has thick lava that solidifies and plugs the magma conduit, causing violent eruptions of ash and cinders

volcanic plug: magma that has solidified in a magma conduit and is now exposed because the rest of the volcano has been eroded

volcanism: tectonic process caused by volcanoes

volcano: a mountain formed from molten rock

wadi: a water course that only has water flowing in it for part of the time; also called intermittent stream

watershed: the total area drained by a river and its tributaries

wave refraction: the wave pattern that occurs when waves bend around headlands; more wave action is concentrated on the headland, eroding it more than sheltered beaches

wave train: a series of waves with the same characteristics

weather: all the conditions that exist in the atmosphere

weathering: the process whereby rocks are worn down or broken up

weathered material: rock particles that have been broken away through the processes of weathering

westerlies: winds in the middle latitudes that blow from the west

wet adiabatic lapse rate: the rate at which rising air becomes saturated to the dew point, creating condensation of vapour into water droplets and releasing latent heat that partly offsets the cooling of the air with height; occurs at a rate of about 6°C per 1000 m

wind duration: the length of time the wind blows from one direction; an important variable for wave formation

wind gap: a dry valley that once had a stream flowing through it

wind shear: the difference between the magnitude and direction of an upper level wind and a lower level wind

wind velocity: wind speed

xerophyte: plant species that are adapted to dry ecosystems

yazoo stream: a tributary that flows alongside a major river but does not join it for several kilometres because it cannot breach the levee separating the two; a feature of old rivers

years before the present (ybp): the unit of measure for geological time periods

zero tillage: a method of planting crop seeds that does not require ploughing

zone of ablation: the part of a glacier that melts faster than it grows

CREDITS

Illustrations

The Canadian Oxford School Atlas, 6th Edition (Oxford University Press Canada, 1992): pages 6, 9, 45, 271

Dunlop, Stewart and Michael Jackson, *Understanding Our Environment* (Oxford University Press Canada, 1991): page 310

Energy, Mines, and Resources Canada, Surveys and Mapping Branch: pages 10, 132, 220, 238, 240, 243, 254, 255, 304

Environment Canada, Atmospheric Environment Service: pages 77, 79, 84

French and Squire, *Modern Certificate Geography: A Man-Land Approach* (Oxford University Press China, 1993): pages 49, 50, 51, 58, 63, 88, 170, 182, 186, 200, 203, 209, 218, 241, 250-51, 293, 294

Fundamentals of Physical Geography, Third Edition, by Trewartha, Robinson and Hammond. © McGraw-Hill, Inc. 1977. Reprinted by permission: page 116.

Geomatics Canada, Canada Centre for Mapping: Page 11

Grenyer, Neville, *Investigating Physical Geography* (Oxford University Press, 1985): pages 27, 48, 91, 184

Institut Géographique National, Paris, France: page 177

Reprinted by permission of the James Bay Corporation: page 312

Keung Yiu Ming, *Contemporary Geography 3* (Oxford University Press China, 1993): pages 166, 335

National Aeronautics and Space Administration: page 245

Ontario Ministry of Transportation: page 12

People and Their Environment by Graves, Lidstone, et al, Heinemann Publishers (Oxford) Ltd., 1987. Reprinted by permission: page 60 (top)

Porter, John, *Physical Environment and Human Activities* (Oxford University Press, 1989): pages 174, 180, 237

Stanford, Quentin, *Geography: A Study of Its Physical Elements* (Oxford University Press Canada, 1988): pages 47, 60 (bottom), 72

Trails Illustrated, P.O. Box 3610, Evergreen, CO 80439-3425, map courtesy of: page 187

UNESCO, *The Human Impact on the Natural Environment* (The United Nations, 1992): page 322

United States Geological Survey: pages 178, 191, 232, 256

Photos

Abril Imagens: Oscar Cabral, page 299

Agriculture and Agri-Food Canada Research Branch, CLBRR: page 381 (middle and right)

CANAPRESS: pages 94; Andrew Vaughan, 336

Case, Loralee: pages 40, 62, 229, 247, 269, 270 (bottom), 332

Chasmer, Ron: pages 56, 158, 160, 161 (both)

CIDA Photo: Virginia Boyd, page 53 (left); Paul Chiasson, page 319; David Barbour, page 323

COMSTOCK: S. Vidler, page 295

Cowan, Maxine: pages 210, 266 (middle left)

The Field Museum: CK23Tc, page 148; CK30Tc, page 151

FIRST LIGHT: D. & J. Heaton, page 77; B. Milne, page 291; Jerry Kobalenko, page 302; Larry J. MacDougal, page 337

Geological Survey of Canada: page 197

Geophysical Institute, University of Alaska Fairbanks: R. Overmyer, page 73

GEOPIC®, Earth Satellite Corporation: pages 4, 26, 179, 219 (top), 227, 252

Harvey, Al: pages 53 (top right), 108 (top and bottom), 242, 266 (top left), 279 (left), 306 (top left), (middle left), (middle right), (bottom), 330 (left)

Hong Kong Government Information Services: page 208

Last, Victor, Geographical Visual Aids, Wiarton, ON: pages 70, 146, 159, 202, 203, 206, 221, 223, 229 (left), 258, 262, 263, 268 (top), 269 (top), 270 (top), 288, 306 (top right), 314, and background photo for pages 3, 15, 69, 145, 199, 257, 287

Miller, Ron/*The Globe and Mail*: page 138

Ministry of Northern Development and Mines: © Queen's Printer for Ontario, 1994. Reproduced with permission: page 160 (both)

Morgan, Vince: page 286

NASA: page 320

Penner, Arlene: pages 16, 103, 106 (both), 157, 181 (top), 219 (bottom) 268, 279 (right), 315, 330 (right)

PONOPRESSE INTERNATIONALE INC./REX: page 181 (middle)

PUBLIPHOTO: K. Krafft/Explorer, page 22; 74; P. Dannic/DIAF, page 100; J.C. Labbe/SYGMA, page 115; M. Renaudeau, page 119 (left); Denis-HUOT, page 119 (right); J.C. Hurni, page 126; G. Gayrard/Explorer, page 130; Warren Faidley, page 135; T. Campion/SYGMA, page 192; Jaffre, page 229 (right); A. Keler/SYGMA, page 260; P. Quittemelle, page 266 (top right); M. Lowan, page 266 (middle left); M. Petit, page 266 (middle right); Y. Hamel, page 266 (bottom); Paolo Da Silva/SYGMA, page 296; A. Allstock, page 298

Ritchie, Frances, pages 233, 325

Royal Tyrrell Museum/Alberta Community Development: page 150 (top), J. Sovak, page 150 (bottom)

Soils of Canada: pages 274 (both), page 318 (No. 1)

Photo-Swiss National Tourist Office, pages 66, 175

TONY STONE IMAGES: Jean-Marc Truchet, page 252; Jacques Jangoux, page 292; Steve Taylor, page 292; M. Sabliere, page 300

Tilsley, Jim: pages 163, 168

U.S. National Park Service: page 181

INDEX

acid rain 310-311
aeolian landscapes 228-231
agriculture. See farming
air. See atmosphere
air masses 85-86
air pressure 32-33
   humidity and 47-48
   measurement 81
   temperature and 32-35
   weather and 71
   wind and 34, 35, 36
albedo 17-18, 322
alfisols 283, 317-318, 326-327
alluvial fans 228, 231
alpine glaciers 234-235, 241-244
Al Qattarah Depression 311
altitude 32-33
Alvarez, Dr. Walter 138-139
American Soil Conservation
   Service 282-283
anemometer 82
aneroid barometer 81
angle of repose 207
Antarctic ice sheet 234
anticyclones 89
aquaculture 337
aquatic ecosystems 300
   acid shock 310-311
aquifers 328, 329-30
Aral Sea 265
Archaeon crust 168
Archaeozoic Era 148
aridisols 277, 282, 326-327
asthenosphere 153, 155
atmosphere 5, 18, 26, 70-73
   composition 21
   energy systems 27-39
   gases 71
   greenhouse effect 59
   layers 73-73
   pressure 32, 34-45
   properties of 71-73
   radiant energy 27
   solar exposure 33
   temperature 30-33
   wind systems 36-39
Atmospheric Environment
   Service (AES) 99, 141
atolls 253
Aurora Borealis (Northern
   Lights) 73
avalanches 207

barchans 228-229

barometric pressure 81
barrier reef 253
barrier waters 249
basalt 155
base level 201-205, 247
bayous 218
Beaufort Windscale 81
Bering Glacier 235
Bermuda Triangle 96
big bang theory 23
biogenic sedimentary rocks
   160, 161
biomass 63
biosphere 5
   chemical energy 58-63
   climate 106-144
   development 23-24
   ecosystems 58, 62-64
   energy 20-21
   kinetic energy 56-57
   potential energy 63-64
   solar energy 58-59
blizzards 103-104
   global warming and 135
block mountains 179
boreal forests 129
   animals 303
   climate 303
   drainage systems 305
   glacial deposits 303-305
   indigenous peoples 305
   landforms 303-305
   plants 303
   resource management 305
   soils 303-305

Cambrian Period 150
Canadian Geologic Survey 152
Canadian Shield 156, 327
   exfoliation (sheeting) 203
   glacial erosion 236
   lakes 304
   soil 276
canyons 219, 231
carbon 24, 25, 168
carbon dating 139
carbon dioxide 59, 70, 71, 140
   in atmosphere 71
   global warming and
      135-136, 142
   mining 167
Carboniferous Period 150,
   167
cattle ranching

   in temperate grasslands
      330-331
   in tropical rainforests 300
ceilometer 80
Cenozoic Era 150-151
CFCs. See chlorofluorocarbons
chaos theory 89
chemical energy 17, 25
   in biosphere 58-62
   in fossil fuels 25
   in lithosphere 25
   in plants 58-59
chemical sedimentary rocks
   160, 161
chemical weathering 203-204
Chicxulub, Mexico 138-139
chinook winds 51
chlorofluorocarbons (CFCs)
   136, 143-144
circle of illumination 27
cirques 241, 242
clastic sedimentary rocks 160
climate 75-76, 106-144
   boreal forests 303
   classification 107-113
   forecasting 141
   global warming and 133-144
   maritimes 333
   microclimate 132-133
   regions 117-130
   statistics 110
   temperate grasslands 326
   tropical grasslands 315
climatologists 133
climax vegetation 264-265
clouds 71
   development 84
   formations 70
   fronts and 86, 87
coal 167, 168
coastal landforms 247-256
coastlines
   coral formations 253
   emergent 249
   headlands 249
   ria 253
   submergent 253-256
cold desert climate 112
cold steppe climate 112
comets 138-139
compass rose 8
condensation 47-48
continental climate 112, 127-128
continental drift 170-174

continental glaciers *234-240*
contour mapping *188-189*
convection *19*
convection currents *35*
    and water *41, 42-45*
    and deserts *123*
coral formations *253*
coriolis effect *36-38*,
    ocean currents *44*
    cyclones *89*
    hurricanes *96*
    tornadoes *100*
Cretaceous Period *150*
cyclones *51, 88-89, 93, 128*

Davis, William *215*
Death Valley, California *232*
decomposers *62*
deforestation *62, 285*
deltas *218*
deposition *205*
deserts
    aeolian landscapes *228-232*
    climate *108, 111, 112, 121-124*
    ecosystem *275, 322*
    natural vegetation *270*
    pavement *231*
    soil *317-318*
desertification *319, 322*
Devonian Period *150*
dewpoint *47*
diamonds *163, 168-169*
dinosaurs *71, 149, 150, 259*
    extinction *137, 138-139*
doldrums *36, 38*
dolines *221*
drumlins *239*
dry adiabatic lapse rate *48, 50, 51*
dry adiabatic winds *50-51*
dry climate *109*

earth
    age of *152*
    core *153-154*
    crust *155-156, 157*
    curvature *30*
    daylight *27*
    elements *155, 157*
    layers *154-156*
    major plates *172*
    mantle *153, 154-155*
    origins *23-24*
    rotation *24, 154*
    seasons *27*
    tilt *27*
Earth Resources Survey Program *244*

earth-sun relations
    circle of illumination *27*
    daylight *27*
    equinox *27*
    seasons *27*
earthquakes *24, 155, 173, 179, 192-197*
    measurement *195-197*
    Mexico City *192-194*
    forecasting *195-196*
    Richter Scale *195*
    seismographs *196-197*
ecosphere *258-286*
ecosystems *5, 58, 62-64, 259-260*
    animal adaptations to *259-260*
    aquatic *300, 310-311*
    biomass *63*
    boreal forests *302-312*
    deserts *265-266*
    elements *157-158*
    emergent coastline *249*
    energy flow in *58-59*
    human impact on *260-261, 264-267, 288-289, 321-323*
    hydroelectric development and *312*
    marine *323-337*
    monocultures and *299*
    pollution *260-261*
    sustainable development *289*
    temperate grasslands *326-331*
    tropical grasslands *314-323*
    tropical rainforests *292-301*
    types *64*
El Niño *90-91*
emergent coastlines *249*
entisols *283, 318*
environmental lapse rate *33*
equinox *27, 31*
ergs *228*
erosion *204, 298*
    glacial *236-237, 241*
    slope *207*
    soil *286, 321-222*
    wave *248-249, 250-251*
    wind *228-231*
esker *239*
estuary *219*
extractive reserves *300-301*

farming
    fertilizers *328*
    intercropping *323*
    irrigation *329*
    pastoral *321-322*
    politics and *323*

    slash and burn *296, 297*
    subsistence *297-298, 322-323*
    tropical rainforests *296*
faulting *179-183*
fault patterns *173*
ferrous metals *165*
fertilizers *328*
finger lakes *236*
fish depletion *334-337*
fishing areas of the world *335*
fishing maritimes *334-337*
fishing moratorium *335*
fjords *241*
floods *218-219*
flow velocity *205*
fold mountains *174-176*
food chain *259-260*
food pyramid *63*
forecasting *140*
    climate *141*
    tornadoes *99-100*
forester *308-309*
forestry
    boreal forests *305-308*
forests *268-269*
fossil correlation *148*
fossil fuels *24-25, 142*
    chemical energy *25*
    coal *167*
    natural gas *166*
    oil *166-167*
    global warming and *136, 142*
fossils
    continental drift theory and *171, 173*
    formation *148*
Frank Slide *206*
frigid zone *108*
fringing reefs *253*
frontal precipitation *51*
fronts *86, 87, 88*
frost shattering *202*

gazetteer *9*
general circulation models (GCMs) *141*
geological time scale *147, 148-151*
geologists *164*
geomorphic cycle *215-219*
geothermal energy *19*
glacial ice *234*
glacial landforms *232, 237-240, 241-242, 249*
glaciation
    alpine *234-235, 241-244*
    cirques *241, 242*

continental 236-240
land use patterns 242
Pleistocene 234
theory 140-141
glaciers 232-246
   deposition 237-239, 242, 303-305
   erosion 236-237, 241
   formation 234
   mass balance 235
   movement 234-235
   surges 234-235
   types 234
global cooling 140
global warming 95, 133-144
   fossil fuels and 136, 142
   ice age and 137
   oceans and 137
   projections in Canada 143
   storms and 95, 135
gradational processes 200-205
   base level 201-205
   glaciers 232-246
   gravity 206-209
   principles 201-205
   rivers 210-222
   waves 247-256
   wind 227-232
Grand Canyon 219
Grand Carajas Project 300
grasslands 269-270
Great Barrier Reef 253
Great Lakes 236-237
greenhouse effect 18, 25, 133, 135, 139
greenhouse gases 135-137, 140-141
   reduction 142-144
Gulf Stream 44

Hadley, George 36
hail 49-50
hanging valleys 241
harmattan winds 47
headlands 249
Hildebrand, Dr. Alan 138-139
histosols 283
hot desert climate 108, 112
hot steppe climate 112
humidity 47-48
humid subtropical climate 126-127
humus 274-275
hurricanes 9, 94-98
   destruction 97
   determining speed and direction 96
   global warming and 135
   increasing frequency 135
   total number by month 98
Hurricane Andrew 97, 103, 133
Hurricane Gilbert 97
Hurricane Hazel 95
Hutton, James 148
hydroelectric development 237
   adverse effects on ecosystem 312
   Al Qattarah Depression 311
   boreal forests 311-312
   Grande Carajas Project 300
   James Bay Project 237, 311-312
   James Bay II Power Project 312
hydrologic cycle 46-47
hydrolysis 203-304
hydrosphere 5, 40-55
   climate and 106-107
   composition 21
   solar energy 41
   temperature 41
   thermal energy 41

ice age 140-141, 234
   global warming and 137
ice cap climate 111, 112, 130
ice sheet
   Antarctic 234
igneous intrusions 165, 168-169
igneous ore bodies 165
igneous rocks 159
inceptisols 283, 318
indigenous peoples
   in boreal forests 305
   in maritimes 334
   in temperate grasslands 327
   in tropical grasslands 319-321
   in tropical rainforests 296
inselbergs 231
intercropping 323
interfluve 217
ionosphere 71, 73
irrigation 265
   aquifers 329-330
   Aral Sea 265
isostasy 24-25, 155-156, 237

James Bay Project 237, 311-312
James Bay II Power Project 312
Jari Project 299
Jurassic Period 150

Kalahari Desert 228, 265-266
karst topography 221-222
katabatic winds 35
kerogen 166
kimberlite pipes 169
kinetic energy 17
   in atmosphere 34
   in biosphere 56-57
   in hydrosphere 42-45
   in lithosphere 24-25
Koppen Climate Classification System 108-114
Kotewall Road Landslide 208

lacustrine soils 214
lahar 183
landforms
   boreal forests 303-305
   coastal 247-56
   glacial 232, 237-40, 241-242, 249
   karst 221-222
   mapping techniques 188-189
   maritimes 333-34
   temperate grasslands 326-327
   tropical grasslands 318
   tropical rainforests 295
   volcanic 183-187
   weathering and 201-204
   wind 228-231
landsat images 244-45
landscape
   formation of 227-231
landscape systems
   alpine 65-66
   techniques for studying 67
landslides 206-209
latitude 8, 30-32, 39
levees 217
lithosphere 5
   composition 21
   elements 106
   formation 146-152
   isostacy 24
   kinetic energy 24-25
   nuclear energy 23
   structure 153-156
longitude 8
Lorenz, Edward 89
lumbering
   clear cutting 307-308
   Jari project 299
   selective cutting 298-299, 307-308
   strip cutting 308
   tropical rainforests 298299

magnetosphere *71, 73, 154*
mantle *153, 154-155*
mapping techniques *188-189*
    models *189*
    profiles *189*
    three-dimensional diagrams *189*
    triangulation *195-196*
maps
    bearings *8*
    direction *7*
    grids *8-9*
    legends *9*
    monitoring hurricanes using *96*
    projections *9, 11, 13*
    relief *188-189*
    scales *7*
marine west coast climate *112, 126*
    ecosystems *323-337*
maritimes
    animals *333*
    climate *333*
    ecosystem *323-337*
    fishing *334-337*
    indigenous peoples *334*
    landforms *333-334*
    plants *333*
    resource development *334-337*
    soil *333*
mass wasting *206-209*
    angle of repose *207*
    avalanches *207*
    Frank Slide *206*
    landslides *208*
    rock slides *207*
    solifluction *209*
    tectonic stability and *208-209*
meanders *215, 216, 218*
meander scars *218*
mediterranean climate *112, 124-125*
Mediterranean Sea *265*
Mendes, Chico *301*
mesas *231*
mesopause *73*
mesoscale eddies *46*
mesosphere *71,73*
Mesozoic Era *150*
metals *165*
metamorphoric rocks *159, 161-162*
meteorites *137, 152*
meteorologists *91-92, 96, 99*
meteorology *84-93*
    chaos theory *89*
methane *136, 140*
Mexico City, Mexico

earthquakes *192-194*
microclimate *132-133*
minerals *157-158, 163-169, 214*
    in boreal forests *309-311*
    ferrous metals *165*
    non-ferrous metals *165*
    precious metals *165*
mining
    boreal forests *309-311*
    carbon dioxide
    Grande Carajas Project *300*
    open-pit *309-310*
    pollution *310-311*
mollisols *282, 326-327*
monsoon climate *112, 121*
moraines *239, 242*
moratorium on fishing *335*
Mt. St. Helen's *147, 183*
Mt. Pinatubo *27*
Mt. Vesuvius *183*
mountains *155-156, 175-182*

NAFTA. *See* North American Free Trade Agreement
natural gas *166*
natural resources. *See also* resource development
    exploitation *334*
    non-renewable *289*
    renewable *289*
    sustainable development *289, 337*
natural selection *259*
natural vegetation
    classification *268-271*
    forest *268-269*
    grasslands *269-270*
    local varieties *270-271*
Nevada del Ruiz *183*
nitrogen *71*
North American Free Trade Agreement (NAFTA) *236-237*
Northern Lights (Aurora Borealis) *73*
nuclear energy *23-24, 154*
nuée ardente *183*

oceans *41, 44, 46*
    atmosphere and *137*
    basins *44*
    climate and *137*
    crust *153*
    currents *43-44, 45*
    El Niño *90-91*
    erosion *249*
    evaporation *42-43*

    global warming and *137*
    movement under *174*
    precipitation *51*
    salinity *42*
    storms *95*
Ogallala aquifer *329*
oil *166-167*
open-pit mining *309-310*
orographic precipitation *50-51*
Ordovician Period *150*
outwash fans *239*
oxbow lakes *218*
oxidals *277*
oxidation *25*
oxisols *282*
oxygen *24, 58-59, 70, 71*
ozone *72, 136-137*
ozone layer *17*
ozonometer *81*

Palaeozoic Era *150*
pampas. *See* temperate grasslands
Pangaea *171*
Permian Period *150*
phosphorous cycle *61*
photosynthesis *24, 70, 71, 139*
    global warming and *136*
    process *58-59*
plants
    hydrologic cycle *62*
    needs *263-64*
    nutrients *59, 62*
plains. *See* temperate grasslands
plate tectonics *24, 155*
    continental drift *170-174*
    earthquakes *173*
    fault patterns *173*
    folding *179-180, 182*
    volcanoes *183-185*
playa *228*
Pleistocene Epoch *234*
polar climate *109, 130*
    frost shattering in *202*
pollution *133, 260-61*
    acid rain *310-311*
    chlorofluorocarbons (CFCs) *136, 143-144*
    farming and *328-329*
    fossil fuels *136, 142*
potential energy *24, 63*
    in biosphere *63*
    fossil fuels *24-25*
    in lithosphere *24*
prairies. *See* temperate grasslands
Pre-Cambrian Era *150*
precious metals *165*

precipitation *108, 111*
   acid *310-311*
   convection *49-50*
   frontal *51*
   hail *49-50*
   orographic *50-51*
   relief *50-51*
prevailing winds *36-39, 44*
Prime Meridian *8*
Proterozoic Era *148-150*
protoplanet *23*
psychrometer *80*

Quaternary Period *151*

radiant energy *27-29, 31*
radiocarbon dating *151*
radiometric dating *151*
rainforest. *See* tropical rainforests
rainshadow *50*
rainshadow deserts *122*
recycling *143-144*
reforestation *299*
relief precipitation *50-51*
remote sensing *244*
resource development
   aquaculture *337*
   cattle ranching *300*
   extractive reserves *300-301*
   farming *328-330*
   industrial development *299-300*
   in boreal forests *305-312*
   in maritimes *334-337*
   in temperate grasslands *327-331*
   in tropical grasslands *321-323*
   in tropical rainforests *297-301*
   lumbering *298-299*
   mining *309-311*
Rhine Rift Valley *182*
Richter Scale *195*
rift valley *179, 182*
river basins *211*
   tropical rainforests *295*
   watersheds *211*
river capture *211*
river deposition
   karst topography *221*
rivers
   deltas *218*
   drainage patterns *211-212*
   flood plains *217*
   flow velocity *213, 214, 225*
   geomorphic cycle *215-219*
   interfluve *217*
   levees *216, 217*

meanders *215*
meander scars *218*
measuring *223-226*
oxbow lakes *218*
rejuvenation *219*
soil *214*
systems *211*
transport *214, 215*
underground *221-222*
watershed *211*
rock classes *159-162*
rock cycle *161*
rock slide *207*
rubber trees *300-301*

Sahara Desert
   ecosystem *322*
   expansion or growth *319, 320*
   soil *317-318*
San Andreas fault *182*
sand transport *228-230*
sand dunes
   transport by wind *228-230*
   types *230*
satellite images *93, 96*
   of hurricanes *94*
   landsat images *244-245*
   measure desertification *319*
savanna climate *112, 118-120, 315*
scrub vegetation *270-271*
sea level *249*
seamount *253*
seasons *27, 31*
sedimentary ore bodies *165-166*
sedimentary rocks *159, 160-161*
seismic waves *153, 196-197*
seismographs *153, 196-197*
shaft mining *310-311*
sial *153*
Silurian Period *150*
sima *153*
skerries *242*
slope *207*
snow line *234*
soil
   boreal forests *303-305*
   classification *282-283*
   climate and *277-278*
   composition *274*
   development *275*
   erosion *286, 298*
   farming *276-279, 285*
   fertilization *278-279*
   horizons *274*
   humus *274-75*
   leaching *106*

maritimes *333*
moisture *276-277*
parent material *276*
properties *274-277*
taxonomy *282-283*
temperate grasslands *326-327*
texture *276*
translocation *277*
transport *228-231*
tropical grasslands *317-318*
tropical rainforests *295*
United Nations Soil Survey *285*
vegetation succession and *264-65*
weathering and *62*
soil creep *207*
soil management *278-279*
soil moisture *276-277*
soil profiles *274, 281*
soil types
   inseptisols *318*
   lacustrine *214*
   sentisols *318*
   vertisols *318*
solar energy *17-19, 26, 58-59, 70*
solar exposure *33*
solar system *23*
solstice *31*
spillways *239*
spodosols *283, 310*
stalactites *221*
stalagmites *222*
steppe regions
   aeolian landscapes *228-32*
   climate *110-111, 112, 121-124, 315*
storms
   blizzards *103-104*
   cyclones *51, 88-89, 93, 128*
   El Niño *90*
   fronts and *86-87*
   global warming and *135*
   hurricanes *95-98*
   tornadoes *98-102*
stratosphere *71-72*
subarctic climate *112, 129*
submergent coastlines *253-256*
subsistence farming *297-298*
subtropical climate *112, 126-127*
sun exposure *33*
sustainable resources *289, 337*
   aquaculture *337*

tectonic processes *170-174*
   creation of submergent coastline *253*

continental drift *170-174*
   earthquakes *173*
   fault patterns *173*
   folding *179-180, 182*
   volcanoes *183-185*
tectonic stability *208-209*
temperate cold winter climate *109*
temperate grasslands
   animals *326*
   cattle ranching *330-331*
   climate *326*
   ecosystem *326-331*
   farming *328-330*
   indigenous peoples *327*
   irrigation *329-330*
   landforms *326-327*
   plants *326*
   resource management *327-331*
   soil *326-27, 328*
temperate mild winter climate *109*
temperate zone *108*
temperature *108, 111*
temperature zones *107-113*
Tertiary Period *151*
thermal energy *17-18, 30-33, 46-55*
   in atmosphere *30-33*
   in hydrosphere *41*
   storms *95*
thermal equator *39, 120*
thermal expansion *202*
thermometer *80*
thermosphere *71, 73*
thunderstorms *49-50, 96*
tides *44*
tornadoes *98-102*
   Coriolis effect *100*
   detection *99*
   forecasting *99-100*
   formation *99*
   frequency *101*
   in Ontario *100*
   necessary conditions *99*
   speed *98-99*
   weather watchers *99*
torrid zone *108*
trade winds *38*
transitional zones *122, 124*
transpiration *62*
triangulation mapping *195-196*

Triassic Period *150*
tropical climate *109*
tropical cyclones. *See* hurricanes
tropical grasslands
   climate *315*
   ecosystem *314-323*
   farming *321-323*
   indigenous peoples *319-321*
   landforms *318*
   plants *315-317*
   resource management *321-323*
   soil *317-318*
tropical rainforests *112, 117-118*
   animals *293-294*
   cattle ranching *300*
   climate *293*
   extractive reserves *300-301*
   farming *296-298*
   forest layers *293-295*
   Grande Carajas Project *300*
   hydroelectricity *300*
   indigenous peoples *296*
   industrial development *299-300*
   landforms *295*
   plants *293-294*
   reforestation *299*
   resource development *297-301*
   soil *295*
tropopause *71*
troposphere *71-72, 85*
tsunami *248*
tundra climate *111, 112, 130*
typhooon. *See* hurricanes

ultisols *283*
United Nations Soil Survey *285*
u-shaped valleys *239*

vegetation *264-265*
   classification *268-271*
   forest *268-269*
   grasslands *269-270*
   local varieties *270-271*
veld. *See* temperate grasslands
ventifacts *229*
vertisols *282, 318*
volcanic activity *149, 186*
volcanic landforms *183-187*
volcanic rock *159*
volcanic soil *295*

volcanoes *22, 155, 253, 295*
   formation *175, 183-187*
   Hawaii *184-185*
   Iceland *184*
   Mt. St. Helen's *147, 183*
   Mt. Vesuvius *183*
   Nevada del Ruiz *183*
   tectonic plates and *172-173*
   Vesuvian *183-184*

wadis *228*
waves *43, 247-256*
   deposition *249*
   erosion *248-249, 250-251*
   landforms *250-251*
   refraction *249*
   size *248*
weather *74-76, 94-105*
   balloons *78-79*
   forecasting *78, 84, 89-91, 96*
   instruments *80*
   maps *90*
   myths *78, 83*
   observation stations *78*
   radar *79*
   satellites *79*
weather stations *77*
weather watchers *99*
weathering *201-204*
   chemical *203-204*
   freeze-thaw action *203*
   mechanical *202-203*
Wegener, Alfred *171*
wet adiabatic lapse rate *48*
willy-willies. *See* hurricanes
wind *34-39, 227-232*
   atmospheric systems *36-39*
   Beaufort Windscale *81*
   chinook *51*
   deposits *228-230*
   duration *248*
   erosion *228-231*
   fetch *248*
   flow velocity *228-229*
   harmattan *47*
   katabatic *35*
   velocity *248*

yazoo stream *218*

SOUTH HILL
EDUCATION CENTRE
6010 Fraser Street
Vancouver, B.C. V5W 2Z7
Ph: 324-2414 Fax: 324-9476